Atmospheric Greenhouse Gases: The Hungarian Perspective

László Haszpra
Editor

Atmospheric Greenhouse Gases: The Hungarian Perspective

Springer

Editor
László Haszpra
Hungarian Meteorological Service
H-1675 Budapest
P.O. Box 39
Hungary
haszpra.l@met.hu

ISBN 978-90-481-9949-5 e-ISBN 978-90-481-9950-1
DOI 10.1007/978-90-481-9950-1
Springer Dordrecht Heidelberg London New York

Library of Congress Control Number: 2010937996

Printed on acid-free paper

Springer is part of Springer Science+Business Media (www.springer.com)

Contents

Contributors

Giorgio Alberti
Department of Agricultural and Environmental Sciences, University of Udine, I-33100 Udine via delle Scienze 208, Italy

János Balogh
Plant Ecology Research Group of Hungarian Academy of Sciences, H-2103 Gödöllő, Páter K. u. 1., Hungary
and
Department of Botany and Plant Physiology, Szent István University, H-2103 Gödöllő, Páter K. u. 1., Hungary

Zoltán Barcza
Department of Meteorology, Eötvös Loránd University, H-1117 Budapest, Pázmány P. s. 1/A, Hungary

Márta Birkás
Institute of Crop Production, Szent István University, H-2103 Gödöllő, Páter K. u. 1., Hungary

Alberte Bondeau
Potsdam Institute for Climate Impact Research, PIK, D-14412 Potsdam, Telegrafenberg, P.O. Box 601203, Germany

György Borka
Research Institute for Animal Breeding and Nutrition, H-2053 Herceghalom, Gesztenyes utca 1., Hungary

László Bottlik
Institute of Crop Production, Szent István University, H-2103 Gödöllő, Páter K. u. 1., Hungary

Galina Churkina
Leibniz-Centre for Agricultural Landscape Research,
D-15374 Müncheberg, Eberswalder Strasse 84, Germany

Philippe Ciais
Laboratoire des Sciences du Climat et de l'Environnement, CEA CNRS UVSQ,
F-91191 Gif sur Yvette, France

Szilárd Czóbel
Institute of Botany and Ecophysiology, Szent István University,
H-2103 Gödöllő, Páter K. u. 1., Hungary

Kenneth J. Davis
Department of Meteorology, The Pennsylvania State University, University Park,
PA 16802, USA

Ed Dlugokencky
National Oceanic and Atmospheric Administration, Earth System Research
Laboratory, 325 Broadway, Boulder, CO 80305, USA

Csilla Farkas
Bioforsk, Norwegian Institute for Agricultural and Environmental Research,
Frederik A. Dahlsvei 20, Ås 1432, Norway

Ernő Führer
Department of Ecology and Silviculture, Hungarian Forest Research Institute,
H-9400 Sopron, Paprét 17., Hungary

Györgyi Gelybó
Department of Meteorology, Eötvös Loránd University,
H-1117 Budapest, Pázmány P. s. 1/A, Hungary

Balázs Grosz
Department of General and Inorganic Chemistry, Eötvös Loránd University,
H-1117 Budapest, Pázmány P. s. 1/A, Hungary

Andrea Hagyó
Research Institute for Soil Science and Agricultural Chemistry of the Hungarian
Academy of Sciences, H-1022 Budapest, Herman O. út 15., Hungary

László Haszpra
Hungarian Meteorological Service, H-1675 Budapest, P.O. Box 39, Hungary

Dóra Hidy
Hungarian Meteorological Service, H-1675 Budapest, P.O. Box 39, Hungary

László Horváth
Hungarian Meteorological Service, H-1675 Budapest, P.O. Box 39, Hungary

Anikó Jagodics
Department of Ecology and Silviculture, Hungarian Forest Research Institute, H-9400, Sopron, Paprét 17., Hungary

Noémi Káposztás
Department of Meteorology, Eötvös Loránd University, H-1117 Budapest, Pázmány P. s. 1/A, Hungary

Anikó Kern
Adaptation to Climate Change Research Group, Hungarian Academy of Sciences, H-1117 Budapest, Pázmány P. s. 1/A, Hungary
and
Department of Meteorology, Eötvös Loránd University, H-1117 Budapest, Pázmány P. sétány. 1/A, Hungary

Gábor Kis-Kovács
Hungarian Meteorological Service, H-1525 Budapest, P.O. Box 38, Hungary

Natascha Kljun
Department of Geography, Swansea University, Singleton Park, Swansea SA2 8PP, UK

Sándor Koós
Research Institute for Soil Science and Agricultural Chemistry of the Hungarian Academy of Sciences, H-1022 Budapest, Herman O. út. 15., Hungary

Györgyi Kovács
Karcag Research Intsitute, Research Institutes and Study Farm, Centre for Agricultural and Applied Economic Sciences, University of Debrecen, H-5300 Karcag, Kisújszállási út 166., Hungary

Katalin Lovas
Hungarian Meteorological Service, H-1525 Budapest, P.O. Box 38, Hungary

Attila Machon
Hungarian Meteorological Service, H-1675 Budapest, P.O. Box 39, Hungary
and
Center for Environmental Science, Eötvös Loránd University, H-1117 Budapest, Pázmány P. s. 1/A, Hungary
and
Department of Botany and Plant Physiology, Szent István University, H-2103 Gödöllő, Páter K. u. 1., Hungary

Hrvoje Marjanović
Croatian Forest Research Institute, Cvjetno naselje 41, HR-10450 Jastrebarsko, Croatia

Edit Nagy
Hungarian Meteorological Service, H-1525 Budapest, P.O. Box 38, Hungary

Zoltán Nagy
Department of Botany and Plant Physiology, Szent István University, H-2103 Gödöllő, Páter K. u. 1., Hungary
and
Plant Ecology Research Group of Hungarian Academy of Sciences, H-2103 Gödöllő, Páter K. u. 1., Hungary

Maša Zorana Ostrogović
Croatian Forest Research Institute, HR-10450 Jastrebarsko, Cvjetno naselje 41, Croatia

László Pásztor
Research Institute for Soil Science and Agricultural Chemistry of the Hungarian Academy of Sciences, H-1022 Budapest, Herman O. út 15., Hungary

Alessandro Peressotti
Department of Agricultural and Environmental Sciences, University of Udine, I-33100, Udine via delle Scienze 208, Italy

Krisztina Pintér
Plant Ecology Research Group of Hungarian Academy of Sciences, H-2103 Gödöllő, Páter K. u. 1., Hungary
and
Department of Botany and Plant Physiology, Szent István University, H-2103 Gödöllő, Páter K. u. 1., Hungary

Zoltán Somogyi
Hungarian Forest Research Institute, H-1142 Budapest, Stefánia út 14., Hungary

Attila Stingli
Institute of Crop Production, Szent István University, H-2103 Gödöllő, Páter K. u. 1., Hungary

István Szilágyi
Chemical Research Center of the Hungarian Academy of Sciences, H-1025 Budapest, Pusztaszeri út 59–67., Hungary

Pieter Tans
National Oceanic and Atmospheric Administration, Earth System Research Laboratory, 325 Broadway, Boulder, CO 80305, USA

Klára Tarczay
Hungarian Meteorological Service, H-1525 Budapest, P.O. Box 38, Hungary

Eszter Tóth
Research Institute for Soil Science and Agricultural Chemistry of the Hungarian Academy of Sciences, H-1022 Budapest, Herman O. út 15., Hungary

Kristof Van Oost
Georges Lemaître Centre for Earth and Climate Research (TECLIM), Earth and Life Institute (ELI), Université catholique de Louvain, 3, Place Louis Pasteur, B-1348 Louvain-la-Neuve, Belgium

József Zsembeli
Karcag Research Intsitute, Research Institutes and Study Farm, Centre for Agricultural and Applied Economic Sciences, University of Debrecen, H-5300 Karcag, Kisújszállási út 166., Hungary

Chapter 1
Introduction

László Haszpra

Earth, our planet, would be a rather cold place if the solar energy absorbed by the surface could be radiated back to space without any obstruction. The equilibrium global average surface temperature would be about −19°C instead of the current much higher temperature of approximately +14°C (Le Treut et al. 2007). Already Fourier, the famous French mathematician-physicist guessed that the atmosphere somehow "slows down" the departure of energy from the Earth (Fourier 1827).

Fourier's hypothesis was supported by Tyndall (1861) when he discovered that water vapor and carbon dioxide, basic components of the atmosphere, absorb part of the energy that departs from the surface in the form of infrared radiation. Then, the absorbed energy is partly reradiated back to the surface increasing its temperature.

In 1896, based on the previous works of Tyndall, Pouillet, Langley, and others, Arrhenius studied the effect of the amount of atmospheric greenhouse gases on the global climate (Arrhenius 1896). He realized that decrease in the carbon dioxide (CO_2) content of the atmosphere could lead to the formation of ice ages, while its increase could cause global warming. Although he knew that a huge amount of carbon dioxide was released into the atmosphere from anthropogenic sources, he supposed that the geochemical processes (like chemical weathering and uptake by oceans) largely balanced this input, and thus warming would only be the problem of the far future. However, Callendar, gathering the available carbon dioxide measurements, argued already in 1938 that carbon dioxide was demonstrably accumulating in the atmosphere, and warming of Earth was in progress (Callendar 1938). His study based on sporadic, often uncertain measurements, did not convince the scientific community believing that the oceans could uptake the anthropogenic CO_2 emission at a rate of its release.

Finally, the radiocarbon measurements of Suess (1955) and Craig (1957) stimulated Roger Revell, director of Scripps Institution of Oceanography, to clarify how much of anthropogenic carbon dioxide actually remained in the atmosphere

L. Haszpra (✉)
Hungarian Meteorological Service, H-1675 Budapest, P.O. Box 39, Hungary
e-mail: haszpra.l@met.hu

L. Haszpra (ed.), *Atmospheric Greenhouse Gases: The Hungarian Perspective*,
DOI 10.1007/978-90-481-9950-1_1,

(Revelle and Suess 1957). He contracted Charles D. Keeling to develop a high-precision method to monitor the atmospheric carbon dioxide concentration. The first instruments were deployed at South Pole and at the Mauna Loa Observatory, Hawaii, in 1957–1958, in the framework of the International Geophysical Year (Keeling 1978). These measurements quickly proved that the atmospheric carbon dioxide concentration was increasing at a high rate (Brown and Keeling 1965; Pales and Keeling 1965).

Recognizing the potential social, economical, and overall environmental risk of a human-induced global climate change, the World Meteorological Organization (WMO) introduced the monitoring of atmospheric carbon dioxide to its forming global Background Air Pollution Monitoring Network (BAPMoN – its successor is the present Global Atmosphere Watch [GAW] network) in the late 1960s. As the cyclic behavior of the vegetation (photosynthesis-respiration, seasonal variation in the biological activity) strongly influences the local carbon dioxide concentration, its monitoring was recommended at the most remote places, far away from any anthropogenic and biological activity (polar or desert regions, high mountains, isolated islands, etc.) (WMO 1974).

The increased number of concentration measurements and the development of the global circulation models allowed to make the first attempts to allocate the CO_2 sources and sinks by means of the so-called inverse models in the late 1980s and early 1990s (Tans et al. 1989). These calculations indicated that the continental biosphere, avoided by the monitoring programs before, plays a crucial role in the global carbon dioxide budget (Tans et al. 1990). As it had been supposed earlier that the biosphere had been in equilibrium, this recognition turned the attention to the existing continental monitoring sites and encouraged the expansion of monitoring and research in these regions (Tans 1991). In addition to the monitoring of the atmospheric concentration, significant effort has been invested to understand the role and behavior of the biosphere. Measuring programs and mathematical models of different complexities have been developed to follow and describe the exchange processes between the vegetation and the atmosphere, to estimate the contribution of the biosphere to the greenhouse gas dynamics of the atmosphere under changing climate conditions, to study the feedbacks and interactions between these two spheres.

Carbon dioxide is not the only atmospheric greenhouse gas affected by human activity. As soon as the monitoring technology made it possible, the monitoring programs and the mathematical models based on the measurements have also been extended to non-CO_2 greenhouse gases.

Monitoring of atmospheric carbon dioxide was started in Hungary in 1981. The monitoring program has been significantly developed since the change in the global monitoring strategy in the early 1990s giving high priority to the continental measurements. The more evident the risk of the global climate change became, the more climate and greenhouse gas related research projects were initiated also in Hungary. Since the 1990s, research projects have been initiated to study the greenhouse gas budget of the different ecological systems, and to study their responses to the increasing atmospheric carbon dioxide concentration and changing climate. Agriculture and forest management can significantly affect the greenhouse gas budget of land; thus, research has been needed to help the development of more environmental-friendly land-use practices. In accordance with relevant international agreements, detailed emission inventories have been compiled that also help the development of the national climate policy.

The political and economic changes in Hungary (and in the neighboring Eastern European countries) in the beginning of the 1990s put these works into special context. The collapse of the heavy industry, and the economical crisis in general, reduced the energy demand and consequently the greenhouse gas emission of the energy sector. The use of fertilizer and stock-raising also fell significantly. New sectors, like biofuel production and use, have appeared. The restructuring of the country's economy has also developed a new greenhouse gas budget structure. The recovery of the economy might increase the greenhouse gas emission again that needs special attention. The application of renewable energy, energy conservation, and training of farmers on the best land-use management may contribute to the worldwide efforts to mitigate climate change.

The large-scale climate and greenhouse gas budget models are necessarily of coarse spatial resolution for technical reasons, and thus they cannot give fine spatial details. They use generalized parameters, which is another reason why their results may not be appropriate for local decision making. They cannot provide accurate results on the scale of a relatively small country like Hungary. To feed the smaller scale (regional) models we need area-specific information with respect to how the vegetation, the technologies used, and the climatic conditions interact, and how they may be affected by climate change. All this information can only be provided by local investigations involving scientists working on spot, knowing the characteristics of the region best.

This book covers a coherent subset of the Hungarian climate change oriented research that is directly related to greenhouse gases. Topics discussed in the book range from the monitoring of the concentrations and fluxes of atmospheric greenhouse gases, through the modeling of atmosphere–biosphere interaction and greenhouse gas exchange processes, to the review of the anthropogenic contribution of Hungary to the greenhouse gas budget of the atmosphere. The monitoring sites performing different measurement programs referred in this book are presented in Fig. 1.1.

The increase in the atmospheric amount of greenhouse gases and the potentially induced climate change are rather global than local issues. We hope that, in addition to their local use, the recent results of the Hungarian research projects presented in this book will contribute to the understanding of the global processes, to the interpretation of the measurements, to the development of mitigation strategies, and they can stimulate further research to save our planet for the next generations.

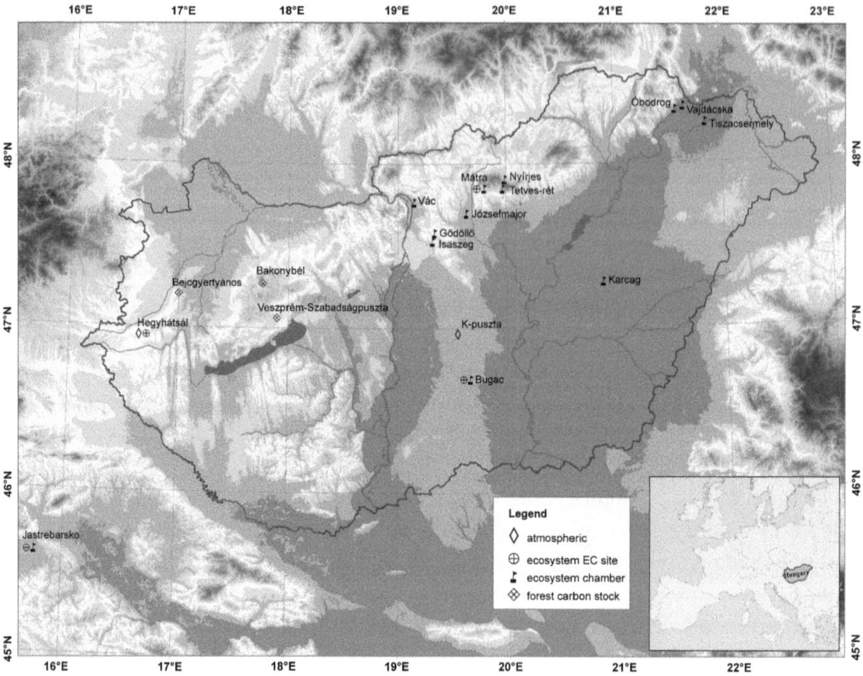

Fig. 1.1 Locations of the monitoring stations and measurement sites referred in this book (EC – eddy covariance) (prepared by L. Pásztor, Research Institute of Soil Science and Agricultural Chemistry of the Hungarian Academy of Sciences)

References

Arrhenius S (1896) On the influence of carbonic acid in the air upon the temperature of the ground. Philos Mag Ser 5(41):237–276

Brown CW, Keeling CD (1965) The concentration of atmospheric carbon dioxide in Antarctica. J Geophys Res 70:6077–6085

Callendar GS (1938) The artifical production of carbon dioxide and its influence on temperature. Quart J R Meteorol Soc 64:223–240

Craig H (1957) The natural distribution of radiocarbon and the exchange time of carbon dioxide between atmosphere and sea. Tellus 9:1–17

Fourier JBJ (1827) Mémoires sur les températures du globe terrestre et des espaces planétaires. Mémoires de l'Académie Royale des Sciences de l'Institut de France VII:570–604

Keeling CD (1978) The influence of Mauna Loa Observatory on the development of atmospheric CO_2 research. In: Miller, J (ed) Mauna Loa Observatory – a 20th Anniversary Report . NOAA Special Report, Sept 1978 (http://www.mlo.noaa.gov/pdf/maunaloa20thanniversary/maunaloa 20th1-50zip7.0later.pdf)

Le Treut H, Somerville R, Cubasch U, Ding Y, Mauritzen C, Mokssit A, Peterson T, Prather M (2007) Historical overview of climate change. In: Solomon S, Qin D, Manning M, Chen Z, Marquis M, Averyt KB, Tignor M, Miller HL (eds) Climate Change 2007: The physical science basis. Contribution of Working Group I to the fourth assessment report of the Intergovernmental Panel on Climate Change. Cambridge University Press, Cambridge, United Kingdom and New York, NY, USA

Pales JC, Keeling CD (1965) The concentration of atmospheric carbon dioxide in Hawaii. J Geophys Res 70:6053–6076

Revelle R, Suess HE (1957) Carbon dioxide exchange between atmosphere and ocean and the question of an increase of atmospheric CO_2 during the past decades. Tellus 9:18–27

Suess HE (1955) Radiocarbon concentration in modern wood. Science 122:415–417

Tans PP (1991) An observational strategy for assessing the role of terrestrial ecosystems in the global carbon cycle: scaling down to regional levels. In: Ehleringer J, Field C (eds) Scaling processes between leaf and landscape levels. Academic, San Diego, CA

Tans PP, Conway TJ, Nakazawa T (1989) Latitudinal distribution of the sources and sinks of atmospheric carbon dioxide derived from surface observations and an atmospheric transport model. J Geophys Res 94D:5151–5172

Tans PP, Fung IY, Takahashi T (1990) Observational constrains on the global atmospheric CO_2 budget. Science 247:1431–1438

Tyndall J (1861) On the absorption and radiation of heat by gases and vapours, and on the physical connexion of radiation, absorption, and conduction. Philos Mag Ser 4(22):169–194, 277–302

WMO (1974) WMO operational manual for sampling and analysis techniques for chemical constituents in air and precipitation. WMO No. 299

Part I
Atmospheric Trends and Fluctuations

Chapter 2
History and Sites of Atmospheric Greenhouse Gas Monitoring in Hungary*

László Haszpra, Zoltán Barcza, and István Szilágyi

Abstract This chapter introduces the Hungarian atmospheric greenhouse gas monitoring stations, their environment, monitoring program, and instrumentation. Quality of the measurements is discussed using the results of interlaboratory comparisons and independent parallel measurements. Study on the spatial representativeness of the measurements leads to the revelation that only the early afternoon measurements can be used for studies other than investigation of diurnal variations. The early afternoon carbon dioxide data may be representative for 200,000–300,000 km^2, and they can characterize the average boundary layer mixing ratio with a reasonable accuracy. Finally, the chapter lists the major international projects, in which the Hungarian monitoring sites participate, the databases, where their data are accessible, and the projects, which support or supported the monitoring activity.

Keywords Hegyhátsál • Interlaboratory comparison • K-puszta • Monitoring system • Quality control • Representativeness

---

*Citation: Haszpra, L., Barcza, Z., Szilágyi, I., 2010: Atmospheric trends and fluctuations – History and sites of atmospheric greenhouse gas monitoring in Hungary. In: Atmospheric Greenhouse Gases: The Hungarian Perspective (Ed.: Haszpra, L.), pp. 9–27.

L. Haszpra (✉)
Hungarian Meteorological Service, H-1675 Budapest, P.O. Box 39, Hungary
e-mail: haszpra.l@met.hu

Z. Barcza
Department of Meteorology, Eötvös Loránd University, H-1117 Budapest, Pázmány P. s. 1/A, Hungary

I. Szilágyi
Chemical Research Center of the Hungarian Academy of Sciences, H-1025 Budapest, Pusztaszeri út 59–67., Hungary

L. Haszpra (ed.), *Atmospheric Greenhouse Gases: The Hungarian Perspective*,
DOI 10.1007/978-90-481-9950-1_2, © Springer Science+Business Media B.V. 2011

2.1 Introduction

Keeling's measurements performed at South Pole and Hawaii made the accumulation of anthropogenic carbon dioxide (CO_2) in the atmosphere evident (Brown and Keeling 1965; Pales and Keeling 1965). The recognition of the potential social, economical, and overall environmental risks of an anthropogenically induced global climate change resulted by the strengthening atmospheric greenhouse effect stimulated the World Meteorological Organization (WMO) to initiate the monitoring of atmospheric CO_2 on global level. WMO was organizing its Background Air Pollution Monitoring Network (BAPMoN) in the late 1960s, and CO_2 concentration measurement was also included in the monitoring program of the BAPMoN stations (WMO 1974). This measurement was primarily expected from the so-called baseline stations located at very remote places, far away from any local anthropogenic source and extended vegetated area (source and sink for CO_2). WMO contracted the Hungarian Meteorological Service (HMS) to host training courses for the staff of BAPMoN (from 1989: Global Atmosphere Watch, GAW) stations between 1978 and 1992 (Mészáros 1978). For these courses, demonstration instruments were also provided by WMO. For the training on atmospheric CO_2 measurements, a Siemens Ultramat 3 nondispersive infrared (NDIR) gas analyzer was donated to HMS. This instrument was deployed at the newly established regional background air pollution monitoring station of HMS (K-puszta) in 1981 to demonstrate this measurement to the participants of the training courses in real-world environment rather than in a classroom. The operative use of the instrument at a regular monitoring station also gave an insight into the variation of atmospheric CO_2 concentration in a region where it had never been studied before, although the location of the station (middle of a highly industrialized, densely populated continent, surrounding vegetation) did not comply with the siting criteria prescribed by WMO.

The measurements performed at K-puszta attracted little scientific attention because of the improper location. The situation changed significantly at the end of 1980s when the model calculations revealed that the northern hemispheric continental biosphere played a crucial role in the global carbon budget. Any data from this previously avoided region became valuable. It was the first time when the Hungarian atmospheric CO_2 measurements could be published in the international scientific literature (Haszpra 1995).

The change in the focus of the global carbon budget research required the expansion of the monitoring networks to the continents, vegetated regions. Tans (1991) suggested the use of tall towers in the monitoring to increase the spatial representativeness of the measurements in the regions where the local biosphere generates a significant "noise" in the CO_2 signal. In 1992, the US–Hungarian Scientific and Technological Joint Fund offered a unique opportunity to both the Hungarian scientists, who wanted to develop the measurements and integrate them into the international efforts, and National Oceanic and Atmospheric Administration (NOAA), U.S.A., who was seeking for new monitoring sites in the critical continental regions. In the framework of this cooperation, building on the available Hungarian measurement tradition and on NOAA's experience with the first tall tower monitoring site in North Carolina, U.S.A. (Bakwin et al. 1995), one of the first European tall tower monitoring sites dedicated to greenhouse gas measurements was put into operation at Hegyhátsál, Hungary, 220 km to the west from K-puszta, where no tall tower was available. It was not scientifically reasonable and economically feasible to operate two CO_2 monitoring sites so close to each other; therefore, the outdated, worn-out system at K-puszta was switched off in 1999.

The state-of-the-art tall tower monitoring site attracted further international projects since the late 1990s that allowed the development and expansion of the monitoring program. After the signature of the Kyoto Protocol, the European Commission supported several greenhouse gas oriented research projects in the framework of its R&D Framework Programmes. Hegyhátsál station, and the scientists working there, joined AEROCARB, CHIOTTO, CarboEurope-IP, IMECC, GEOMON, etc., projects (also involving Hungarian financial support), which allowed the introduction of non-CO_2 greenhouse gas measurements, vertical CO_2 flux measurements, aircraft measurements, carbon and oxygen isotope measurements, as well as the installation of a state-of-the-art, remotely controllable data acquisition and real-time data reporting system.

2.2 K-puszta

2.2.1 Location and Environment

K-puszta, the regional background air pollution monitoring station of the HMS, the member of the WMO GAW and the European EMEP[1] networks, is located at 46°58′N, 19°33′E, 125 m above the sea level, on the Hungarian Great Plain, in the middle of the Carpathian Basin (Fig. 1.1). The station is located in a large forest clearing, where the immediate surroundings of the station are kept undisturbed. The dominant tree species in the region are Scots Pine and Black Pine with acacia patches. The organic layer of the sandy soil is thin. The nearest public road carried approximately 700–800 vehicles per day during the time of CO_2 measurements (1981–1999), and there is a small settlement (370 inhabitants in 2001) about 3 km to the east/southeast off the station. The nearest big city is the dominantly agricultural Kecskemét (85,000 inhabitants in 2001). It is located approximately 20 km to the southeast, opposite to the prevailing wind direction. In the direction of the prevailing northwesterly wind, the nearest remarkable anthropogenic source is Budapest (1.9 million inhabitants in 2001) located about 80 km off the station. In general, K-puszta is as free from direct anthropogenic pollution as it is possible in this geographical region.

[1]Co-operative Programme for Monitoring and Evaluation of the Long-range Transmissions of Air Pollutants in Europe (European Monitoring and Evaluation Programme – EMEP).

2.2.2 Carbon Dioxide Monitoring System

At K-puszta, the HMS was operating a Siemens Ultramat 3 CO_2 NDIR analyzer between July 1981, and June 1999. The air inlet of the instrument was mounted at 10 m above the grass-covered sandy ground. The data acquisition system recorded the 30-min average mixing ratio. For the calibration, three CO_2-in-nitrogen standards prepared by the Scripps Institution of Oceanography, La Jolla, California, U.S.A., were used until 1992, when they were replaced by four CO_2-in-air standards prepared and certified by NOAA Climate Monitoring and Diagnostics Laboratory (CMDL), Boulder, Colorado, U.S.A. The pre-1992 mixing ratio data have been corrected for the pressure broadening effect, and they are expressed on the WMO X2007 scale in this book and in the WMO World Data Centre for Greenhouse Gases (http://gaw.kishou.go.jp/wdcgg/) at the time of writing.

In the case of the Siemens Ultramat 3 CO_2 analyzer, the water vapor content of air to be analyzed influences the output CO_2 signal. Usually, water vapor is removed by a freezer to avoid this interference, but this method requires a powerful gas cooler taking into account the flow rate of 1 L min^{-1}. At K-puszta, Köhler's method (Köhler 1974) was applied to handle the water vapor interference: both air to be analyzed and standard gases were saturated with water vapor at room temperature, then cooled down to +4°C on the way to the analyzer (Fig. 2.1). In this way the water vapor content of the gases to be analyzed was kept approximately constant (saturated at +4°C), and the variable water vapor content of atmospheric air did not influence the CO_2 measurements any more. Köhler's method is not perfect, among others, because of the low frequency noise of relatively large amplitude caused by

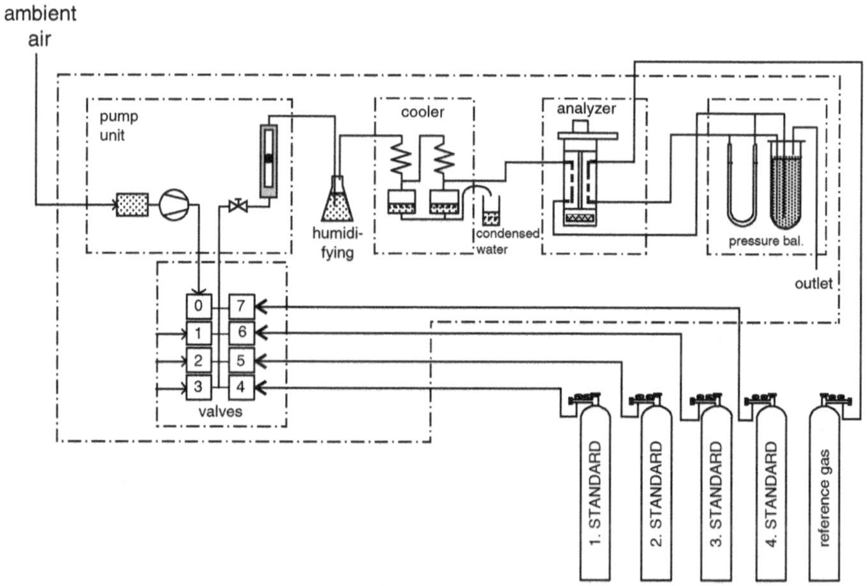

Fig. 2.1 Schematic of the CO_2 monitoring system used at K-puszta between 1981 and 1999

the switch on/off hysteresis of the gas cooler, but it was used at a few sites until the 1990s (e.g. Hensen et al. 1997). However, it should be mentioned that the low frequency noise caused by the improper removal of water vapor is largely eliminated by the longer time averaging, and thus the hourly and longer term averages may be fairly accurate. The outline of the measuring system can be seen in Fig. 2.1.

2.3 Hegyhátsál

2.3.1 Location and Environment

At Hegyhátsál, a 117 m tall, free-standing TV/radio transmitter tower owned by Antenna Hungária Corp. is equipped with intake tubes, meteorological sensors, and other instrumentation. The lower 56 m of the tower is a 7.8-m diameter cylinder made of reinforced concrete, while the upper 61 m is a steel cylinder of 1.8 m diameter. The tower is located in a fairly flat region of western Hungary (46°57′N, 16°39′E), at an altitude of 248 m above the sea level (Fig. 1.1). The tower is surrounded by agricultural fields (mostly crops and fodder of annually changing types) and forest patches. The distribution of vegetation types (53% arable land, 35.5% forest, 11.5% other [vineyard, settlements, etc.]) within 10 km of the tower is not greatly different from the average for the Western Hungarian Landscape Unit (7,300 km^2) or the whole country (93,030 km^2) (48.4% arable land, 19.7% is forest, 11% grassland, 17.7% other; see Chapter 13). The soil type in the region of the tower is lessivated brown forest soil (Alfisol, according to USDA system). These soils have clay migration and moderate acidity as well as the more widespread humification, leaching, and clay formation (Stefanovits 1971). The upper layer is generally 10–20 cm thick, and its organic matter content is 5–8%.

Human habitations within 10 km of the tower are only small villages (100–400 inhabitants). The nearest village is Hegyhátsál (170 inhabitants) about 1 km to the northwest. The nearest city is Körmend (13,000 inhabitants), 9 km to the northwest of the station. There is no notable industrial activity in this dominantly agricultural region. Local roads have mostly low levels of traffic. One of the few main roads of the region, which carries approximately 3,600 vehicles per day on average, passes approximately 400 m to the southwest of the tower. The monitoring site is fairly free from direct anthropogenic pollution as it is shown by sulfur hexafluoride measurements (see Chapter 3).

Further details can be found on the website of the monitoring station: http://nimbus.elte.hu/hhs/.

2.3.2 Monitoring of Atmospheric Carbon Dioxide Mixing Ratio

At Hegyhátsál tall tower site, CO_2 mixing ratio has been continuously monitored at four elevation levels along the tower (10, 48, 82, and 115 m above the ground) since

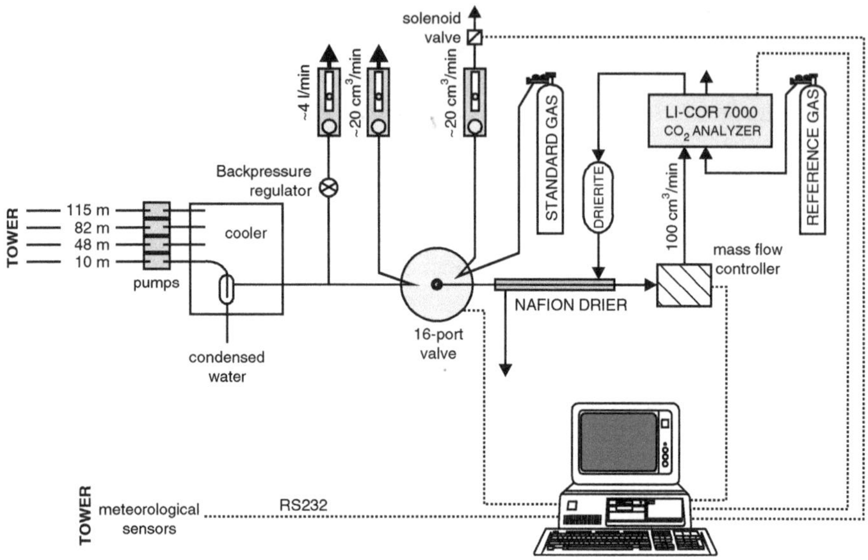

Fig. 2.2 Schematic of the CO_2 monitoring system used at Hegyhátsál tall tower site. For clarity, only one standard gas and one sampling line are presented

September 1994. The schematic diagram of the CO_2 mixing ratio profile monitoring system is shown in Fig. 2.2.

Air is pumped through 9.5 mm diameter tubes (Dekoron Type 1300) from the intake levels on the tower to the CO_2 analyzer located on the ground floor of the transmitter building. 47 mm diameter (Whatman EPM) particle filters are applied in each intake tube to prevent the pumps, valves, and the analyzer from dust. The setup for CO_2 analysis is very similar to that described by Zhao et al. (1997). Diaphragm pumps (1994–2005: KNF Neuberger type UN73MVP; from 2005: KNF Neuberger type N 811 KN.18) are used to draw air continuously through each of the tubes from the four monitoring levels at a flow rate of about 2 L min^{-1} (4 L min^{-1} since 2005). After the pump, the air at 40 kPa overpressure enters a glass trap for liquid water, which is cooled in a regular household refrigerator to dry the air to a dew point of 3–4°C. Liquid water is forced out through an orifice at the bottom of each trap.

The four inlet tubes and the standard gases are connected to a computer-controlled 16-position valve (VICI AG, Valco Europe) that selects which monitoring level or standard gas is sampled by the analyzer. The valve head is protected by 7 μm in-line filters. Ambient air flows continuously through the multiport valve so that the system is constantly flushed. The (expensive) standard gases are shut off when they are not in use by means of computer-controlled solenoid valves. The air leaving the multiport valve through its common outlet is further dried to a dew point below −25°C by passage through a 182 cm long Nafion drier (Permapure, type MD-110-72P), so that the water vapor interference and dilution effect are less than 0.1 μmol mol^{-1} equivalent CO_2 (Zhao et al. 1997). The Nafion drier is purged in a counterflow (100 cm^3

min^{-1}) arrangement using the waste sample air that has been further dried by passage through anhydrous $CaSO_4$ (W.A. Hammond Drierite Co. Ltd.).

CO_2 analysis is carried out using an infrared gas analyzer (1994–2007: Li-Cor Inc. model LI-6251; from 2007: Li-Cor Inc. model LI-7000). A constant sample flow rate of 100 cm^3 min^{-1} is maintained by a mass flow controller (Tylan, model FC-260). The reference cell of the CO_2 analyzer is continuously flushed at a flow rate of 5–10 cm^3 min^{-1} with a compressed reference gas of 330–350 μmol mol^{-1} CO_2 in synthetic air (Messer Hungarogáz). Calibration of the analyzer is carried out using four standards spanning 330–440 μmol mol^{-1} CO_2 prepared by NOAA CMDL (from 2007: NOAA Earth Systems Research Laboratory, ESRL) (Kitzis et al. 1999). Both the CO_2 mixing ratio in the reference gas and the calibration range have been slightly shifted upward since the beginning of the measurements to follow the changes in the atmosphere.

The basic measuring cycle is 2 min, consisting of 1 min flushing and 1 min signal integration. (Note, as the intake tubes are permanently flushed, only the analyzer and the short 1/8 in. ID tube from the selector valve to the analyzer have to be flushed during this relatively short time.) Each 1-min average and standard deviation is based on 6–7 individual measurements. The multiport valve steps through the four monitoring levels in 8 min. Every 32 min, after four 8-min measuring cycles, the standard gas with the lowest CO_2 mixing ratio is selected and analyzed, and this measurement is termed "zero." After every sixth cycle (every 202 min), a full four-point calibration is carried out. The reference and sample cells of the CO_2 analyzer are not pressure or temperature controlled. The "zero" measurements are used to account for any short-term drift of the analyzer due to changes in ambient pressure or temperature. A quadratic response function is fit to each set of calibration gas measurements. The "zero" offsets and response functions are linearly interpolated in time to obtain values appropriate to calculate CO_2 mixing ratio from the instrument response.

2.3.3 *Monitoring of the Atmospheric Mixing Ratio of non-CO_2 Greenhouse Gases*

At Hegyhátsál tall tower site, methane (CH_4), nitrous oxide (N_2O), and sulfur hexafluoride (SF_6) mixing ratios have been monitored since January 2006. The sampling elevation is at 96 m above the ground. For the monitoring of the mixing ratios, a measuring system built on an Agilent 6890N gas chromatograph is used. Its design is similar to those at other present European sites (Thompson et al. 2009; van der Laan et al. 2009). The schematic diagram of the system is shown in Fig. 2.3.

The gas chromatograph is equipped with two analysis lines. One of them is used for methane and carbon monoxide (CO – not discussed here) measurements. It consists of a 4 ft Unibeads 1S precolumn followed by a 6 ft Molecular Sieve 5A analytical column and a flame ionization detector (FID). The other line consists of a 4-ft Hayesep Q precolumn followed by a 6-ft Hayesep Q analytical column and

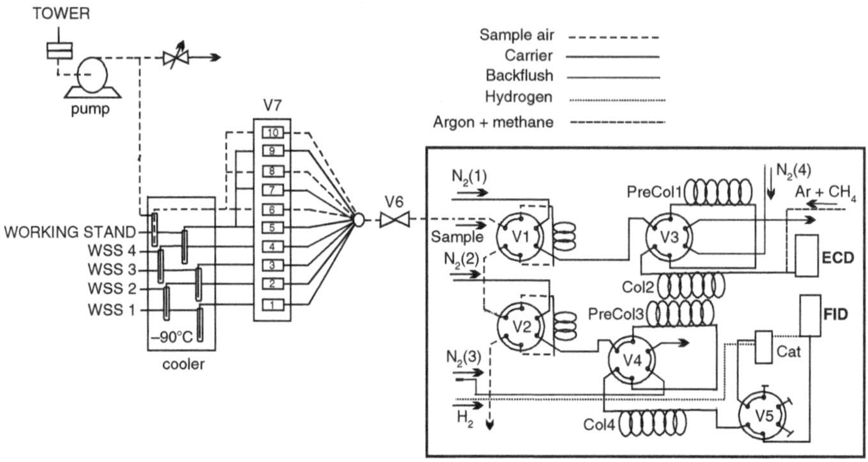

Fig. 2.3 Schematic of the gas chromatographic system used at Hegyhátsál tall tower site for the monitoring of non-CO_2 greenhouse gases

an electron capture detector (ECD), and it is capable to measure the nitrous oxide and sulfur hexafluoride mixing ratios. Nitrogen with 5.0 purity (Linde Gáz, Hungary), further purified by a Supelco 23801 high-capacity gas purifier, is used as carrier gas for both lines. Hydrogen for FID and methanizer nickel catalyst (for CO analysis only) is produced by a Parker–Balston hydrogen generator (model H2PEM-100), while air is supplied by a Jun-Air oil-less compressor (model OF-302) through a Parker Zero Air Generator. For the ECD, 5% methane/95% argon make-up gas is used.

Air to be analyzed is continuously pumped by a diaphragm pump (KNF Neuberger Type N86) through a 9.5 mm diameter tube (Dekoron Type 1300) from the intake level on the tower, at 96 m above the ground, at a rate of about 6 L min^{-1} to the measuring system located on the ground floor of the transmitter building. A 47 mm diameter (Whatman EPM) particle filter is applied to prevent the system from dust. The inlet tube, the four standard gases used for the calibration of the system, and a so-called working standard with approximately ambient mixing ratios used to check the stability of the system are connected to a computer-controlled, 10-position valve (VICI AG, Model: DCSD10MWE) through separated glass water vapor traps held at –90°C in a freezer (FTS System, Model VT490). The 10-position valve selects a standard, the working standard or the ambient air to be analyzed to forward it to the gas chromatograph.

At the beginning of an analysis the gas, selected by the 10-position valve (V7 in Fig. 2.3), flows through the loops of the two gas sampling valves (Valco, DC6UWE, V1 and V2 in Fig. 2.3). After 40 s flushing time, the on/off valves (Numatics S series, V6 in Fig. 2.3) between the 10-position valve and the gas sampling valves closes, and the pressure in the loops equilibrate to atmosphere pressure. The two sampling valves simultaneously inject the sample to the precolumns. For the separation of methane and carbon monoxide, a 0.9 m × 1/8 in. Unibeads S1 (Alltech) precolumn and a 1.2 m × 1/8 in. Molecular Sieve 5A (Alltech) separating column (PreCol3 and Col4, respectively, in Fig. 2.3) are used. When the air, methane, and carbon monoxide have left the precolumn and entered the main column, the backflush valve (Valco DC6WE, V4 in Fig. 2.3) changes the direction of flow in the precolumn, eliminating the heavier compounds to enter the Molecular Sieve column. When air/oxygen has left the main column, valve 5 (V5 in Fig. 2.3) switches and directs the separated CH_4 and CO to the nickel catalyst converting carbon monoxide to methane. Methane sweeps through the catalyst unchanged. The analyses are executed at 60°C isothermally. The FID is held at 250°C.

The separation of nitrous oxide and sulfur hexafluoride is executed on a Hayesep Q (Alltech) columns of 1/8 in. diameter. The 1.2 m precolumn (PreCol1 in Fig. 2.3) separates the heavier substances from the two compounds to be measured, and it is backflushed (V3 in Fig. 2.3). N_2O and SF_6 are separated on a 1.8 m long main separating column (Col2). ECD is held at 370°C.

We use a 12-h long analysis program started at midnight and noon. Each 12-h long sequence begins with the calibration of the system using the four standard gases (WSS1-WSS4) prepared by Max-Planck-Institute for Biogeochemistry (Jena, Germany) in the framework of the European Commission funded CHIOTTO project (http://www.chiotto.org). The scales are traceable to the WMO GAW Central Calibration Laboratory for these gases operated by NOAA. During this initial calibration the mixing ratios in the working standard (WS) are also determined. Later in the sequence the ambient air and the WS are analyzed alternately (WS-air–WS-air–WS-air–...) to monitor and correct the potential drift in the scale. An analysis takes 10 min; thus ambient air data is available in every 20th minute.

2.3.4 Ancillary Measurements

For the interpretation of the mixing ratio data, meteorological and other measurements are indispensable. The basic meteorological measurements (wind speed, wind direction, air temperature, relative humidity) are available from the same elevation levels (10, 48, 82, 115 m) as the CO_2 mixing ratio data. The distance of the sensors from the tower is 4.4 m at 82 and 115 m elevations, 2.5 m at 48 m elevation, while the sensors of 10 m are mounted on the top of a freestanding mast. The flow distortion caused by the tower influences the measured wind speed at 48, 82, and 115 m. We correct the measured wind speeds based on a theoretical laminar flow pattern around a cylindrical body using wind direction information (Barcza 2001). Especially in sunny summer days, when the temperature sensors are on the lee side of the tower, the warmed body of the tower may influence the temperature measurements. This effect is also checked and corrected (Haszpra et al. 2001). The station is also equipped with several other sensors (global solar radiation sensor, radiation balance sensor, photosynthetically active radiation sensor, soil temperature, soil moisture, and soil heat flux sensors), which are primarily used in the surface–atmosphere gas exchange studies (see Chapters 6 and 8).

The biosphere plays a crucial role in the formation of the atmospheric concentrations. To have a better insight into the processes and to help in the interpretation of the mixing ratio measurements, two independent vertical CO_2 flux measuring systems based on eddy covariance technique are in operation at Hegyhátsál. One of them is mounted on the tall tower at 82 m above the ground. Its large footprint covers the neighboring agricultural fields and it gives information about the CO_2 exchange of the typical Hungarian agricultural lands in general. Further details on this measuring system can be found in Chapter 8. The other eddy covariance system is mounted at 3 m elevation and it monitors the CO_2 flux between a seminatural grassland and the atmosphere. This monitoring system and its measurements are discussed in Chapter 6.

Hegyhátsál tall tower site has been a member of NOAA global cooperative air sampling network for greenhouse gases (station code: HUN) since March 1993. Every week, two flasks (2.5 L each) have been flushed and pressurized in series with air in the early afternoon hours, when the vertical mixing of the atmosphere is the most vigorous. The sampling height is 96 m above the ground. The samples are analyzed by NOAA ESRL Carbon Cycle Greenhouse Gases group (Boulder, Colorado, U.S.A.) for greenhouse gases and by the Institute for Arctic and Alpine Research (University of Colorado, Boulder, Colorado, U.S.A.) for carbon and oxygen stable isotope composition of CO_2. Carbon dioxide content of the samples is measured by means of an NDIR analyzer, while for CH_4, N_2O, and SF_6, a gas chromatograph equipped with FID and ECD is used. For details, see http://www.esrl.noaa.gov/gmd/ccgg/flask.html. The parallel in situ and flask measurements offer a unique opportunity to monitor the quality of the measurements regularly.

In the framework of the European Commission supported AEROCARB project and CarboEurope Integrated Project (IP) regular aircraft measurements of greenhouse gases were also performed over the tower between 2001 and 2008. Flask air samples were taken between 200 and 3000 m elevation above the ground 8–16 times a year. The samples were analyzed by Laboratoire des Sciences du Climat et l'Environnement, Gif-sur-Yvette, France.

2.4 Quality Assurance and Quality Control of the Measurements

Comparability of the measurements made at different monitoring sites or by different laboratories is a basic requirement in any scientifically sound study. One of the prerequisites to the acceptable comparability is the use of standards traceable to the same etalon. Since 1992, both K-puszta and Hegyhátsál have been using CO_2 standards prepared and certified by the Central Calibration Laboratory (CCL) of the World Meteorological Organization (NOAA ESRL, Boulder, Colorado, U.S.A.). The N_2O, CH_4, and SF_6 standards are also traceable to the WMO standards maintained by the CCLs of these gases, also operated by NOAA ESRL.

The monitoring sites should regularly participate in interlaboratory comparisons when air samples of unknown composition are distributed among the participating laboratories to control their scales. WMO requests the interlaboratory comparability in case of CO_2, CH_4, N_2O, and SF_6 within ±0.1 μmol mol^{-1}, ±2 nmol mol^{-1}, ±0.1 nmol mol^{-1}, and ±0.02 pmol mol^{-1}, respectively (WMO 2009). WMO organizes an interlaboratory comparison approximately in every fourth year. In the framework of the European Commission supported CarboEurope-IP, a more frequent interlaboratory comparison exercise was initiated in Europe, which has been continued after the end of CarboEurope-IP at the end of 2008 (http://www.cucumbers.uea.ac.uk/).

In the first two WMO interlaboratory comparisons (1992 and 1996) the measuring system at K-puszta was used. In 1992, CO_2-in-nitrogen was used as standards yet, so the average −2.47 μmol mol^{-1} offset (http://gaw.kishou.go.jp/wcc/co2/co2comparison.html) was quite understandable. In 1996, the average offset was −1.02 μmol mol^{-1} indicating that the Köhler's method for the removal of water vapor interference (Köhler 1974) could not provide the requested accuracy. The Köhler's method caused at least ±0.5 μmol mol^{-1} low-frequency noise that strongly influenced the momentary measurements, like the measurement of a sample of unknown mixing ratio from WMO, while this noise was largely smoothed out from the longer term averages. Consequently, the accuracy of the longer term averages (hourly, daily, monthly, etc.) used in different studies was significantly higher. Nevertheless, the potential inaccuracy of the values in a few tenth of μmol mol^{-1} range should be considered in the evaluations.

In the next three WMO interlaboratory CO_2 comparisons (1999, 2003, 2005) already the measuring system at Hegyhátsál participated. The more developed system provided significantly better results (0.06, 0.07, and −0.38 μmol mol^{-1}, respectively – unpublished results). In the CarboEurope-IP interlaboratory CO_2 comparisons (2008–2009) the new LI-7000 analyzer was already used, the stability of which exceeded that of the previous model. The results of the interlaboratory comparison were within the required range in all five exercises performed so far (overall mean offset: 0.06 μmol mol^{-1}, stability of offset: ±0.03 μmol mol^{-1} – Manning et al. (2009) and results unpublished yet).

As the monitoring system for methane, nitrous oxide, and sulfur hexafluoride was installed in the beginning of 2006, we could not participate in WMO interlaboratory comparison yet. However, the interlaboratory comparisons organized by CarboEurope-IP indicate that the accuracy of the methane measurements satisfies the WMO requirement (overall mean offset: 1.3 nmol mol^{-1}, stability of offset: ±0.9 nmol mol^{-1}, unpublished project results). The bias of the SF_6 measurements ranges from 0.09 to 0.14 pmol mol^{-1}, which is out of the range requested by WMO. SF_6 measurements are not easy as it is indicated by the fact that meeting the accuracy requirement causes problems even to some of the regional reference laboratories (Manning et al. 2009). There are more serious problems with the nitrous oxide measurements. The scale at Hegyhátsál is biased from the reference scale by 0.8–1.1 nmol mol^{-1} without obvious reason. That is why, these data are used only for the characterization of the local daily variation.

Interlaboratory comparisons by means of circulation of air samples of unknown composition offer only rare, occasional possibility to check the performance of the monitoring systems, and they cannot simulate the normal atmospheric sampling. The parallel, independent measurements give a more rigorous control on data quality.

As it was mentioned in the previous section, Hegyhátsál tall tower site is a member of NOAA global cooperative air sampling network for greenhouse gases, where every week two flasks (2.5 L each) are filled with ambient air. The samples are analyzed by NOAA ESRL Carbon Cycle Greenhouse Gases group (Boulder, Colorado, U.S.A.) for all those greenhouse gases that are also measured in situ at the station. In the case of methane, nitrous oxide, and sulfur hexafluoride, the sampling elevation of the flasks and the in situ measurements coincides (96 m above the ground); however, CO_2 mixing ratio is measured at 82 and 115 m by the in situ system. While the flasks contain momentary air samples, in situ non-CO_2 measurements are available only in every 20th minute, and in situ CO_2 measurements are usually performed only in every 8th minute at a given sampling elevation. The dislocation of the intakes of the NOAA sampler unit and that of the in situ CO_2 analyzer, the time synchronization problem, the different residence time of air in the sampling systems cause methodological problems in the comparison.

For the comparison, in the case of CO_2 measurements, the arithmetic mean of the consecutive in situ measurements carried out at 115 and 82 m are calculated. For the comparison, those in situ measurements are taken into account, and averaged, which fall in a ±10 min time-window around the recorded NOAA sampling time. Usually, three in situ measurement pairs (82 and 115 m) can be used to calculate a data to compare with the flask measurements. In the case of unfortunate timing, zeroing or calibration in the time-window, there might be only two or one in situ data pairs for the comparison. In the case of the other greenhouse gases, in situ data from a ±20 min time-window around the recorded NOAA sampling time are used. In most cases, the time-window includes two in situ measurements. The temporal and spatial dislocations of the samples in the comparison cause significant scatter in the flask–in situ deviation, especially if significant short-term fluctuation can be observed in the atmospheric mixing ratios. These periods and a few extreme outliers (potential sampling or instrument errors) are rejected from the data series.

After data filtering mentioned above, the atmospheric CO_2 mixing ratios measured from the flask samples and by the in situ analyzer deviate by −0.03 μmol mol^{-1} on average (standard deviation is 1.50 μmol mol^{-1}, correlation coefficient is 0.99 – based on 407 flask–in situ data pairs), which is much better than expected taking into account the absolute independence of the methods and the methodological difficulties. The flask–in situ difference does not show any recognizable temporal variation or tendency (Fig. 2.4).

For methane, nitrous oxide, and sulfur hexafluoride, shorter data series are available (63–71 valid data pairs). The atmospheric mixing ratios measured from the flask samples and by the in situ gas chromatograph deviate by −4.71 nmol mol^{-1}, −0.79 nmol mol^{-1}, and −0.02 pmol mol^{-1} for CH_4, N_2O, and SF_6, respectively (standard deviations are 9.35 nmol mol^{-1}, 0.77 nmol mol^{-1}, and 0.10 pmol mol^{-1}, respectively, while the correlation coefficients are 0.98, 0.74, and 0.93). Temporal variations in the biases do not show any trend or regular variation (Fig. 2.4). In the case of methane, the bias indicated by this comparison is somewhat higher than it could be expected from the results of the interlaboratory comparisons (1.3 and 4.71 nmol mol^{-1}), while the present comparison confirms that the SF_6 measurements are reasonably

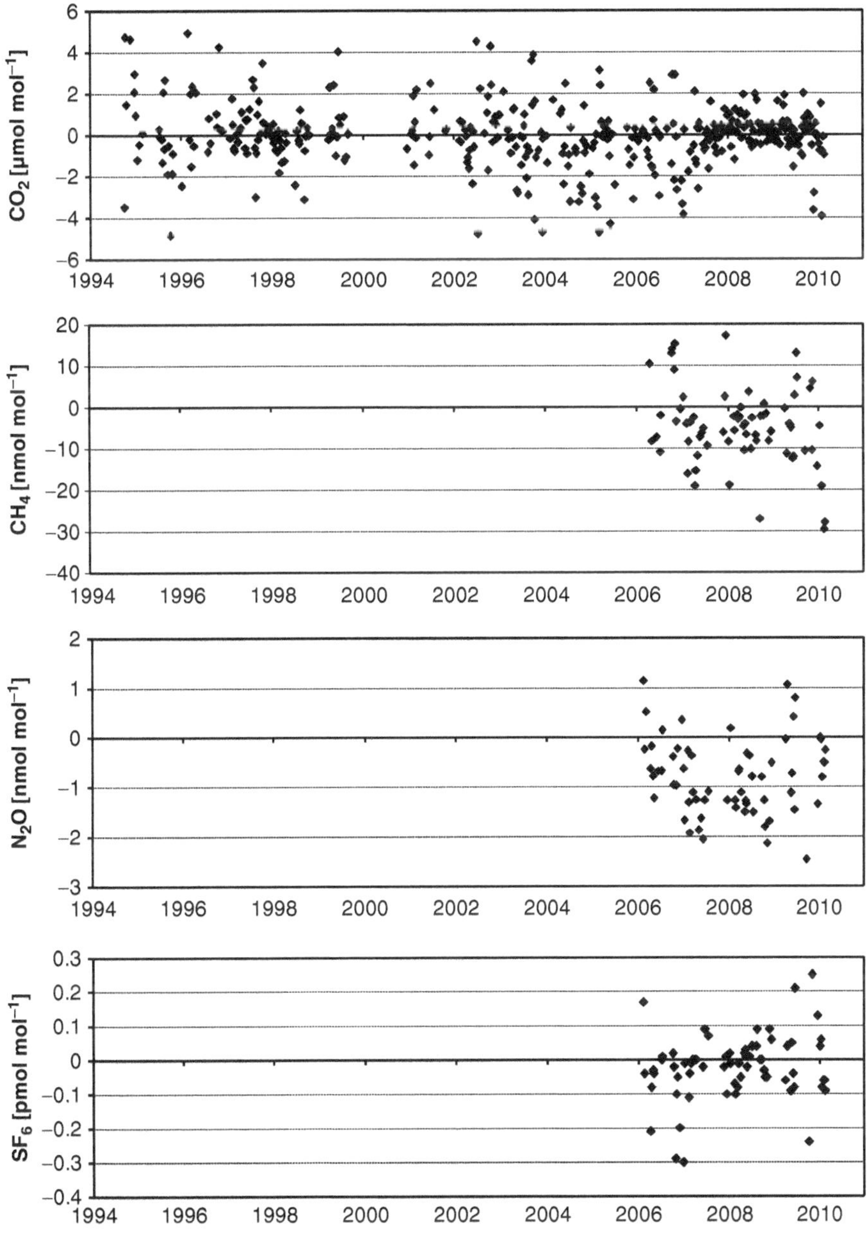

Fig. 2.4 Temporal variations of flask – in situ difference (flask minus in situ mixing ratio) for the different greenhouse gases measured at Hegyhátsál

accurate, although their accuracy does not satisfy the WMO requirement yet. The flask–in situ difference in the nitrous oxide measurements call the attention again to a possible scale problem, to the overestimation of the actual mixing ratio by the in situ system.

At several monitoring sites a clean air sector is defined. Data measured in the air masses arriving from the complement sector are rejected because of the potential contamination from anthropogenic sources. In the case of the Hungarian monitoring stations no clean air sector can be defined because the stations are located in the middle of a highly industrialized, densely populated continent where the natural and anthropogenic sources (and natural sinks in the case of CO_2) are distributed more or less evenly around them. The topography of the Carpathian Basin also helps the mixing of air masses arriving from different directions. The sectoral distribution of the atmospheric CO_2 mixing ratio does not show any characteristic feature at either monitoring sites (Haszpra 1999b). At several monitoring stations, minimum wind speed criteria are applied to avoid the dominance of local sources. Both Hungarian measuring sites are located in climatologically calm regions, where the annual average wind speeds are 3.5 m s^{-1} (Hegyhátsál) and 1.7 m s^{-1} (K-puszta), respectively. Therefore, a data selection based on wind speed is not introduced either. Under the given environmental conditions, any algorithmic data selection procedure might lead to some sort of bias. Therefore, the data measured are manually checked using data visualization software, and only the technically false data are removed from the data series.

2.5 Representativeness of the Monitoring Sites

Representativeness is the extent to which a set of measurements taken in a space-time domain reflects the actual conditions in the same or different space-time domain taken on a scale appropriate for a specific application (Nappo et al. 1982). Knowledge of the representativeness of the measurements is essential in the interpretation of the data. The parallel CO_2 concentration measurements performed at Hegyhátsál and K-puszta between 1994 and 1999 offered a possibility to get an insight into the horizontal representativeness of these measurements.

For the estimation of the horizontal representativeness of the measurements and its temporal variation, the average differences between the mixing ratios measured at the two stations were calculated for every hour for summer (June–August) and winter (December–February) separately (Fig. 2.5). In summer daytime hours (12–16 h local standard time, LST), the difference between the two stations is negligible (0.06 μmol mol^{-1}). In winter and in the transition seasons (not presented in Fig. 2.5), the mixing ratio at Hegyhátsál is higher by 1.84 and 1.55 μmol mol^{-1}, respectively, on average. It is suspected to be the consequence of the higher anthropogenic source density in Western Europe closer to Hegyhátsál than to K-puszta in the prevailing wind direction and the more limited atmospheric mixing in the Carpathian Basin, especially in winter.

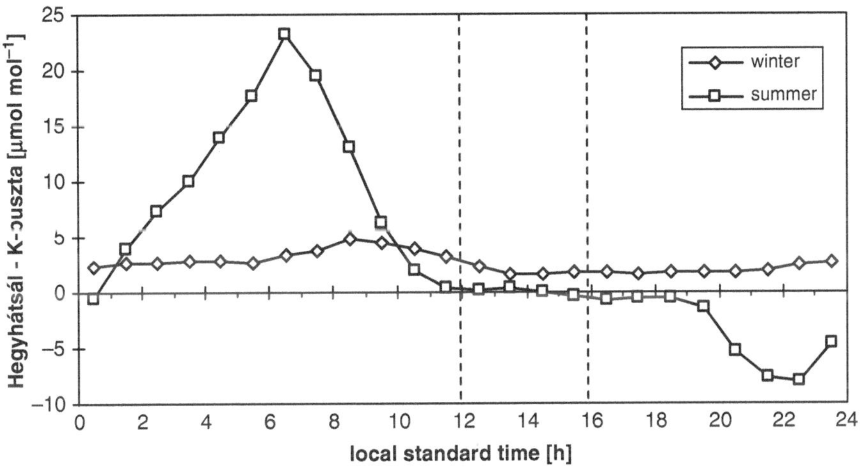

Fig. 2.5 Daily variations of the average differences between the mixing ratios measured at K-puszta and Hegyhátsál at 10 m elevation above the ground in winter and summer

During summer nights, the difference between the two monitoring sites is remarkable; it is above 20 μmol mol^{-1} in the early morning hours. In the evening, the mixing ratio is higher at K-puszta. This region is climatologically calmer than the other, and a stable boundary layer may form here earlier in the evening in which CO_2 starts to accumulate. Later, the intensity of the soil/vegetation respiration takes over the control on the CO_2 content of the boundary layer. The respiration at Hegyhátsál is more intensive than at the sandy K-puszta. As a consequence of this process, the mixing ratio becomes higher at Hegyhátsál, and the difference is the most remarkable in summer. The smallest and almost constant difference between the two sites can be observed in the early afternoon hours.

Figure 2.6 shows the average vertical CO_2 mixing ratio profiles measured at the Hegyhátsál tall tower site for different seasons and for different periods of the day. In Fig. 2.7, the daily variation of the correlation between the mixing ratios measured at 10 and 115 m is presented. Both figures suggest that the representativeness of the early afternoon measurements is high enough to allow the use of the data in regional scale studies. Therefore, in all studies, except for those focusing on the diurnal variation, the data measured between 12 and 16 o'clock LST are used, and this period is defined as "early afternoon" in our studies.

Gloor et al. (2001) developed a simple method for the estimation of the horizontal representativeness of CO_2 concentration measurements based on the tall tower measurements in Wisconsin, U.S.A. Supposing that the parameters estimated for that region are also valid for our region, the area from where the signal contribution to the measured mixing ratio is greater than 2.5% is 200,000–300,000 km^2 depending on the season. The influencing area is not concentric around the station. Due to the prevailing northwesterly wind direction, it is somewhat more extended to the northwest.

Figure 2.6 shows that due to the intensive vertical mixing of the atmosphere in summer, the mixing ratio at 115 m above the ground is almost the same as at 10 m.

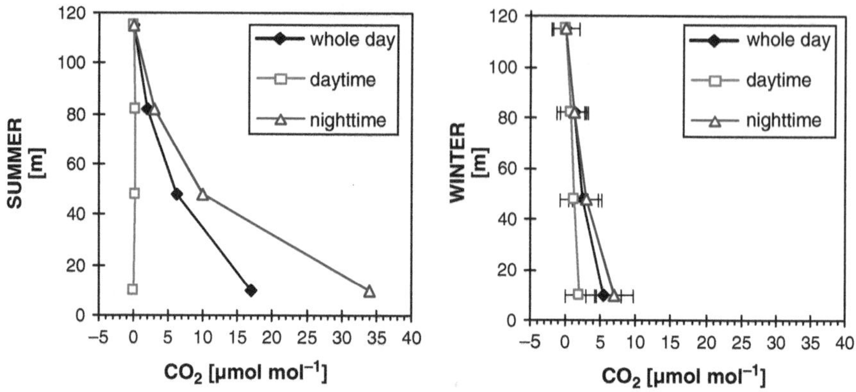

Fig. 2.6 Mean vertical CO_2 mixing ratio profiles in different seasons and in different periods of the day at Hegyhátsál, relative to the values measured at 115 m

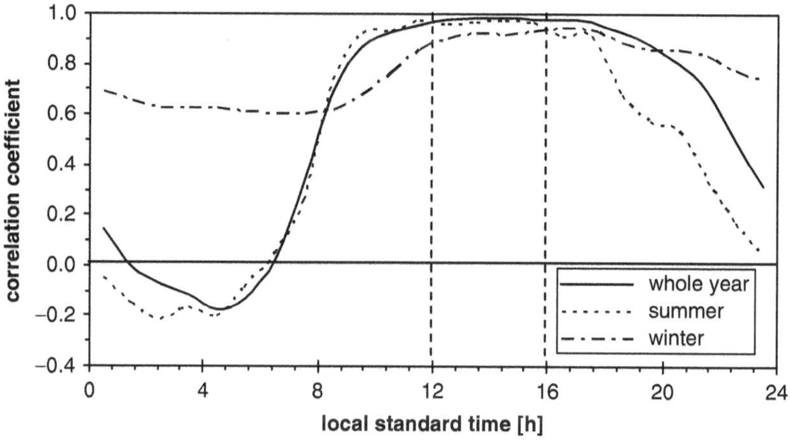

Fig. 2.7 Daily variations of the correlation coefficient between the mixing ratios measured at 10 and 115 m

The measurements presented by Bakwin et al. (1995) suggest that the mixing ratio measured at about 100 m above the ground may not differ from the characteristic value of the planetary boundary layer (PBL) by more than a few tenths of μmol mol^{-1}. Therefore, it can also be assumed that during summer, the measurements carried out at 10 m elevation do not underestimate the regional CO_2 mixing ratio of the boundary layer by much more. Following the same reasoning, it can also be assumed that in winter time, the measurements close to the surface (10 m) in a low-elevation temperate continental region do not overestimate the mixing ratio of the PBL by more than 2–3 μmol mol^{-1}. Combining the CO_2 mixing ratio data from air samples taken on board of a small aircraft over the tower and those measured in situ on the tower, as well as knowing the height of the PBL (MARS archive,

European Centre for Medium-Range Weather Forecasts), the average mixing ratio of the PBL can be calculated. In the summer half year (April–September), this value is underestimated by the measurements at 115 m above the ground on the tower by 0.84 μmol mol^{-1} (standard deviation: 1.79 μmol mol^{-1}), indicating the intensive CO_2 uptake at the surface by the vegetation. In the winter half year (October–March), the top monitoring level on the tower overestimates the PBL average by 1.51 μmol mol^{-1} (standard deviation: 3.68 μmol mol^{-1}) on average. This average deviation would be reduced to 0.61 μmol mol^{-1} if we rejected those three profiles from the available 38 where the deviation is bigger than +10 μmol mol^{-1}. These profiles indicate either internal stratification in the PBL or problems in the determination of the PBL height. (It should be noted that some sort of "fair weather bias" cannot be excluded in the case of the above data because air samples could only be taken when the weather was suitable for flying).

The time scale is a critical issue in the evaluation of the representativeness. The longer the averaging period the more extended the represented area. As we will see in Chapter 3, the annual growth rate of the CO_2 mixing ratio calculated from the measurements performed at K-puszta/Hegyhátsál correlates well with the global growth rate calculated from the data of the most remote monitoring sites. In this sense, the CO_2 data of the Hungarian monitoring sites are globally representative.

The data series are too short yet to estimate the representativeness of the non-CO_2 measurements directly. Taking into account the long atmospheric residence time of those gases and the less intensive interaction with the surface, we can suppose that the discussion above is also more or less valid for those measurements.

2.6 Summary

Monitoring of atmospheric greenhouse gases was started in Hungary as early as 1981 with the measurements of CO_2 at K-puszta. During the past almost 3 decades, several publications presented the data to the scientific community (e.g. Haszpra 1995, 1999a, b; Haszpra et al. 2008), and they have infiltrated into regional and global studies (e.g. Bakwin et al. 2004; Geels et al. 2007; Ramonet et al. 2010). K-puszta and Hegyhátál are regional GAW stations. The results of the atmospheric CO_2 measurements are publicly available at the World Data Centre for Greenhouse Gases of WMO (WMO WDCGG – http://gaw.kishou.go.jp/wdcgg/). The data from Hegyhátsál are also used for the preparation of NOAA's GLOBALVIEW-CO2 data product (http://www.esrl.noaa.gov/gmd/ccgg/globalview/). In addition to WMO WDCGG, the monthly mean CO_2 mixing ratios are reported to the Carbon Dioxide Information Analysis Center, Oak Ridge, U.S.A. (http://cdiac.esd.ornl.gov/). The data measured by NOAA from the air samples taken at Hegyhátsál are publicly available from NOAA (http://www.esrl.noaa.gov/gmd/ccgg/iadv/) and WMO WDCGG. Hegyhátsál also reports its measurements to the CarboEurope database (http://ce-atmosphere.lsce.ipsl.fr/database/index_database.html). In situ non-CO_2 greenhouse gas data are available only there yet. Hegyhátsál has its own website (http://nimbus.elte.hu/hhs/) where, in addition to other information, quasi-real-time data are also accessible.

Hegyhátsál tall tower site participated and participates in several international research projects. Since 2000, it has been a permanent participant (full-right partner or subcontractor) in several European projects in the field of greenhouse gas research (AEROCARB – http://aerocarb.lsce.ipsl.fr/, CHIOTTO – http://www.chiotto.org/, CarboEurope-IP – http://www.carboeurope.org/, IMECC – http://imecc.ipsl.jussieu.fr/, GEOMON – http://geomon.ipsl.jussieu.fr/, etc.).

In addition to the continuous work for WMO and NOAA, Hegyhátsál is also a potential node for the forming unified European greenhouse gas observation system, ICOS (http://www.icos-infrastructure.eu/).

Acknowledgments During the years, Hungarian atmospheric greenhouse gas monitoring received significant financial support from the US–Hungarian Scientific and Technological Joint Fund (J.F. 162, J.F. 504), the European Commission (AEROCARB – EVK2-CT-1999-00013, CHIOTTO – EVK2-CT-2002-00163, CarboEurope-IP – GOCE-CT-2003-505572, IMECC – RII3 026188), the European Union INTERREG IIIB CADSES program (5D038), as well as from Hungarian funding organizations (Hungarian Scientific Research Fund – T7282, T042941, CK77550, Hungarian Ministry of Economy and Transport – GVOP-3.2.1.-2004-04-0107/3.0, Hungarian Ministry for Environment – 027739-01/2001, K0441482001, K-36-02-00010H). The authors thank Pat Lang, Tom Conway, and Molly Heller at NOAA ESRL Carbon Cycle and Greenhouse Gases group for the analyses and logistics of the flask samples. The authors also thank Michel Ramonet, Martina Schmidt, and their colleagues at Laboratoire des Sciences du Climat et l'Environnement, Gif-sur-Yvette, France, for the analyses of the aircraft air samples. In addition to original material, this chapter also contains, mostly updated, sections and figures from Haszpra (1999b) and Haszpra et al. (2001, 2008).

References

Bakwin PS, Tans PP, Zhao C, Ussler W III, Quesnell E (1995) Measurements of carbon dioxide on a very tall tower. Tellus 47B:535–549

Bakwin PS, Davis KJ, Yi C, Wofsy SC, Munger JW, Haszpra L, Barcza Z (2004) Regional carbon dioxide fluxes from mixing ratio data. Tellus 56B:301–311

Barcza Z (2001) Long term atmosphere/biosphere exchange of CO_2 in Hungary. Ph.D. Thesis, Eötvös Loránd University, Department of Meteorology, Budapest. http://nimbus.elte.hu/~bzoli/thesis/

Brown CW, Keeling CD (1965) The concentration of atmospheric carbon dioxide in Antarctica. J Geophys Res 70:6077–6085

Geels C, Gloor M, Ciais P, Bousquet P, Peylin P, Vermeulen AT, Dargaville R, Aalto T, Brandt J, Christensen JH, Frohn LM, Haszpra L, Karstens U, Rödenbeck C, Ramonet M, Carboni G, Santaguida R (2007) Comparing atmospheric transport models for future regional inversions over Europe – Part 1: mapping the atmospheric CO_2 signals. Atmos Chem Phys 7:3461–3479

Gloor M, Bakwin P, Hurst D, Lock L, Draxler R, Tans P (2001) What is the concentration footprint of a tall tower? J Geophys Res 106D:17831–17840

Haszpra L (1995) Carbon dioxide concentration measurements at a rural site in Hungary. Tellus 47B:14–22

Haszpra L (1999a) Measurement of atmospheric carbon dioxide at a low elevation rural site in Central Europe. Időjárás 103:93–106

Haszpra L (1999b) On the representativeness of carbon dioxide measurements. J Geophys Res 104D:26953–26960

Haszpra L, Barcza Z, Bakwin PS, Berger BW, Davis KJ, Weidinger T (2001) Measuring system for the long-term monitoring of biosphere/atmosphere exchange of carbon dioxide. J Geophys Res 106D:3057–3070

Haszpra L, Barcza Z, Hidy D, Szilágyi I, Dlugokencky E, Tans P (2008) Trends and temporal variations of major greenhouse gases at a rural site in Central Europe. Atmos Environ 42:8707–8716

Hensen A, van den Bulk WCM, Vermeulen AT, Wyers GP (1997) CO_2 exchange between grassland and the atmosphere. Report ECN-C--97-032

Kitzis DC, Zhao C (1999) CMDL/Carbon Cycle Gases Groupstandards preparation and stability. Technical Memorandum ERL CMDL-14, National Oceanic and Atmospheric Administration, Boulder, CO. http://www.esrl.noaa.gov/gmd/ccgg/refgases/airstandard.html

Köhler A (1974) Bemerkungen über die Messung der CO_2-Konzentration in Gegenwart von Wasserdampf. Staub-Reinhalt. Luft 34:305–307

Manning AC, Jordan A, Levin I, Schmidt M, Neubert REM, Etchells A, Steinberg B, Ciais P, Aalto T, Apadula F, Brand WA, Delmotte M, Giorgio di Sarra A, Hall B, Haszpra L, Huang L, Kitzis D, van der Laan S, Langenfelds RL, Leuenberger M, Lindroth A, Machida T, Meinhardt F, Moncrieff J, Morguí JA, Necki J, Patecki M, Popa E, Ries L, Rozanski K, Santaguida R, Steele LP, Strom J, Tohjima Y, Thompson RL, Vermeulen A, Vogel F, Worth D (2009). Final report on CarboEurope 'Cucumber' intercomparison programme. http://cucumbers.webapp2.uea.ac.uk/documents/CucumberFinalReport_Final.pdf

Mészáros E (1978) Background air pollution monitoring. Prepared for the World Meteorological Organization and the United Nations Environment Programme training courses. Meteorological Service of Hungary

Nappo CJ, Caneill JY, Furman RW, Gifford FA, Kaimal JC, Kramer ML, Lockhart TJ, Pendergast MM, Pielke RA, Randerson D, Shreffler JH, Wyngaard JC (1982) Workshop on the representativeness of meteorological observations. Bull Am Meteorol Soc 63:761–764, June 1981, Boulder, CO

Pales JC, Keeling CD (1965) The concentration of atmospheric carbon dioxide in Hawaii. J Geophys Res 70:6053–6076

Ramonet M, Ciais P, Aalto T, Aulagnier C, Chevallier F, Cipriano D, Conway TJ, Haszpra L, Kazan V, Meinhardt F, Paris JD, Schmidt M, Simmonds P, Xueref-Rémy I, Necki JN (2010) A recent build-up of atmospheric CO_2 over Europe. Part 1: observed signals and possible explanations. Tellus 62B:1–13

Stefanovits P (1971) Brown forest soils of Hungary. Akadémiai Kiadó, Budapest

Tans PP (1991) An observational strategy for assessing the role of terrestrial ecosystems in the global carbon cycle: scaling down to regional levels. In: Ehleringer J, Field C (eds) Scaling processes between leaf and landscape levels. Academic, San Diego, CA

Thompson RL, Manning AC, Gloor E, Schultz U, Seifert T, Hänsel F, Jordan A, Heimann M (2009) In-situ measurements of oxygen, carbon monoxide and greenhouse gases from Ochsenkopf tall tower in Germany. Atmos Meas Tech 2:573–591

van der Laan S, Neubert REM, Meijer HAJ (2009) A single gas chromatograph for accurate atmospheric mixing ratio measurements of CO_2, CH_4, N_2O, SF_6 and CO. Atmos Meas Tech 2:549–559

WMO (1974) WMO operational manual for sampling and analysis techniques for chemical constituents in air and precipitation. WMO No. 299

WMO (2009) 14th WMO/IAEA meeting of experts on carbon dioxide, other Greenhouse gases and related tracers measurement techniques, WMO GAW Report 186, Helsinki, Finland, 10–13 Sept 2007

Zhao CL, Bakwin PS, Tans PP (1997) A design for unattended monitoring of carbon dioxide on a very tall tower. J Atmos Oceanic Technol 14:1139–1145

Chapter 3
Trends and Temporal Variations of Major Greenhouse Gases at a Rural Site in Central Europe*

László Haszpra, Zoltán Barcza, István Szilágyi, Ed Dlugokencky, and Pieter Tans

Abstract In this chapter, the trends and temporal variations of four major greenhouse gases (carbon dioxide (CO_2), methane (CH_4), nitrous oxide (N_2O), sulfur hexafluoride (SF_6)) measured at Hegyhátsál, Hungary, are analyzed on the basis of in situ and flask measurements. The long-term trends observed closely follow the global tendencies. The relatively small positive offset can be attributed to the European anthropogenic sources. The seasonal cycles are basically governed by that in the atmospheric mixing; however, in the case of CO_2 and N_2O, it is also modulated by the temporal variation in the biological activity. A secondary maximum in SF_6 mixing ratio in summer may indicate the additional contribution of the seasonally changing circulation pattern. The daily cycles are dominated by the diurnal variation in the vertical mixing of the atmosphere. However, in the case of CO_2, the diurnal cycle in the biospheric uptake/release is the governing process, especially in the growing season. The lack of a diurnal cycle in the mixing ratio of the exclusively anthropogenic SF_6 indicates that there is no notable anthropogenic activity in the

---

*Citation: Haszpra, L., Barcza, Z., Szilágyi, I., Dlugokencky, E., Tans, P., 2010: Atmospheric trends and fluctuations – Trends and temporal variations of major greenhouse gases at a rural site in Central Europe. In: Atmospheric Greenhouse Gases: The Hungarian Perspective (Ed.: Haszpra, L.), pp. 29–47.

L. Haszpra (✉)
Hungarian Meteorological Service, H-1675 Budapest, P.O. Box 39, Hungary
e-mail: haszpra.l@met.hu

Z. Barcza
Department of Meteorology, Eötvös Loránd University, H-1117 Budapest, Pázmány P. s. 1/A, Hungary

I. Szilágyi
Chemical Research Center of the Hungarian Academy of Sciences, H-1025 Budapest, Pusztaszeri út. 59–67., Hungary

E. Dlugokencky and P. Tans
National Oceanic and Atmospheric Administration, Earth System Research Laboratory, 325 Broadway, Boulder, CO 80305, USA

L. Haszpra (ed.), *Atmospheric Greenhouse Gases: The Hungarian Perspective*,
DOI 10.1007/978-90-481-9950-1_3,

influence area of the station, which supports that Hegyhátsál tall tower can be considered to be a rural monitoring site as free from direct anthropogenic pollution as it is possible in Central Europe.

Keywords Carbon dioxide • Methane • Nitrous oxide • Sulfur hexafluoride • Trend • Temporal variation

3.1 Introduction

The atmosphere of the Earth contains trace constituents absorbing radiation in the range of the thermal radiation of the planet. The energy absorbed in the atmosphere in this way is radiated partly back increasing the temperature of the surface. The major natural trace constituents warming the planet are water vapor (H_2O), carbon dioxide (CO_2), methane (CH_4), nitrous oxide (N_2O), and ozone (O_3), and they are commonly called as greenhouse gases (GHGs). Anthropogenic activity can directly influence the concentration of CO_2, CH_4, N_2O; it releases GHGs of exclusively anthropogenic origin (e.g. halogenated compounds like CFC, HCFC, PFC, etc.), and it may also modify O_3 and H_2O content of the atmosphere indirectly through physical and chemical processes. Any change in the atmospheric amount of GHGs results in the redistribution of the energy in the atmosphere–surface system and leads to a change in the surface temperature. It causes global climate change statutorily; however, the relation between the concentration of GHGs and the climate is extremely complex due to the high number, and partly poorly known, feedbacks and interactions.

The amount of water vapor in the atmosphere is too high and turns over too rapidly to be alterable directly by anthropogenic activity. However, ice core data and contemporary direct atmospheric measurements show increasing and significant anthropogenic influence on the amount of the other GHGs (Jansen et al. 2007). The risks of the potential global climate change generated by this (and expected future) increase in the amount of atmospheric GHGs are the biggest challenge humankind faces today.

The atmospheric residence times of CO_2, CH_4, N_2O, and halogen-containing GHGs are rather long that allow them to distribute more or less evenly in the global atmosphere with little regional differences on a longer time scale. Therefore, it is understandable why most of the studies analyze only the global tendencies, including the large spatial scale (e.g. continental or hemispheric) differences at best. There are only a few studies attempting to compare measurements of individual monitoring sites with the global tendencies and to reveal possible local/regional features, seasonal or diurnal variations, especially in the case of non-CO_2 GHGs. In this chapter, we analyze the CO_2, CH_4, N_2O, and sulfur hexafluoride (SF_6) measurements carried out at Hegyhátsál tall tower GHG monitoring station, a rural site in Hungary, partly in cooperation with the National Oceanic and Atmospheric Administration (NOAA), USA. We also include the CO_2 measurements from K-puszta, another Hungarian monitoring site going back as early as 1981. The detailed description of the monitoring sites and their instrumentation can be found in Chapter 2.

3.2 Carbon Dioxide

Carbon dioxide is one of the most important components for life on Earth. Plants build up their organic material from CO_2 absorbed from the atmosphere through photosynthesis. In addition to photosynthesis, another way to produce energy for living is oxidation when the organism oxidizes its organic material releasing CO_2 into the atmosphere (respiration). The total amount of CO_2 exchanged between the terrestrial biosphere and the atmosphere through photosynthesis/respiration and oxidative decomposition of dead organic material is about 120 Gt carbon per year (Denman et al. 2007).

Carbon dioxide is a soluble gas. The surface layer of the ocean, as a solution, is in approximate equilibrium with the atmosphere. It can take up or release CO_2 depending on the actual conditions. Ocean currents transport the dissolved CO_2 across the globe while the surface tends to be in equilibrium with the atmosphere. While in certain geographical region the ocean is a net source of CO_2, it is a net sink at others. The total amount of CO_2 exchange between the atmosphere and the ocean is about 90 Gt carbon per year (Denman et al. 2007).

There are minor fluxes between the atmosphere and the other reservoirs through chemical weathering and volcanism. These processes have played a determining role in the formation of the contemporary atmospheric CO_2 level, but they are effective only in the very long term, during geological eras. These fluxes are well below 1 Gt carbon per year, and thus they do not influence the atmospheric CO_2 content in a short term (Prentice et al. 2001; Denman et al. 2007).

During most of the Holocene era, the photosynthesis/respiration, solution/dissolution were so finely balanced (on longer time scale) that the change in the atmospheric CO_2 content was only about 40 Gt expressed in carbon amount during 10,000 years (~20 μmol mol^{-1} increase – Jansen et al. (2007)). This finely balanced system has been essentially disturbed by the anthropogenic release of CO_2, especially since the beginning of the industrial era. The quickly growing use of fossil fuel, certain industrial processes, deforestation, and other land use change have released about 450 Gt carbon into the atmosphere in the form of CO_2 since the middle of the 18th century. About half of this amount (~200 Gt carbon) has remained in the atmosphere (estimated on the base of Sabine et al. (2004) and Denman et al. (2007)) increasing the atmospheric CO_2 mixing ratio from ~280 to 385 μmol mol^{-1} by the end of 2008 (Denman et al. 2007; WMO 2009). The rest has been absorbed by the ocean, which accommodates to the atmospheric concentration as a solution (Sabine et al. 2004), and by the terrestrial biosphere, for which the explanation is not as obvious as that (Denman et al. (2007) and references therein). The present imbalance in the atmospheric CO_2 budget is 3.2–4.1 Gt year^{-1} (Denman et al. 2007), expressed in carbon mass, orders of magnitude larger than that in the undisturbed atmosphere used to be.

Figure 3.1 shows the diurnal variation in the CO_2 mixing ratio in the different seasons of the year at Hegyhátsál tall tower monitoring site. The atmospheric concentration clearly indicates the cyclic photosynthesis–respiration process that results

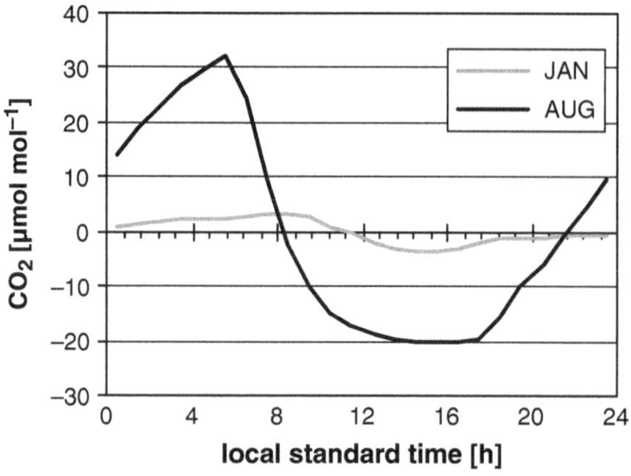

Fig. 3.1 Monthly mean diurnal variation in CO_2 mixing ratio at Hegyhátsál (2000–2009), at 10 m above the ground, in January (*gray line*) and August (*black line*), relative to the daily average

in high daily amplitude, especially in summer. The remarkable difference in the atmospheric mixing depth (typically 50–200 m during night and 1,500–2,000 m during summer afternoon) also contributes to the formation of the high daily amplitude. Such a diurnal variation in the mixing ratio can hardly be seen at remote oceanic or arctic sites where there is no surrounding vegetation near the monitoring site and the variation in the mixing depth is smaller. As the sources and sinks are located at the surface, the amplitude of the diurnal cycle attenuates with height. While the mean diurnal amplitude is 52 μmol mol^{-1} at 10 m in August, it is only 16 μmol mol^{-1} at 115 m above the ground. In winter, the amplitudes are proportionally lower, it is only 7 μmol mol^{-1} on average in December–January at 10 m (see also Fig. 2.6).

As it was demonstrated in Chapter 2, the spatial representativeness of the night-time measurements is rather low due to the shallow boundary layer and the modest atmospheric mixing; therefore, in the further data analysis, only the early afternoon (12–16 h local standard time (LST)) values are used.

Hegyhátsál is located in the temperate zone, where the activity of the vegetation also exhibits an annual cycle. The resulting annual cycle in the atmospheric CO_2 mixing ratio can be seen in Fig. 3.2, but it is also reflected in the seasonally dependent daily concentration amplitude in Fig. 3.1. Although the vegetation becomes a net CO_2 sink only in late March, early April, the mixing ratio decreases from January. This phenomenon is explained by the more vigorous vertical mixing in the late winter/early spring months than in the middle of the winter injecting relatively clean, CO_2-depleted free tropospheric air into the boundary layer (for further details on this phenomenon read Chapter 4).

Chapter 2 also showed that the spatial representativeness of the early afternoon mixing ratio values are large enough to allow the combination of the data from the two Hungarian CO_2 monitoring stations (K-puszta and Hegyhátsál). As atmospheric CO_2 monitoring was started at K-puszta in the middle of 1981 (see Chapter 2), by the time of writing a 29-year long data series are available for the evaluation.

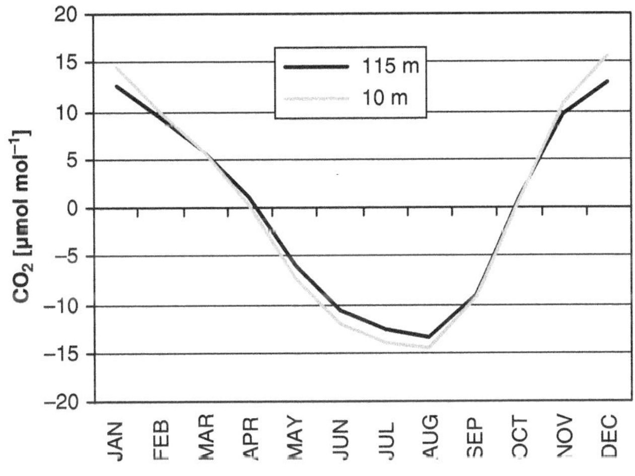

Fig. 3.2 Mean seasonal variation in CO_2 mixing ratio at Hegyhátsál, at 10 and 115 m above the ground, relative to the annual average

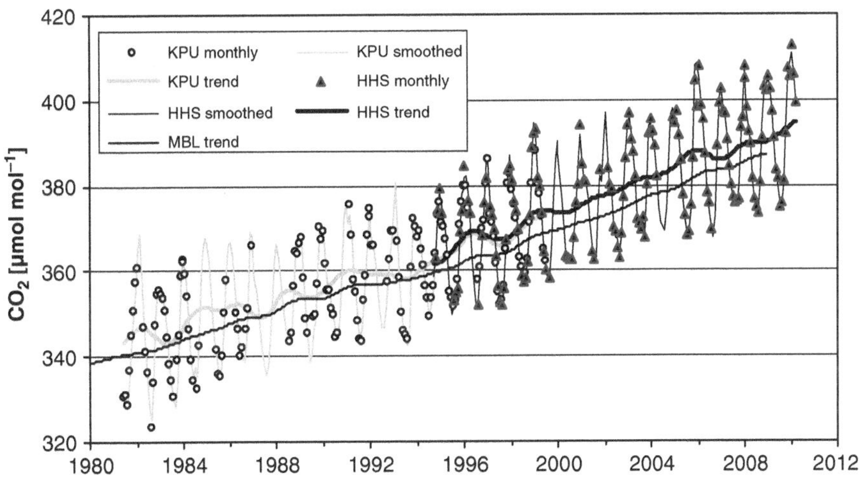

Fig. 3.3 The long-term trend in the atmospheric CO_2 mixing ratio based on the combined data series measured at the same elevation (10 m above the ground) at the two Hungarian stations (K-puszta – KPU, Hegyhátsál – HHS). The figure also shows the marine boundary layer reference mixing ratio for Hegyhátsál (MBL trend) (GLOBALVIEW-CO2 2009). Trends and smoothed curves are calculated using Thoning's method (Thoning et al. 1989)

Figure 3.3 presents the two data series. From the beginning of the measurements in summer, 1981, until spring, 2010, the CO_2 mixing ratio increased from 343 to 395 µmol mol^{-1} at a fluctuating but generally positive growth rate. Taking into account the meridional gradient in the background CO_2 mixing ratio (e.g. WDCGG 2009), it is more reasonable to compare the Hungarian measurements with the zonally representative marine boundary layer mixing ratio (GLOBALVIEW-CO2 2009)

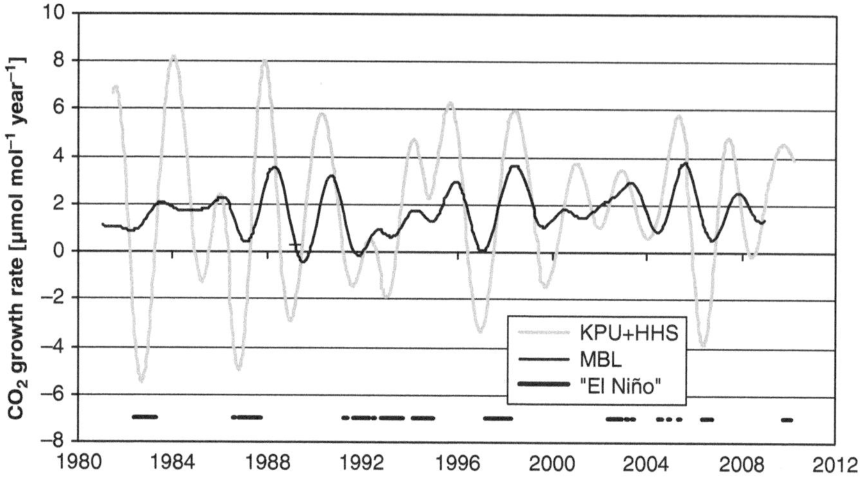

Fig. 3.4 Covariance between the growth rate of the atmospheric CO_2 mixing ratio at K-puszta–Hegyhátsál (KPU + HHS) and in the characteristic background atmosphere represented by the marine boundary layer reference mixing ratio (MBL) (GLOBALVIEW-CO2 2009). At the bottom of the figure, the El Niño periods are also indicated

than with an overall global background average. On average, the Hungarian measurements exceed this reference value by 3.0 and 4.3 µmol mol^{-1} at K-puszta (1981–1999) and Hegyhátsál (1994–2009), respectively. Europe is mostly dominated by the westerly winds and thus the offset must be the contribution of the Western European anthropogenic sources (Chevillard et al. 2002; Ramonet et al. 2010). The measurements are also compared with those at three European coastal NOAA sampling sites located in different directions from Hungary: Mace Head (MHD – Atlantic Ocean, 53°19′N, 9°54′W), Baltic Sea (BAL – 55°21′N, 17°13′E), and Black Sea (BSC – 44°10′N, 28°40′W). While Mace Head station on the west coast of the continent measures 4.0 µmol mol^{-1} lower values than Hegyhátsál on 14-year (1995–2008) average, the sampling site on Black Sea coast, southeast off Hegyhátsál, reports values higher by 4.3 µmol mol^{-1}. Baltic Sea coastal station, located north off the Hungarian site, shows a negligible surplus (0.1 µmol mol^{-1}).

The course of the growth rate is not a monotonic function of time, it shows significant interannual variability, typically between −4 and +6 µmol mol^{-1} year^{-1} (Fig. 3.4). The growth rate observed at the stations correlates reasonably well with that measured at remote sites but the range of the fluctuation is wider here, in the middle of Europe. It is known for quite some time that the fluctuation in the growth rate is primarily generated by the El Niño/Southern Oscillation through its effect on the biosphere (Bacastow 1976; Jones et al. 2001). The more intense fluctuation indicates that the Central European biosphere plays an important role in the global fluctuation in spite of the fact that it is located on the globe almost opposite to the El Niño region. It takes some time for the biosphere and the atmospheric CO_2 mixing ratio to react to the El Niño phenomenon. Figure 3.5 shows the correlation coefficient between the growth rates measured and the Southern Oscillation Index

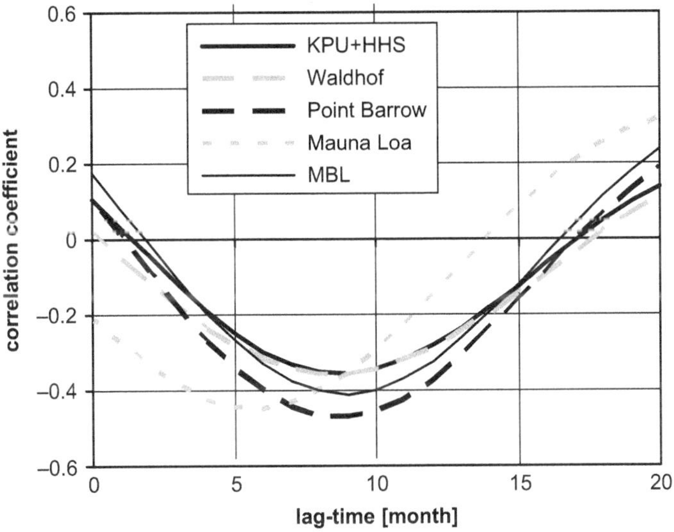

Fig. 3.5 Correlation between the growth rate of the atmospheric CO_2 mixing ratio measured at K-puszta–Hegyhátsál (KPU + HHS) and the SOI as the function of lag-time. The figure also shows the correlation function for selected northern hemispheric stations (WMO WDCGG – http://gaw.kishou.go.jp/wdcgg/) and for the characteristic background atmosphere represented by the marine boundary layer reference data (MBL) (GLOBALVIEW-CO2 2009)

(SOI) as a function of lag-time for a subtropical Pacific station (Mauna Loa, Hawaii, 19°32′N, 155°35′W, 3,397 m asl), for an Arctic station north off the Pacific Ocean (Point Barrow, Alaska, 71°19′N, 156°36′W, 11 m asl), for a Central European station not far from Hegyhátsál (Waldhof, Germany, 52°48′N, 10°46′E, 74 m asl), for Hegyhátsál, and for the marine boundary layer reference mixing ratio characteristic for the Hungarian monitoring site (data from the foreign monitoring sites are from WMO WDCGG; marine boundary layer reference is from GLOBALVIEW-CO2 (2009)). The growth rate responds to the change in SOI the fastest at Mauna Loa being the closest to the region directly affected by the El Niño phenomenon. The lag-time at the other stations being farther from the Equatorial Pacific, including Hegyhátsál, is 8–9 months.

The trend in the atmospheric CO_2 mixing ratio also shows shorter-term variability that is the result of the regional climate fluctuation. These variations will be discussed in detail in Chapter 4.

3.3 Methane

Most of CH_4 of natural origin is produced by the anaerobic decomposition of organic material. The typical source sites are wetlands, marshes, and tidal regions. The contribution of termites and ruminant wild animals are also notable. A small amount can be attributed to wildfires, geological sources, and to the ocean (Denman et al. 2007).

The growing human population has been increasing the food demand to be satisfied by extending agricultural activity and stock-raising. Both rice paddies (like marshes) and ruminant animals (e.g. cattle) are huge CH_4 sources. The growing human population has also produced increasing amount of organic waste the anaerobic decomposition of which can release CH_4 into the atmosphere. Abiotic anthropogenic CH_4 sources are biomass burning, coal mining, gas and oil industry. Although the estimations for the individual source categories are rather uncertain, it can be assumed that nowadays 60–70% of the total CH_4 emission is of anthropogenic origin. For numerical data, see e.g. Denman et al. (2007).

The two major CH_4 sinks, atmospheric oxidation (CH_4 + OH) and dry deposition, could not balance the increasing source yield, and the excess CH_4 has increased the atmospheric CH_4 mixing ratio from the pre-industrial ~700 nmol mol^{-1} to the present (end of 2008) 1797 nmol mol^{-1} global average (WMO 2009).

There were no CH_4 measurements at K-puszta; thus, our analyses can only be based on the flask and in situ measurements performed at Hegyhátsál. As the in situ measurements were started at the station only in 2006, the seasonal variation and long-term trend calculations are based on the flask air sample measurements carried out by NOAA Earth System Research Laboratory, Carbon Cycle Greenhouse Gases Group (ESRL CCGG).

Figure 3.6 shows the long-term trend in the atmospheric CH_4 mixing ratio. The decreasing growth rate from the 1980s is observed worldwide (see e.g. Dlugokencky et al. (2003)). One of the possible explanations for the slowing down is the stabilization of the sources and the formation of a steady-state concentration corresponding to the stabilized source yield and the available sink capacity (Denman et al. 2007;

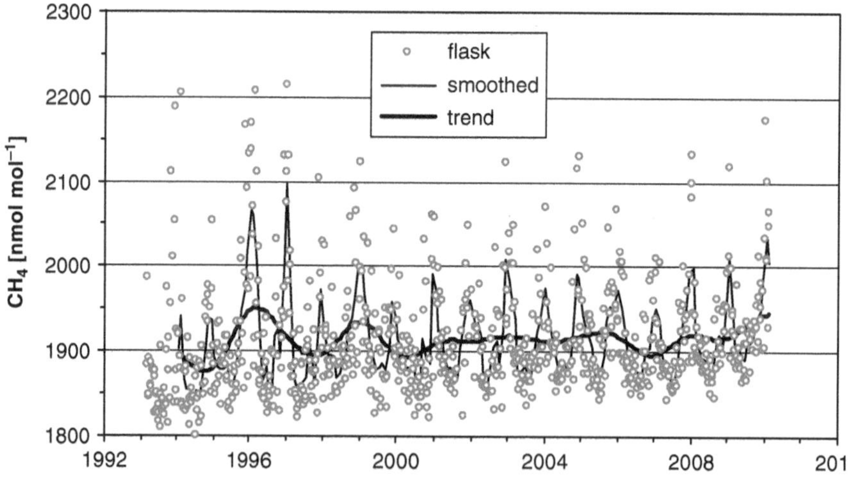

Fig. 3.6 Trend in atmospheric CH_4 mixing ratio measured at Hegyhátsál by NOAA ESRL CCGG and calculated using Thoning's method (Thoning et al. 1989) (Data for 2009 and 2010 are preliminary)

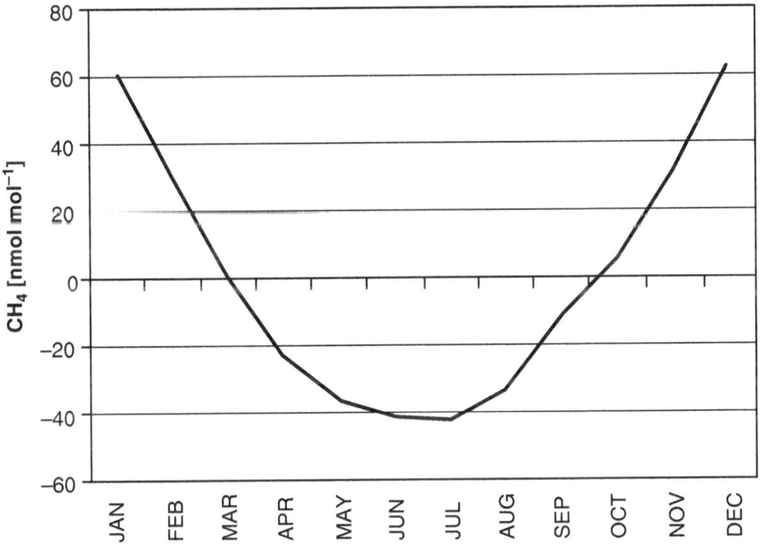

Fig. 3.7 Mean seasonal cycle of atmospheric CH_4 mixing ratio at Hegyhátsál relative to the annual average

Khalil et al. 2007). An equilibrium concentration can be formed much faster than in the case of CO_2 because there are no notable slow systems in the CH_4 cycle like the oceans in the case of CO_2. Since 2007, the growth rate has seemed to increase again. Extreme warm arctic summers and the increased tropical rainfall might be the reasons for the increased CH_4 emission (Dlugokencky et al. 2009). The recent increase in the atmospheric CH_4 mixing ratio can also be observed at Hegyhátsál indicating that the station has a global representativeness of a certain kind. The mixing ratio has increased by almost 50 nmol mol^{-1} during the past 3 years (2007–2009), but the lowest CH_4 mixing ratios of the past decade were measured at Hegyhátsál in 2006–2007. It is hard to say yet if we see the beginning of a new period of high growth rate or just fluctuation.

The CH_4 concentration measured at Hegyhátsál exceeds the latitudinal background represented by the marine boundary layer reference mixing ratio (GLOBALVIEW-CH4 2009) by 74 nmol mol^{-1} on average (1994–2008). The offset from Mace Head is somewhat lower (60 nmol mol^{-1}). The positive offset is certainly the result of the mid-continental location of the Hungarian station, which is surrounded with anthropogenic sources. It is supported by the fact that the average CH_4 mixing ratio at BSC, southeast off Hegyhátsál, is 42 nmol mol^{-1} higher than that at the Hungarian monitoring station.

The seasonal cycle shows a summer minimum and a winter maximum with an average amplitude of 104 nmol mol^{-1} (Fig. 3.7). Both the anaerobic decomposition rate (one of the major sources) and the chemical destruction (the major sink), as well as the atmospheric dilution have their maxima in summer. As the intensity of the daytime vertical mixing changes little between May and August (see Fig. 4.5), the biogenic activity should produce a second maximum in summer if it is not compensated by the sink. In the mixing ratio, no increase can be observed in summer. Without an appropriate atmospheric chemical transport model and a detailed, at least continental scale emission inventory, it is only supposed that the temperature-dependent biogenic processes play only a moderate role in the formation of the regional atmospheric CH_4 concentration; they are (over)compensated by the photochemical destruction process. In fact, Central Europe is free from marshes, swamps, or tidal regions, which can emit large amount of CH_4. Rice paddies are also not typical in this geographical region. As it can be read in Part III of this book, the Hungarian ecosystems are mostly rather weak net CH_4 sinks than sources.

The emission from less temperature-dependent sources (like most of the anthropogenic sources including the ruminants) is dispersed in a much deeper boundary layer in summer than in winter, and consequently, the input causes less concentration increase above the background in summer than in winter. The higher chemical destruction rate, due to the higher photochemical production of OH, acts in the same direction. The average annual amplitude at Hegyhátsál higher than those observed at other monitoring stations (e.g. Tsutsumi et al. (2006); Tyler et al. (2007)) may be the results of the significant difference between the winter and summer vertical mixing.

Figure 3.8 presents the daily variation of the CH_4 mixing ratio on the basis of the in situ GC measurements. The early morning maximum also indicates that the more intensive daytime mixing and chemistry overwhelms the effect of the higher temperature stimulating the anaerobic microbiological activity. The daily amplitude is about 36 nmol mol^{-1} in summer and 22 nmol mol^{-1} in winter. The later formation of the daily maximum in winter indicates the contribution of the vertical mixing that gets more vigorous later in winter than in summer.

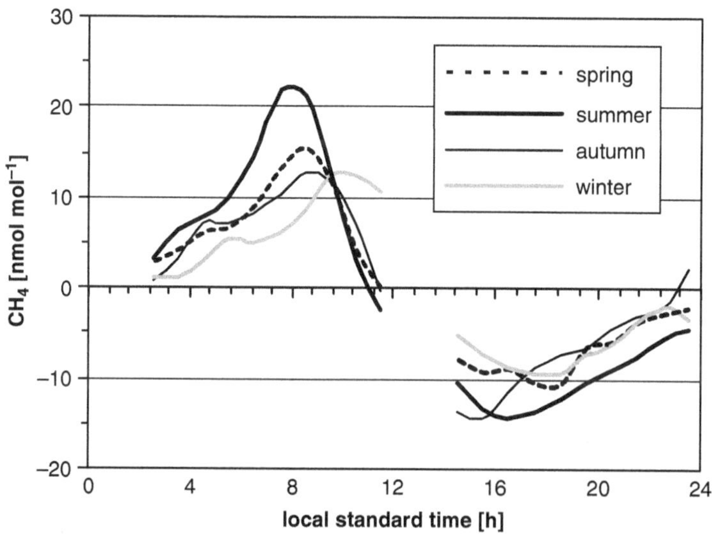

Fig. 3.8 Mean diurnal variations in atmospheric CH_4 mixing ratio at Hegyhátsál relative to the daily average. The gaps at noon and midnight are caused by the regular calibration of the monitoring system

3.4 Nitrous Oxide

The major source of atmospheric N_2O is the microbiological activity in soils and water. N_2O can be formed by both nitrification and denitrification reactions (Firestone and Davidson 1989). These processes may be significantly influenced by the agricultural practice (see also Chapter 13). Land use change, sowing and production of plants, fodder with high nitrogen fixation rate (like legumes) may increase N_2O production. Most agricultural crops are nitrogen limited, and higher yields are achieved by the application of nitrogen fertilizer. The application of nitrogen fertilizer adds fixed nitrogen to the soil, which is released by microbiological activity partly in the form of N_2O. Surface runoff, nitrogen leaching can transport the fixed nitrogen to the rivers, and ultimately to the oceans expanding the area of elevated N_2O emission. Biomass burning, coal combustion, certain industrial processes (e.g. nitric acid and synthetic fertilizer productions), waste management add smaller but not negligible source terms to the N_2O budget (Denman et al. 2007). The growing food demand of the explosively increasing human population, especially during the last 200 years, has resulted in increasing N_2O emission that is not balanced by the photochemical destruction of N_2O, the only sink of N_2O (Mosier and Kroeze 2000). This imbalance leads to a permanent increase in the atmospheric mixing ratio.

NOAA started to analyze N_2O content of the air samples taken at Hegyhátsál in summer, 1997. The past 12 years show an almost linear increase (0.78 nmol mol^{-1} $year^{-1}$) in the atmospheric mixing ratio (Fig. 3.9), close to the global average (Thompson et al. 2004; Jiang et al. 2007) (see also http://www.esrl.noaa.gov/gmd/ccgg/iadv/).

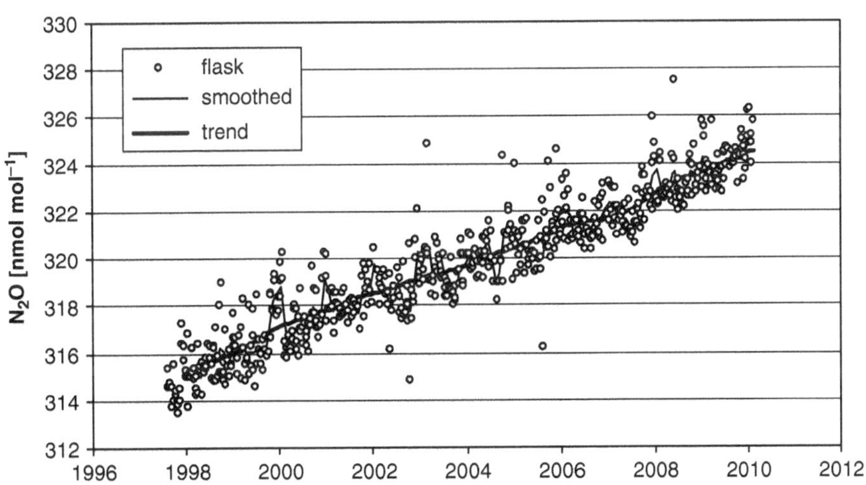

Fig. 3.9 Trend in atmospheric nitrous oxide mixing ratio measured at Hegyhátsál by NOAA ESRL CCGG and calculated using Thoning's method (Thoning et al. 1989) (Data for 2009 and 2010 are preliminary)

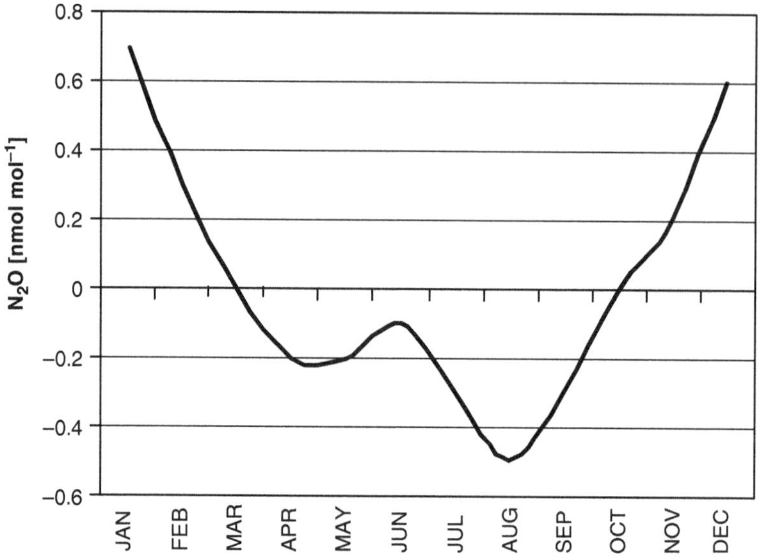

Fig. 3.10 Mean seasonal cycle in atmospheric N_2O mixing ratio at Hegyhátsál relative to the annual average

The mixing ratio is a bit higher than at the remote northern hemispheric oceanic sites (~1.4 nmol mol^{-1} offset from Mauna Loa and 1.3 nmol mol^{-1} from Mace Head on average), but it is closer to the values measured at BAL (0.52 nmol mol^{-1} offset). The mixing ratios at Hegyhátsál and BSC are practically the same on the average (−0.10 nmol mol^{-1} offset).

N_2O mixing ratio shows a weak seasonal signal with a main maximum in January and a smaller one in June (Fig. 3.10). The minimum can be observed in August. The mean range of the annual variation is only 1.2 nmol mol^{-1}, i.e. 0.4%, while the corresponding value for CO_2 and CH_4 is 6–7% of the baseline signal. The annual amplitude is higher than at the stations studied by Liao et al. (2004) and Jiang et al. (2007), but the climatic and ecological location of Hegyhátsál is also different from theirs. Bimodal annual cycle is not observed at those stations. While the winter maximum is certainly due to the limited vertical mixing and the accumulation of the anthropogenic emission in the shallow boundary layer, the somewhat elevated early summer concentration may be the result of the biospheric source (also including anthropogenic contributions like application of mineral fertilizer) having its maximum strength in summer and/or that of the seasonality in the circulation (Levin et al. 2002). The possible influence of the seasonally changing circulation on the trace gas concentrations is also supported by the SF_6 measurements (see below).

The daily amplitude of the mixing ratio is 1.7 nmol mol^{-1} in summer and 0.7 nmol mol^{-1} in winter (Fig. 3.11) with the maximum in the morning hours and minimum in the afternoon. The diurnal variation is supposed to be driven mainly by the daily variation in the atmospheric mixing.

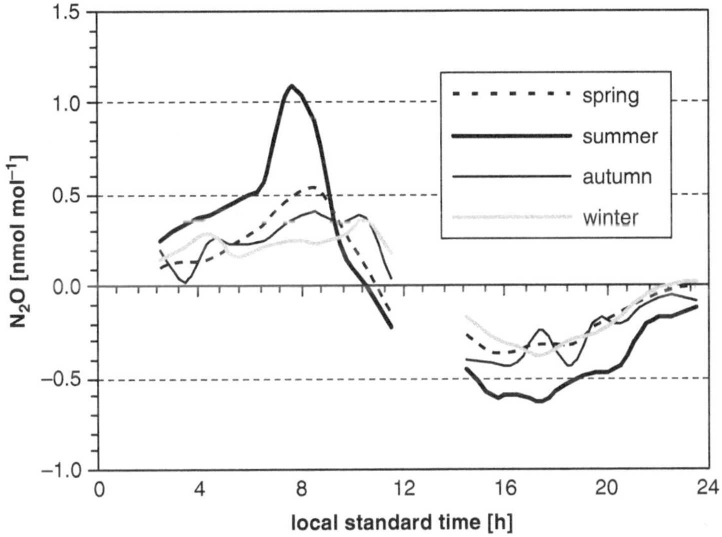

Fig. 3.11 Mean diurnal variations in atmospheric N_2O mixing ratio at Hegyhátsál relative to the daily averages. The gaps at noon and midnight are caused by the regular calibration of the monitoring system

3.5 Sulfur Hexafluoride

Sulfur hexafluoride is an extremely stable gas having practically no natural source at all (Harnisch and Eisenhauer (1998) identified a small natural source from fluorites). Its atmospheric residence time is estimated to be 3,200 years and it decomposes only in the upper atmosphere (Forster et al. 2007). Its unique physicochemical properties make this gas ideally suited for many industrial applications, predominantly in electrical insulation and switching. It is also used in the aluminum industry (Maiss and Brenninkmeijer 1998). As the spatial distribution of its sources reflects well that of the electric network, SF_6 is a good tracer of population density and energy consumption. It can also be used as an indicator of episodic anthropogenic influence on remote monitoring sites.

NOAA started to analyze SF_6 content of the air samples taken at Hegyhátsál, together with N_2O, in 1997. Since then, the mixing ratio has shown an almost linear upward trend (0.22 pmol mol^{-1} $year^{-1}$, Fig. 3.12) similar to the global one (Thompson et al. 2004). The SF_6 mixing ratio at Hegyhátsál is a bit higher than at the most remote northern hemispheric sites (0.38 pmol mol^{-1} offset from Mauna Loa on average). The mid-continental Hungarian station reports ~0.2 pmol mol^{-1} higher values than the other European sites (average offsets from BAL, BSC, and MHD are 0.15, 0.19, and 0.23 pmol mol^{-1}, respectively). The long atmospheric residence time means almost no sink for this trace compound; consequently, almost all emission accumulates in the atmosphere causing a steady increase in its atmospheric mixing ratio.

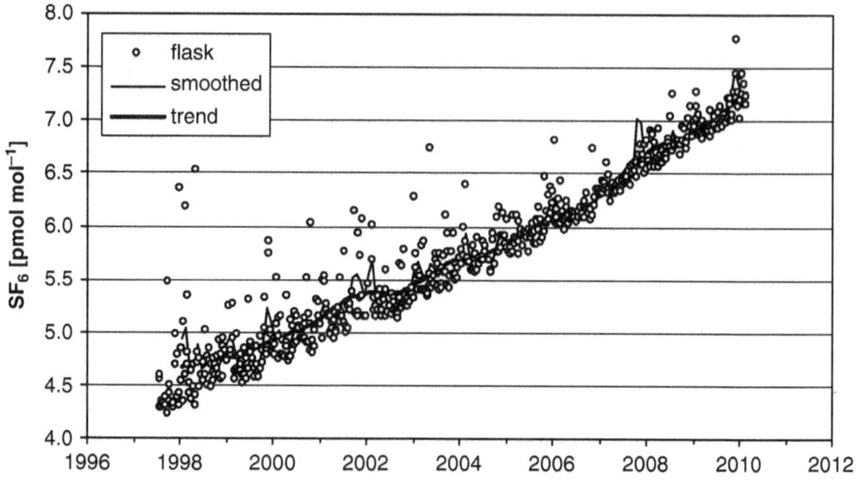

Fig. 3.12 Trend in atmospheric SF_6 mixing ratio measured at Hegyhátsál by NOAA ESRL CCGG and calculated using Thoning's method (Thoning et al. 1989) (Data for 2009 and 2010 are preliminary)

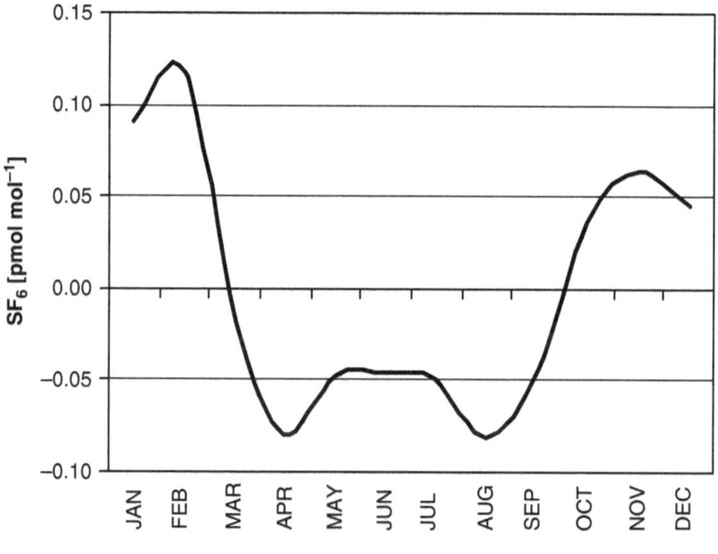

Fig. 3.13 Mean seasonal cycle in atmospheric SF_6 mixing ratio at Hegyhátsál relative to the annual average

The seasonal signal is bimodal at Hegyhátsál having the main maximum in January and a secondary one in May (Fig. 3.13). The range of the annual variation is 0.20 pmol mol^{-1}. Unlike the other GHGs reviewed so far, SF_6 does not show up any clear diurnal cycle (Fig. 3.14). The dominant sources of SF_6 are not supposed to be season-dependent; therefore, the seasonal variation can only be governed by the atmospheric mixing. While the winter primary maximum can be explained by the

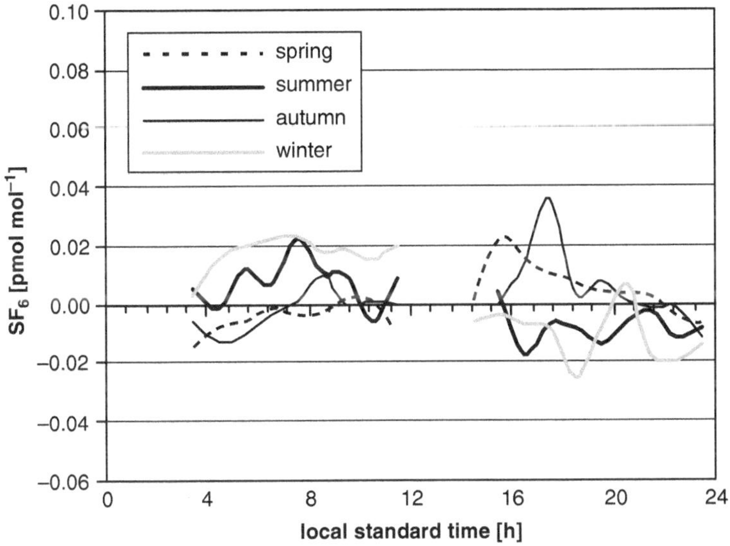

Fig. 3.14 Mean diurnal variations in atmospheric SF_6 mixing ratio at Hegyhátsál relative to the daily averages. The gaps at noon and midnight are caused by the regular calibration of the monitoring system

limited vertical mixing, the secondary summer maximum suggests a seasonal variation in the circulation, which transports more polluted, SF_6-rich air towards the station in early summer than in spring or autumn. The latitudinal difference in the northern hemispheric SF_6 mixing ratio was already presented by Maiss et al. (1996).

The obvious diurnal variation in the atmospheric mixing should produce a concentration fluctuation in the boundary layer if there are surface sources within the influence area of the station. At a mid-continental site, the daytime mixing volume may be 10–20 times bigger than during the night in summer. However, the diurnal variation is within the measurement uncertainty. The lack of daily variation in the SF_6 mixing ratio indicates that there is no remarkable source within the area represented by the measurements.

3.6 Summary and Conclusions

Although the major GHGs studied in this chapter have long residence time, it is widely supposed that no globally representative measurements can be carried out in the middle of a densely populated, highly industrialized continent. The spatial representativeness of a monitoring site strongly depends on the time scale. The results presented in this chapter prove that on annual or longer time scales, the mixing ratio values measured at Hegyhátsál closely follow the global tendencies. This is especially convincing since no special data selection technique other than taking the air samples in early afternoon when the spatial representativeness is the highest is applied.

The positive offsets observed can be attributed to the European anthropogenic sources. It means that monitoring of the global tendencies is possible at rural sites, even in the middle of Europe.

In the climate zone of Hegyhátsál, the biogenic sources of CO_2 (also sink), CH_4, and N_2O show significant seasonal variations, while in the case of anthropogenic sources the variations are modest or small. In general, the seasonal cycle of non-CO_2 GHGs are governed by the seasonal variation in the atmospheric mixing, but there are also signs of the influence of the seasonally changing circulation patterns (see SF_6). In the case of N_2O, the intensive summer biogenic activity may also contribute to the observed secondary maximum in the mixing ratio. Atmospheric transport models governed by high-resolution wind fields are needed for a more detailed interpretation of the measurement data. Receptor-oriented modeling may help the identification of major source areas (Gloor et al. 2001; Aalto et al. 2007). High-resolution anthropogenic emission data together with a transport model and with a realistic boundary layer representation are essential to distinguish between the effect of circulation and anthropogenic/biotic emission (see e.g. Murayama et al. (2004)). The long-term GHG measurements at Hegyhátsál may be a valuable verification/validation tool for atmospheric transport modelers (Geels et al. 2007). The different properties of GHGs (existence or lack of biospheric source/sink, different source locations) could potentially help us to improve the boundary layer representation and horizontal diffusion simulation in the three-dimensional atmospheric transport models.

At most remote, oceanic, high elevation, or arctic stations, no diurnal concentration variation is observed. At Hegyhátsál, the diurnal variation in the CO_2 mixing ratio is basically caused by the photosynthesis–respiration cycle; therefore, the daily amplitude strongly depends on the season following the seasonal change in the activity of the vegetation. Nevertheless, the significant diurnal variation in the atmospheric mixing depth also plays an important role. The nighttime release, comparable with the daytime uptake, is dispersed in a shallow nocturnal boundary layer, while the daytime uptake removes CO_2 from a 10–20 times bigger air volume causing much smaller concentration change in the boundary layer than the nighttime release. In the case of CH_4 and N_2O, there is no significant sink that can influence the temporal variations.

The biogenic emissions of CH_4 and N_2O are temperature-dependent (microbiological activity); consequently the release may be the highest in daytime (note that other factors, like soil water content, etc., may also be important influencing factors, but typically on a longer time scale). The observations show an opposite phase in the atmospheric mixing ratio with the minimum concentration during daytime. It means that the diurnal cycle in the mixing ratio is essentially governed by the variation in the vertical mixing. As the mountain top stations are located mostly above the planetary boundary layer and the diurnal variation in the boundary layer height is low or negligible at remote oceanic, seashore, or arctic sites, at those monitoring stations the diurnal variation in the mixing ratio is not significant or missing. The covariance between the mixing ratios and the vertical mixing at a mid-continental, low elevation site has to be taken into account and properly handled in the transport/dispersion models to avoid artifacts in the results.

Hardly any diurnal variation was observed in SF_6 mixing ratio at Hegyhátsál in spite of the significant variation in the mixing depth. It is only possible in that case if the emission is negligible in the influence area of the station. SF_6, being practically entirely artificial, is a good indicator of anthropogenic activity. As there is no sign of notable SF_6 emission in the influence area of Hegyhátsál, the station can be qualified as a rural monitoring site, as free from direct human influence as it is possible in Central Europe at all.

Acknowledgments The authors thank Pat Lang, Tom Conway, and Molly Heller at NOAA ESRL Carbon Cycle and Greenhouse Gases group for the analyses and logistics of the flask air samples. The authors especially thank Kirk Thoning, also at NOAA ESRL CCGG, for the CCGCRV data analysis software used in this study for the determination of the long-term trends and seasonal variations. During the years, the Hungarian atmospheric GHG monitoring received significant financial support from the US-Hungarian Scientific and Technological Joint Fund (J.F. 162, J.F. 504), from the European Commission (AEROCARB – EVK2-CT-1999-00013, CHIOTTO – EVK2-CT-2002-00163, CarboEurope-IP – GOCE-CT-2003-505572, IMECC – IMECC – RII3 026188), from the European Commission's INTERREG IIIB CADSES program (Carbon-Pro – 5D038), as well as from Hungarian funding organizations (Hungarian Scientific Research Fund – T7282, T042941, CK77550, Hungarian Ministry for Environment – 027739-01/2001, K0441482001, K-36-02-00010H). Data for Mauna Loa, Point Barrow, and Waldhof were retrieved from WMO World Data Centre for Greenhouse Gases (http://gaw.kishou.go.jp/wdcgg/).

In addition to original material, this chapter is essentially based on the updated sections and figures from Haszpra et al. (2008).

References

Aalto T, Hatakka J, Lallo M (2007) Tropospheric methane in northern Finland: seasonal variations, transport patterns and correlations with other trace gases. Tellus 59B:251–259

Bacastow RB (1976) Modulation of atmospheric carbon dioxide by the Southern Oscillation. Nature 261:116–118

Chevillard A, Karstens U, Ciais P, Lafont S, Heimann M (2002) Simulation of atmospheric CO_2 over Europe and western Siberia using the regional scale model REMO. Tellus 54B:872–894

Denman KL, Brasseur G, Chidthaisong A, Ciais P, Cox PM, Dickinson RE, Hauglustaine D, Heinze C, Holland E, Jacob D, Lohmann U, Ramachandran S, da Silva Dias PL, Wofsy SC, Zhang X (2007) Couplings between changes in the climate system and biogeochemistry. In: Solomon S, Qin D, Manning M, Chen Z, Marquis M, Averyt KB, Tignor M, Miller HL (eds) Climate change 2007: The physical science basis. Contribution of Working Group I to the fourth assessment report of the Intergovernmental Panel on Climate Change. Cambridge University Press, Cambridge, UK and New York, USA, pp 499–587

Dlugokencky EJ, Houweling S, Bruhwiler L, Masarie KA, Lang PM, Miller JB, Tans PP (2003) Atmospheric methane levels off: temporary pause or a new steady-state? Geophys Res Lett 30:L1992

Dlugokencky EJ, Bruhwiler L, White JWC, Emmons LK, Novelli PC, Montzka SA, Masarie KA, Lang PM, Crotwell AM, Miller JB, Gatti LV (2009) Observational constraints on recent increases in the atmospheric CH_4 burden. Geophys Res Lett 36:L18803

Firestone MK, Davidson EA (1989) Microbiological basis of NO and nitrous oxide production and consumption in soil. In: Andreae MO, Schimel DS (eds) Exchange of trace gases between terrestrial ecosystems and the atmosphere. Wiley, New York, pp 7–21

Forster P, Ramaswamy V, Artaxo P, Berntsen T, Betts R, Fahey DW, Haywood J, Lean J, Lowe DC, Myhre G, Nganga J, Prinn R, Raga G, Schulz M, Van Dorland R (2007) Changes in atmospheric constituents and in radiative forcing. In: Solomon S, Qin D, Manning M, Chen Z, Marquis M, Averyt KB, Tignor M, Miller HL (eds) Climate change 2007: The physical science basis. Contribution of Working Group I to the fourth assessment report of the Intergovernmental Panel on Climate Change. Cambridge University Press, Cambridge; UK and New York, USA, pp 129–234

Geels C, Gloor M, Ciais P, Bousquet P, Peylin P, Vermeulen AT, Dargaville R, Aalto T, Brandt J, Christensen JH, Frohn LM, Haszpra L, Karstens U, Rödenbeck C, Ramonet M, Carboni G, Santaguida R (2007) Comparing atmospheric transport models for future regional inversions over Europe - Part-1: mapping the atmospheric CO_2 signals. Atmos Chem Phys 7:3461–3479

GLOBALVIEW-CH4 (2009) Cooperative atmospheric data integration project – methane. CD-ROM, NOAA ESRL, Boulder, CO. Also available on Internet via anonymous FTP to ftp.cmdl.noaa.gov, Path: ccg/ch4/GLOBALVIEW

GLOBALVIEW-CO2 (2009) Cooperative atmospheric data integration project – carbon dioxide. CD-ROM, NOAA ESRL, Boulder, CO. Also available on Internet via anonymous FTP to ftp.cmdl.noaa.gov, Path: ccg/co2/GLOBALVIEW

Gloor M, Bakwin P, Hurst D, Lock L, Draxler R, Tans P (2001) What is the concentration footprint of a tall tower? J Geophys Res 106D:17831–17840

Harnisch J, Eisenhauer A (1998) Natural CF_4 and SF_6 on Earth. Geophys Res Lett 25:2401–2404

Haszpra L, Barcza Z, Hidy D, Szilágyi I, Dlugokencky E, Tans P (2008) Trends and temporal variations of major greenhouse gases at a rural site in Central Europe. Atmos Environ 42:8707–8716

Jansen E, Overpeck J, Briffa KR, Duplessy J-C, Joos F, Masson-Delmotte V, Olago D, Otto-Bliesner B, Peltier WR, Rahmstorf S, Ramesh R, Raynaud D, Rind D, Solomina O, Villalba R, Zhang D (2007) Palaeoclimate. In: Solomon S, Qin D, Manning M, Chen Z, Marquis M, Averyt KB, Tignor M, Miller HL (eds) Climate change 2007: The physical science basis. Contribution of Working Group I to the fourth assessment report of the Intergovernmental Panel on Climate Change. Cambridge University Press, Cambridge, UK and New York, USA, pp 433–497

Jiang X, Ku WL, Shia R-L, Li Q, Elkins JW, Prinn RG, Yung YL (2007) Seasonal cycle of N_2O: analysis of data. Glob Biogeochem Cycles 21:GB1006

Jones CD, Collins M, Cox PM, Spall SA (2001) The carbon cycle response to ENSO: A coupled climate – carbon cycle model study. J Climate 14:4113–4129

Khalil MAK, Butenhoff CL, Rasmussen RA (2007) Atmospheric methane: trends and cycles of sources and sinks. Environ Sci Technol 41:2131–2137

Levin I, Ciais P, Langenfelds R, Schmidt M, Ramonet M, Sidorov K, Tchebakova N, Gloor M, Heimann M, Schulze E-D, Vygodskaya NN, Shibistova O, Lloyd J (2002) Three years of trace gas observations over the EuroSiberian domain derived from aircraft sampling – a concerted action. Tellus 54B:696–712

Liao T, Camp CD, Yung YL (2004) The seasonal cycle of N_2O. Geophys Res Lett 31:L17108. doi:10.1029/2004GL020345

Maiss M, Brenninkmeijer CAM (1998) Atmospheric SF_6: trends, sources, and prospects. Environ Sci Technol 32:3077–3086

Maiss M, Steele LP, Francey RJ (1996) Sulfur hexafluoride – a powerful new atmospheric tracer. Atmos Environ 30:1621–1629

Mosier A, Kroeze C (2000) Potential impact on the global atmospheric N_2O budget of the increased nitrogen input required to meet future global food demands. Chemosphere: Glob Change Sci 2:465–473

Murayama S, Taguchi S, Higuchi K (2004) Interannual variation in the atmospheric CO_2 growth rate: Role of atmospheric transport in the Northern Hemisphere. J Geophys Res 109D:D02305. doi:10.1029/2003JD003729

Prentice IC, Farquhar GD, Fasham MJR, Goulden ML, Heimann M, Jaramillo VJ, Kheshgi HS, Le Quéré C, Scholes RJ, Wallace DWR (2001) The carbon cycle and atmospheric carbon dioxide. In: Houghton JT, Ding Y, Griggs DJ, Noguer M, van der Linden PJ, Dai X, Maskell K, Johnson C (eds) Climate change 2001 – the scientific basis, contribution of WG I to the 3rd assessment report of IPCC. Cambridge University Press, Cambridge, NY, pp 183–237

Ramonet M, Ciais P, Aalto T, Aulagnier C, Chevallier F, Cipriano D, Conway TJ, Haszpra L, Kazan V, Meinhardt F, Paris J-D, Schmidt M, Simmonds P, Xueref-Rémy I, Necki JN (2010) A recent build-up of atmospheric CO_2 over Europe. Part 1: observed signals and possible explanations. Tellus 62B:1–13

Sabine CL, Feely RA, Gruber N, Key RM, Lee K, Bullister JL, Wanninkhof R, Wong CS, Wallace DWR, Tilbrook B, Millero FJ, Peng T-H, Kozyr A, Ono T, Rios AF (2004) The oceanic sink for anthropogenic CO_2. Science 305:367–371

Thompson TME, Butler JH, Daube BC, Dutton GS, Elkins JW, Hall BD, Hurst DF, King DB, Kline ES, Lafleur BG, Lind J, Lovitz S, Mondeel DJ, Montzka SA, Moore FL, Nance JD, Neu JL, Romashkin PA, Scheffer A, Snible WJ (2004) Halocarbons and other atmospheric trace species. In: NOAA CMDL Summary Report no. 27, 115–135

Thoning KW, Tans PP, Komhyr WD (1989) Atmospheric carbon dioxide at Mauna Loa Observatory, 2, Analysis of the NOAA GMCC data, 1974–1985. J Geophys Res 94D:8549–8566

Tsutsumi Y, Mori K, Ikegami M, Tashiro T, Tsuboi K (2006) Long-term trends of greenhouse gases in regional and background events observed during 1998–2004 at Yonagunijima located to the east of the Asian continent. Atmos Environ 40:5868–5879

Tyler SC, Rice AL, Ajie HO (2007) Stable isotope ratios in atmospheric CH_4: implications for seasonal sources and sinks. J Geophys Res 112D:D03303

WDCGG (2009) WMO WDCGG data summary: GAW data – Volume IV-Greenhouse gases and other atmospheric gases. Vol. 33. (http://gaw.kishou.go.jp/wdcgg/products/publication.html)

WMO (2009) The state of greenhouse gases in the atmosphere using global observations through 2008. WMO Greenhouse Gas Bulletin No. 5, 23 Nov 2009

Chapter 4
Regional Climate Change and Fluctuations as Reflected in the Atmospheric Carbon Dioxide Concentration*

László Haszpra and Zoltán Barcza

Abstract This chapter analyzes the 15-year long atmospheric carbon dioxide (CO_2) mixing ratio record measured at Hegyhátsál, Hungary, to reveal the effect of regional climate change and fluctuations. While the long-term trend and the temporal fluctuation of the growth rate faithfully follow the global tendencies, the shorter term variations show special features. The authors present the distorted seasonal cycle caused by the seasonality in the atmospheric vertical mixing and the tendentious change in its shape, which is attributed to the gradual warming and the prolongation of the growing season. The decreasing summer diurnal amplitude and the decreasing seasonal amplitude in the mixing ratio and the higher than average summer CO_2 mixing ratio growth rate in the first period of the measurements (1994–2003) with increasing temperature and decreasing precipitation are explained as the consequences of the reduced activity of the biosphere in the influence area of the station and that of the reduced biomass under the increasingly unfavorable environmental conditions. The explanation is supported by the colocated tall tower surface–atmosphere CO_2 exchange measurements and by the crop yield statistics of the dominantly agricultural region around the station.

Keywords Climate fluctuation • Carbon dioxide • Biosphere/atmosphere exchange • Growing season • Seasonal trend • Diurnal variation

*Citation: Haszpra, L., Barcza, Z., 2010: Atmospheric trends and fluctuations Regional climate change and fluctuations as reflected in the atmospheric carbon dioxide concentration. In: Atmospheric Greenhouse Gases: The Hungarian Perspective (Ed.: Haszpra, L.), pp. 49–62.

L. Haszpra (✉)
Hungarian Meteorological Service, H-1675 Budapest, P.O. Box 39, Hungary
e-mail: haszpra.l@met.hu

Z. Barcza
Department of Meteorology, Eötvös Loránd University, H-1117 Budapest, Pázmány P. s. 1/A, Hungary

L. Haszpra (ed.), *Atmospheric Greenhouse Gases: The Hungarian Perspective*,
DOI 10.1007/978-90-481-9950-1_4, © Springer Science+Business Media B.V. 2011

4.1 Introduction

Carbon dioxide (CO_2) is the most important atmospheric greenhouse gas influenced by anthropogenic activity directly (Forster et al. 2007). Most of the studies consider it as a driver of global climate change. Its atmospheric mixing ratio is being monitored to follow the long-term changes, to get information on the carbon budget of the atmosphere, to reveal feedbacks and interactions in the biogeochemical carbon cycle, and ultimately to predict the evolution of the composition of the atmosphere and the resulting climate. In this chapter, it is shown that atmospheric CO_2 records, shorter term variations in the mixing ratio can also indicate regional climate features and climate fluctuations. For the study, CO_2 mixing ratio data measured at the Hungarian rural tall tower greenhouse gas monitoring site (Hegyhátsál) are used. The detailed description of the station and its instrumentation can be found in Chapter 2. For the analysis of the relation between the atmospheric mixing ratio and the activity of the biosphere (including soil), the direct surface–atmosphere carbon dioxide exchange measurements performed by the eddy covariance (EC) system mounted on the tower at 82 m elevation are also used. The technical description of this system can be found in Chapter 8.

4.2 Changes in the Diurnal Variation

Hegyhátsál is a mid-continental, low elevation site in the temperate zone surrounded by vegetation. Photosynthesis/respiration generates a seasonally changing diurnal cycle and a seasonal cycle in the atmospheric CO_2 mixing ratio, as it was presented in Chapter 3. The diurnal cycle is governed by both the photosynthesis/respiration cycle and the diurnal cycle in the intensity of the atmospheric vertical mixing (assuming that anthropogenic emission does not modulate the diurnal cycle substantially). Therefore, the daily maximum can be observed shortly after sunrise, before the vegetation–soil system becomes a net CO_2 sink and the quickening vertical mixing starts to dilute the contaminated shallow nighttime boundary layer (see Fig. 3.1) (Denning et al. 1996; Larson and Volkmer 2008).

Figure 4.1 shows the temporal variation in the monthly mean daily amplitude, where the daily amplitudes are calculated as the difference of the daily maximum and minimum hourly CO_2 mixing ratio. While the wintertime amplitude hardly changes, the summer mean daily amplitudes vary from year to year. There is an almost decade-long period from the beginning of the measurements (1994) to the early 2000s when the summer mean daily amplitudes tended to decrease, then they returned to the level observed in the middle of 1990s.

The diurnal amplitude is determined by the temporally changing anthropogenic emission, natural surface–atmosphere exchange as well as horizontal and vertical mixing of the atmosphere. After the dramatic decrease in anthropogenic CO_2 emission in Eastern Central Europe in the early 1990s (a consequence of the collapse and restructuring of the economy – see also Chapter 14), no further significant

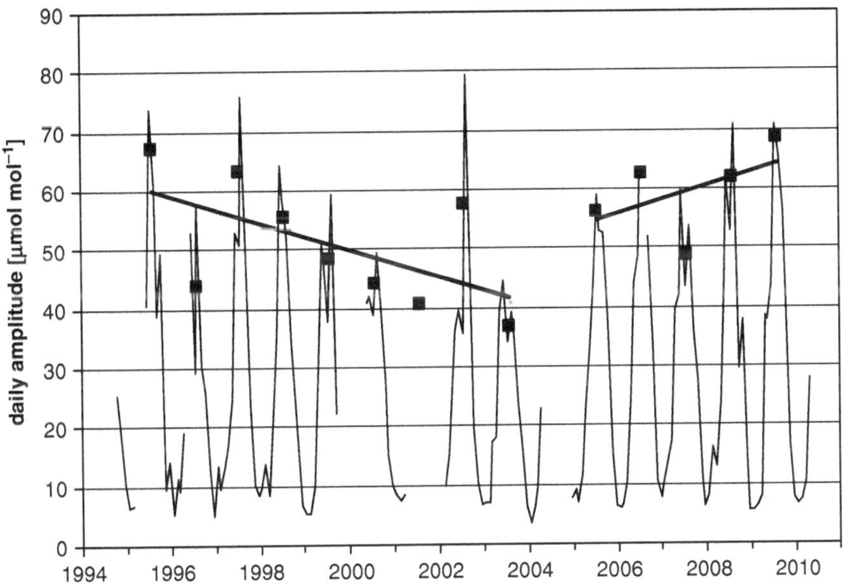

Fig. 4.1 Temporal variation in the monthly mean daily mixing ratio amplitude observed at 10 m above the ground at Hegyhátsál (1994–2009). Straight lines indicate the characteristic tendencies in the July–August mean daily mixing ratio amplitudes (*plotted as black squares*)

change in the anthropogenic emission happened in the influence area of the monitoring station (see EDGAR4 (EC-JRC/PBL 2009) for high-resolution emission data and UNFCCC (http://unfccc.int/) for the total emissions of Hungary and the neighboring countries). There is no indication of systematic circulation changes either.

During a summer day, the convective boundary layer may grow above 2,000 m in this geographical region. Any reasonable change in the surface–atmosphere CO_2 exchange or in the vertical mixing can cause only modest concentration change in this big air volume. The daily amplitude is essentially governed by the nighttime conditions when relatively small absolute changes in the emission or in the mixing height may result in a significant change in the CO_2 mixing ratio in the shallow boundary layer.

Planetary boundary layer (PBL) height data are available from the MARS archive of the European Centre for Medium-Range Weather Forecasts (ECMWF) (Beljaars et al. 2001). For this study, PBL height data were retrieved on the reduced Gaussian grid of the ECMWF operative deterministic model, and the data from the nearest grid point was used. Simulated PBL height data are only available in 3-h time steps from the short-range forecasts started at 00 and 12 UTC.

For the characterization of the CO_2 release at the surface, we can use our own net ecosystem exchange (NEE) measurements carried out at 82 m height above the ground on the tower at Hegyhátsál (see Chapter 8 and Haszpra et al. (2005)). Unlike the traditional short EC towers monitoring only a specific ecosystem in their small footprint area (e.g. FLUXNET network – Baldocchi et al. (2001)), the few tall tower EC systems can provide integrated signal from a much larger area covered by

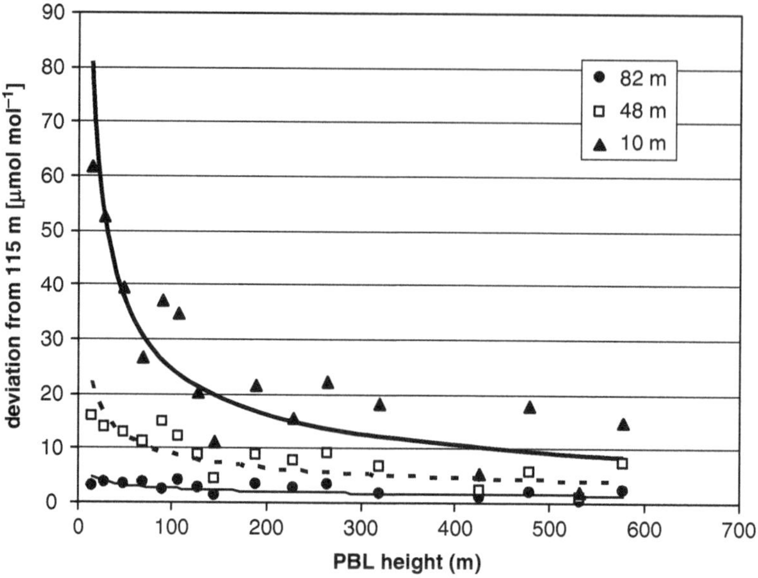

Fig. 4.2 Relationship between the nighttime CO_2 mixing ratio measured at the different monitoring levels on the tall tower at Hegyhátsál relative to the level at 115 m above the ground (0–4 h LST average) and the height of the boundary layer (4 h LST) in August. Data points represent bin averages with resolution of 20 m between 0 and 200 m boundary layer heights, 50 m between 200 and 400 m, and 100 m above 400 m

different ecological systems. So, their results are more comparable with the mixing ratio measurements even if the footprint difference is still large.

As it can be seen in Fig. 4.2, vertical mixing of trace constituents emitted at the surface is rather limited during summer nights. The vertical gradient of the mixing ratio close to the ground hardly depends on the boundary layer height if it is higher than 100–120 m. The nighttime boundary layer is hardly homogeneous unlike the daytime convective boundary layer. Therefore, any reasonable change in the boundary layer height cannot be the primary cause of the variation in the daily amplitude. It is also supported by the low correlation coefficient (–0.07) between the July–August nighttime (4 h local standard time (LST)) PBL height and the daily amplitude. Just because of the limited vertical mixing, CO_2 release influences much more the nighttime concentrations close to the surface.

Figure 4.3 shows the relation of mean nighttime (0–4 h LST) ecosystem CO_2 release and the mean daily amplitude for July–August. The two variables correlate fairly strongly ($r = 0.67$) considering the fact that the influence area of the mixing ratio measurements and that of the NEE measurements are rather different (Gloor et al. 2001; Barcza et al. 2009). Importance of the soil respiration in the formation of the nighttime ground level CO_2 mixing ratio was also indicated by Murayama et al. (2005).

The period from the mid-1990s to the extremely hot and dry 2003 was increasingly warm and dry (Haszpra et al. 2005). For a given ecosystem in a given status,

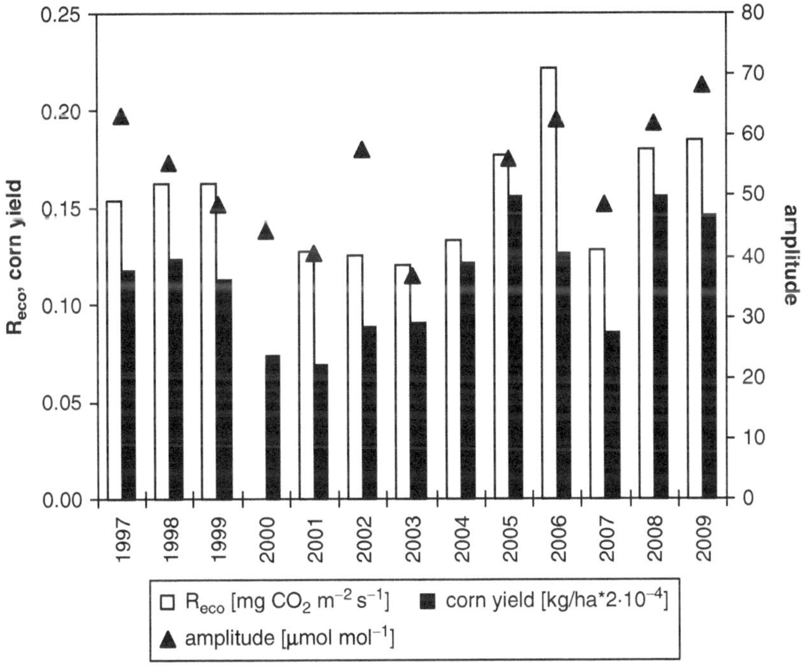

Fig. 4.3 Covarying behavior of mean nighttime (0–4 h LST) ecosystem respiration (R_{eco}) and the mean daily amplitude of CO_2 mixing ratio in July–August, as well as the corn yield in the region of Hegyhátsál. For the time series of average yield of main field crops in the region of Hegyhátsál (Vas County), visit the website of the Hungarian Central Statistical Office (http://ww.ksh.hu) and see the STADAT table 4.1.14 (http://portal.ksh.hu/pls/ksh/docs/eng/xstadat/xstadat_annual/tabl4_01_14iea.html). Corn yield data are scaled to be comparable with R_{eco}

nighttime temperature correlates with the ecosystem respiration (Raich and Schlesinger 1992; Reichstein and Beer 2008), although this relationship is modulated by other environmental variables like soil water content, microbial activity, and leaf area index. The latter, being a proxy of the standing biomass, may characterize the maintenance respiration. The unfavorable environmental conditions reduced the biomass in the region, which triggered a decrease in total ecosystem respiration in spite of the temperature increase as it was also found by Ciais et al. (2005) and Reichstein et al. (2007) on European scale. The assumed biomass reduction is supported by the yield data of summer crops as proxies for biomass and net primary production (Ciais et al. 2005, 2007; Reichstein et al. 2007) (Fig. 4.3).

4.3 Changes in the Seasonal Variation

Life cycle of the vegetation in the temperate and boreal zones generates an annual cycle in the atmospheric carbon dioxide mixing ratio (Fig. 4.4, see also Fig. 3.2 in Chapter 3). During the winter period, the net ecosystem exchange of CO_2 is positive

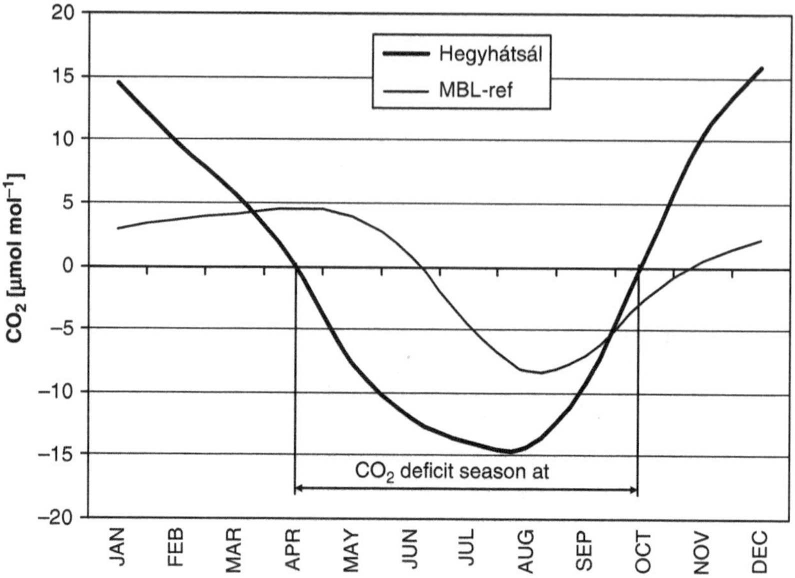

Fig. 4.4 Mean seasonal cycle of the early afternoon (12–16 h LST) CO_2 mixing ratio at Hegyhátsál (10 m) and in the modeled marine boundary layer (GLOBALVIEW-CO2 2009) corresponding to the geographical latitude of Hegyhátsál relative to the annual average mixing ratio

from the point of view of the atmosphere: Ecosystems release more CO_2 than they absorb by photosynthesis. Atmospheric mixing ratio can begin to decrease when NEE becomes negative, typically in March–April, or even later at high latitudes (see e.g. http://www.esrl.noaa.gov/gmd/ccgg/iadv/). However, at Hegyhátsál, CO_2 mixing ratio reaches its annual maximum in December–January (Fig. 4.4) well before the vegetation becomes a net CO_2 sink. A shorter phase-shift could be explained by long-range horizontal transport carrying CO_2-depleted air masses from lower latitudes where growing season starts earlier. However, the seasonal cycle in atmospheric CO_2 mixing ratio leads that of the regional NEE by approximately 12 weeks, which seems to be too long time to consider any advection effect. The strange seasonal cycle observed at Hegyhátsál is essentially explained by the regional feature of the vertical mixing of the atmosphere. From January, following the increasing insolation after the winter solstice, the vertical mixing is getting more vigorous quickly (see Fig. 4.5) injecting more and more relatively clean, CO_2-depleted free tropospheric air into the polluted, CO_2-rich boundary layer. This process overcompensates the contribution of the surface that is a net CO_2 source yet in that time of the year. A similar phase lag was observed by Davis et al. (2003) at the WLEF tall tower site in the continental U.S.A. (Wisconsin), but it was not determined whether the horizontal advection or the vertical mixing is the main contributing factor.

The effect of the changing vertical mixing on the CO_2 mixing ratio in the boundary layer sampled by the monitoring station can be demonstrated even by means of a

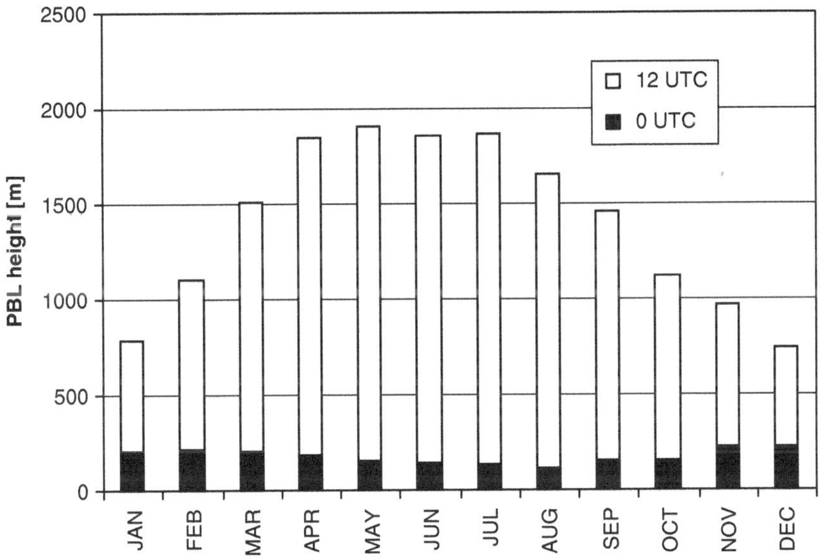

Fig. 4.5 Seasonal variation in the average daytime (13 h LST = 12 UTC) and nighttime (1 h LST = 0 UTC) heights of the PBL at Hegyhátsál (data source: ECMWF MARS archive)

very simple box model. This model, describing only the effects of the vertical mixing and that of the surface exchange, and developed exclusively for this demonstration, is a box model with height following the height of the PBL. When it increases, air not influenced directly by the local surface processes (residual layer air or free tropospheric air – Stull (1988)) is incorporated into the box. To demonstrate the overwhelming effect of the vertical mixing on the CO_2 mixing ratio in the PBL, only the period of November–February is simulated. Thus, photosynthesis can be neglected and a constant 1 g C m^{-2} day^{-1} ecosystem respiration is supposed based on our multiyear measurements (Haszpra et al. 2005). Anthropogenic release is set to 0.112–0.155 g C m^{-2} day^{-1} depending on the year (EC-JRC/PBL 2009). To simulate the contribution of the temporary residual layer of the atmosphere and that of the exchange between the PBL and the free troposphere, a 40%:60% mixture of boundary layer air and free tropospheric air is mixed into the box if its height is increasing. This ratio is adjusted to get the smallest offset from the real atmospheric mixing ratio. CO_2 mixing ratio in the free troposphere is considered identical with that in the marine boundary layer (GLOBALVIEW-CO2 2009) corresponding to the geographical position of the station. In Fig. 4.6, the resulting temporal variation of CO_2 mixing ratio is presented along with the measured values for 2002–2003. The average deviation of the measured values from the simulated ones is -1.26 ± 9.55 μmol mol^{-1} with a correlation coefficient of 0.56. In other years the correlation coefficients range 0.38–0.63.

The seasonal cycle is not only unusual, but it also changes its shape with time. The length of the CO_2 deficit season, when the mixing ratio is lower than the annual

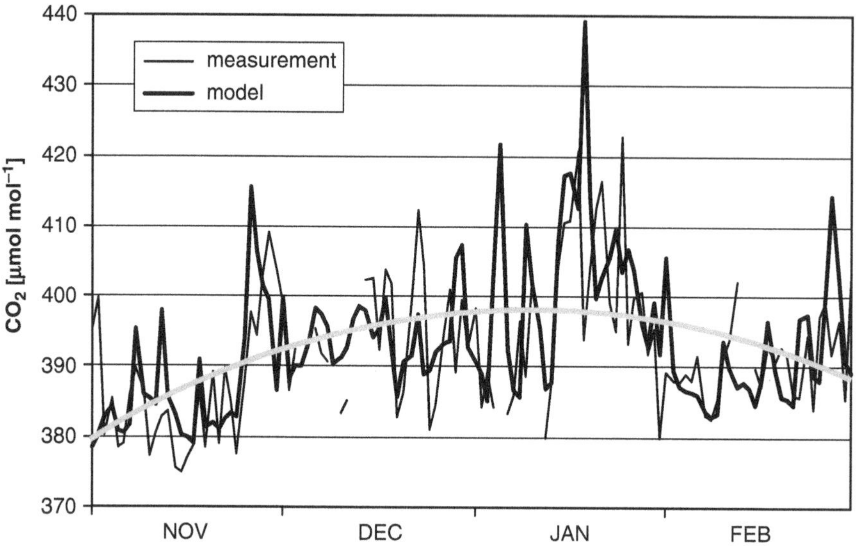

Fig. 4.6 Measured and simulated CO_2 mixing ratios for the period of November 2002–February 2003. The line fitted to the simulated values is a third-order polynomial

average (see Fig. 4.4), is increasing slowly. The beginning of the period is moving to earlier in the year, while its end hardly changes (Fig. 4.7). Supposing a linear trend, the beginning of the CO_2 deficit season shifted almost 9 days earlier between 1995 and 2009. As the seasonal cycle is strongly influenced by that of the vertical mixing at Hegyhátsál, the first idea might be to check this factor. However, based on the ECMWF PBL data, there is not any tendency in the vertical mixing in spring. The earlier start of the CO_2 deficit season may be caused by the generally warming climate, by the earlier start of the growing season, as it has already been indicated by several researches (e.g. Menzel and Fabian 1999; Tucker et al. 2001; Linderholm 2006; Piao et al. 2007; Thum et al. 2009). Although the statistical confidence level of this trend is relatively low (93%), it may indicate a real tendency as the temperature in the critical month (April, day 91–121 of the year) has also been increasing statistically significantly (confidence level >95%).

In principle, warming could also prolong the growing season and move the autumn zero-crossing (the time when CO_2-deficit season turns into CO_2-sufficit season) towards the end of the year. Both international (Linderholm et al. 2008) and Hungarian (Varga-Haszonits 2004) studies showed that warming does not extend the growing season equally towards spring and autumn. The prolongation towards the end of the year is shorter. It may be one of the reasons why no trend can be observed in the autumn zero-crossing of CO_2 mixing ratio. Another contributing factor in this dominantly agricultural region may be the earlier harvest of summer crops, which prevents the biosphere to profit from the prolonged growing season. Furthermore, the increasing respiration caused by the higher autumn temperature can also balance the potentially prolonged autumn carbon uptake (Piao et al. 2008).

In principle, warming may also increase the amplitude of the seasonal cycle as it was discussed by Keeling et al. (1996) and Zimov et al. (1999). The warmer winter

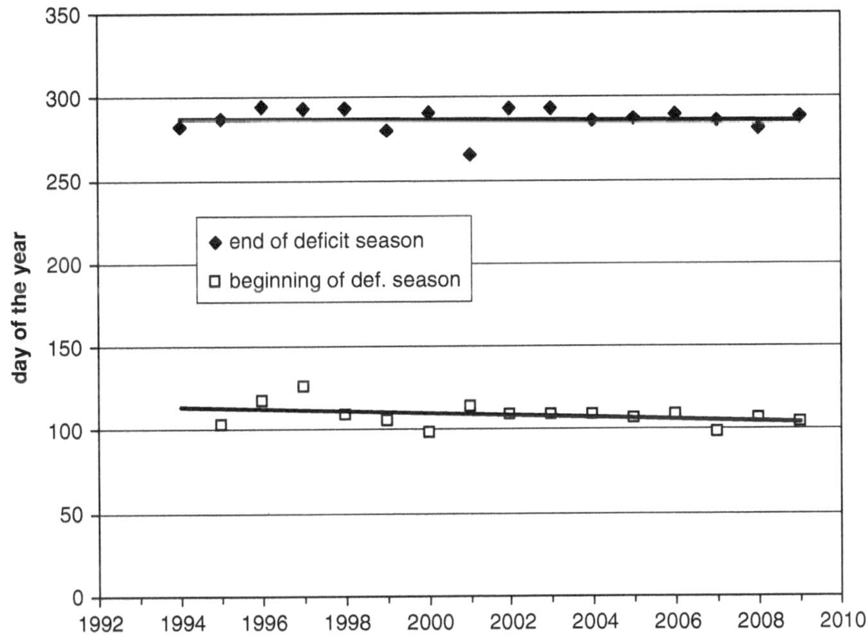

Fig. 4.7 Tendencies of the beginning and end of the CO_2 deficit season (1994–2009)

temperature may intensify the respiration, while the prolonged growing season increases the carbon uptake. In the Hegyhátsál CO_2 record, no obvious trend can be observed in the amplitude of the seasonal cycle. From the beginning of the measurements until 2003, the amplitude of the seasonal cycle decreased by about 20% (estimated from the smoothed time series), then it increased again (Fig. 4.8). The most plausible reason is the same as in the case of the reduced diurnal amplitude: the increasingly unfavorable environmental conditions reduced the standing biomass and it reduced the CO_2 uptake capacity of the biosphere in the influence area of the station. The less-intensive CO_2 uptake in the summer season decreased the annual amplitude of the atmospheric CO_2 mixing ratio. It is in accordance with the higher than average summer growth rate in the mixing ratio discussed in the next section.

4.4 Temporal Variations in the Long-Term Trend

As it was pointed out in Chapter 3, the long-term atmospheric CO_2 mixing ratio trend and its fluctuation faithfully follow the global tendencies in spite of the unfavorable location of the station (middle of a highly populated, industrialized continent, surrounding active vegetation). However, if the long-term trend is decomposed by seasons, interesting tendencies can be revealed. During the period between the beginning of the measurements at Hegyhátsál, 1994 and 2003, the summer growth rate of the CO_2 mixing ratio significantly exceeded that in the other seasons

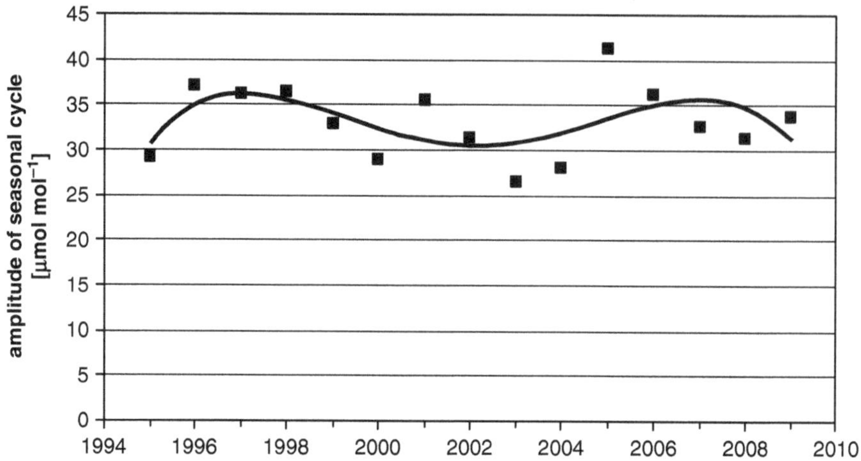

Fig. 4.8 Temporal variation in the amplitude of the seasonal cycle of the early afternoon (12–16 h LST) CO_2 mixing ratio (1996–2009). The fitted line is a fourth-order polynomial

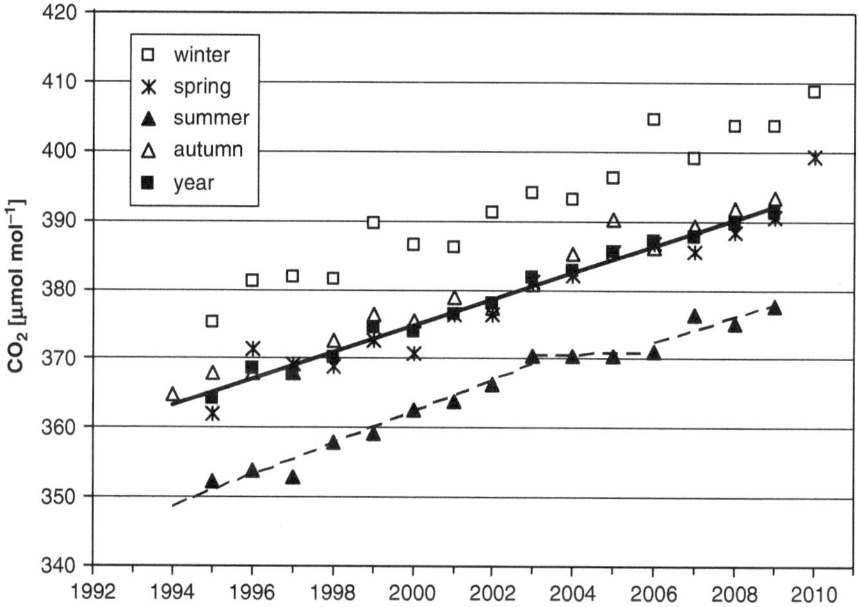

Fig. 4.9 Seasonal trends in the early afternoon (12–16 h LST) atmospheric CO_2 mixing ratio at 10 m elevation above the ground at Hegyhátsál (1994–2009). Lines indicate only the summer trend (*dashed line*) and the overall trend (*thick solid line*)

(Fig. 4.9, Table 4.1). While the mixing ratio grew by 2.66 µmol mol^{-1} year^{-1} in summer during this period, the overall growth rate was only 1.95 µmol mol^{-1} year^{-1}. As it was mentioned before, these years were increasingly warm and dry. The high temperature and the lack of water could significantly reduce the development of

Table 4.1 Seasonal growth rates (μmol mol^{-1} year^{-1}) of early afternoon (12–16 h LST) atmospheric CO_2 mixing ratio at 10 m above the ground at Hegyhátsál for the periods of 1994–2003 and 1994–2009. The bold number emphasizes the extreme summer growth rate between 1994 and 2003 discussed in the text

	1994–2003	1994–2009
Spring	1.96 ± 0.35	1.86 ± 0.13
Summer	**2.66 ± 0.19**	1.94 ± 0.11
Autumn	1.81 ± 0.35	1.96 ± 0.13
Winter	1.89 ± 0.49	1.98 ± 0.15
Year	2.08 ± 0.20	1.95 ± 0.07

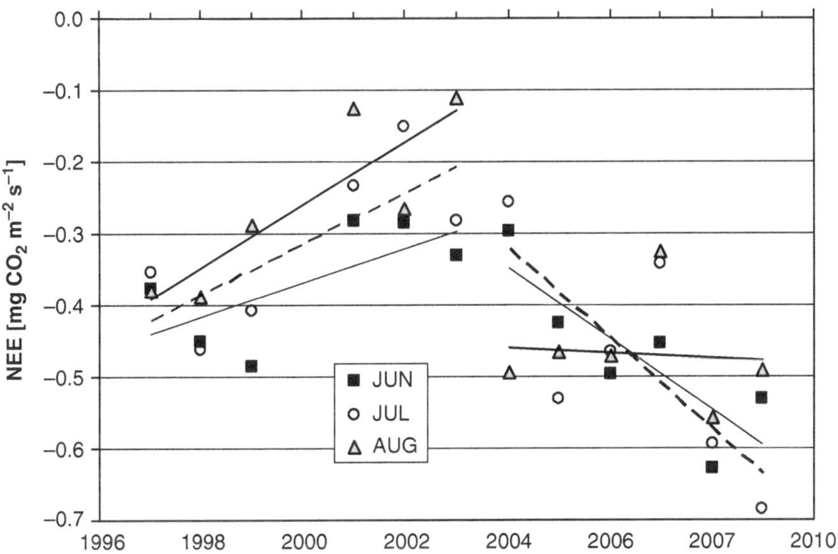

Fig. 4.10 Tendencies in the monthly mean NEE at Hegyhátsál in summer (1997–2008). The lines indicate only the tendencies and do not represent any statistically significant trend (*thin dashed line – June, thick dashed line – July, solid line – August*)

biomass (as it was also referred in the previous sections), which resulted in decreased net CO_2 uptake of the biosphere (Haszpra et al. 2005). From 2004, the climate has returned to normal; it has become relatively cooler and wetter in Eastern Central Europe. It has balanced the seasonal differences on the 15-year time scale (Table 4.1). It can be seen in Fig. 4.9 that the summer mixing ratios have hardly increased for 4 years after 2003.

The effect of the climate anomaly that lasted until 2003 is well presented by the biosphere–atmosphere CO_2 exchange measurements. Somewhat higher biospheric uptake resulted in decreasing net release in March and significantly decreased net uptake in summer, especially in July and August, as can be seen in Fig. 4.10. From 2004, the situation has changed drastically, in parallel with the changing regional climate.

4.5 Summary and Conclusions

In this chapter, we looked at several characteristics of the shorter term variations in the atmospheric CO_2 mixing ratio record measured at a mid-continental, low elevation, rural tall tower site. We could reveal special features like the enhanced diurnal variation, the strongly limited nighttime vertical mixing, and the nontrivial shape of the seasonal cycle. These features may be characteristic only for certain monitoring stations, and this fact should be taken into account in atmospheric modeling. The reduced diurnal and seasonal amplitudes, and the seasonal differences in the longer term trend equally show the joint effect of the warming climate and the developing drought in the first half of the measurements. While the long-term trend in the atmospheric carbon dioxide mixing ratio and its fluctuation measured locally faithfully follow the global tendencies, the shorter term variations reflect the regional climate fluctuations. In a more general approach, due to the complex interactions between the surface/biosphere and the atmosphere, watching the short-term variations in the atmospheric CO_2 mixing ratio records might call the attention to atmospheric/climatic changes in progress.

This study demonstrates the importance and usefulness of colocated, tall tower based CO_2 mixing ratio and NEE measurements as it was initiated by the CHIOTTO project (Vermeulen (2007); http://www.chiotto.org/) and accepted by ICOS, the new pan-European integrated greenhouse gas observation network in preparation (http://www.icos-infrastructure.eu/). The joint use of measurements can highlight important physical processes helping us to better understand the synoptic variability of atmospheric carbon dioxide and can provide modelers with important information on the underlying mechanisms.

Acknowledgments During the years, the Hungarian atmospheric greenhouse gas monitoring received significant financial support from the US–Hungarian Scientific and Technological Joint Fund (J.F. 162, J.F. 504), the European Commission R&D Framework Programmes (AEROCARB – EVK2-CT-1999-00013, CHIOTTO – EVK2-CT-2002-00163, CarboEurope-IP – GOCE-CT-2003-505572, IMECC – RII3 026188), the European Commission's INTERREG IIIB CADSES program (Carbon-Pro – 5D038), as well as the Hungarian funding organizations (Hungarian Scientific Research Fund – T7282, T042941, CK77550; Hungarian Ministry for Environment – 027739-01/2001, K0441482001, K-36-02-00010H).

In addition to original material, this chapter is essentially an updated version of Haszpra and Barcza (2010).

References

Baldocchi D, Falge E, Gu L, Olson R, Hollinger D, Running S, Anthoni P, Bernhofer Ch, Davis K, Evans R, Fuentes J, Goldstein A, Katul G, Law B, Lee X, Malhi Y, Meyers T, Munger W, Oechel W, Paw KT, Pilegaard K, Schmid HP, Valentini R, Verma S, Vesala T, Wilson K, Wofsy S (2001) FLUXNET: a new tool to study the temporal and spatial variability of ecosystem-scale carbon dioxide, water vapor, and energy flux densities. Bull Am Meteorol Soc 82:2415–2434

Barcza Z, Kern A, Haszpra L, Kljun N (2009) Spatial representativeness of tall tower eddy covariance measurements using remote sensing and footprint analysis. Agric For Meteorol 149:795–807

Beljaars A, Jakob C, Morcrette J-J (2001) New physics parameters in the MARS archive. ECMWF Newsl 90:17–21

Ciais P, Reichstein M, Viovy N, Granier A, Ogée J, Allard V, Aubinet M, Buchmann N, Bernhofer C, Carrara A, Chevallier F, De Noblet N, Friend AD, Friedlingstein P, Grünwald T, Heinesch B, Keronen P, Knohl A, Krinner G, Loustau D, Manca G, Miglietta F, Ourcival JM, Papale D, Pilegaard K, Rambal S, Seufert G, Soussana JF, Sanz MJ, Schulze ED, Vesala T, Valentini R (2005) Europe-wide reduction in primary productivity caused by the heat and drought in 2003. Nature 437:529–533

Ciais P, Bousquet P, Freibauer A, Naegler T (2007) Horizontal displacement of carbon associated with agriculture and its impacts on atmospheric CO_2. Glob Biogeochem Cycles 21:GB2014

Davis KJ, Bakwin PS, Yi C, Berger BW, Zhao C, Teclaw RM, Isebrands JG (2003) The annual cycles of CO_2 and H_2O exchange over a northern mixed forest as observed from a very tall tower. Glob Change Biol 9:1278–1293

Denning AS, Randall DA, Collatz GJ, Sellers PJ (1996) Simulations of terrestrial carbon metabolism and atmospheric CO_2 in a general circulation model. Part II.: Simulated CO_2 concentrations. Tellus 48B:543–567

EC-JRC/PBL (European Commission, Joint Research Centre/Netherlands Environmental Assessment Agency) (2009) Emission Database for Global Atmospheric Research (EDGAR), release version 4.0. http://edgar.jrc.ec.europa.eu

Forster P, Ramaswamy V, Artaxo P, Berntsen T, Betts R, Fahey DW, Haywood J, Lean J, Lowe DC, Myhre G, Nganga J, Prinn R, Raga G, Schulz M, Van Dorland R (2007) Changes in atmospheric constituents and in radiative forcing. In: Solomon S, Qin D, Manning M, Chen Z, Marquis M, Averyt KB, Tignor M, Miller HL (eds) Climate change 2007: The physical science basis. Contribution of Working Group I to the fourth assessment report of the Intergovernmental Panel on Climate Change. Cambridge University Press, Cambridge, UK and New York, USA, pp 129–234

GLOBALVIEW-CO2 (2009) Cooperative atmospheric data integration project – carbon dioxide. CD-ROM, NOAA ESRL, Boulder, CO. Also available on Internet via anonymous FTP to ftp.cmdl.noaa.gov, Path: ccg/co2/GLOBALVIEW

Gloor M, Bakwin P, Hurst D, Lock L, Draxler R, Tans P (2001) What is the concentration footprint of a tall tower? J Geophys Res 106D:17831–17840

Haszpra L, Barcza Z (2010) Climate variability as reflected in a regional atmospheric CO_2 record. Tellus, 62B (in print) DOI: 10.1111/j.1600-0889.2010.00505.x

Haszpra L, Barcza Z, Davis KJ, Tarczay K (2005) Long term tall tower carbon dioxide flux monitoring over an area of mixed vegetation. Agric For Meteorol 132:58–77

Keeling CD, Chin JFS, Whorf TP (1996) Increased activity of northern vegetation inferred from atmospheric CO_2 measurements. Nature 382:146–149

Larson VE, Volkmer H (2008) An idealized model of the one-dimensional carbon dioxide rectifier effect. Tellus 60B:525–536

Linderholm HW (2006) Growing season changes in the last century. Agric For Meteorol 137:1–14

Linderholm HW, Walther A, Chen D (2008) Twentieth-century trends in the thermal growing season in the Greater Baltic Area. Clim Change 87:405–419

Menzel A, Fabian P (1999) Growing season extended in Europe. Nature 397:659

Murayama S, Yamamoto S, Saigusa N, Kondo H, Takamura C (2005) Statistical analyses of inter-annual variations in the vertical profile of atmospheric CO_2 mixing ratio and carbon budget in a cool-temperate deciduous forest in Japan. Agric For Meteorol 134:17–26

Piao S, Friedlingstein P, Ciais P, Viovy N, Demarty J (2007) Growing season extension and its impact on terrestrial carbon cycle in the Northern Hemisphere over the past 2 decades. Glob Biogeochem Cycles 21:GB3018

Piao S, Ciais P, Friedlingstein P, Peylin P, Reichstein M, Luyssaert S, Margolis H, Fang J, Barr A, Anping Ch, Grelle A, Hollinger DY, Laurila T, Lindroth A, Richardson AD, Vesala T (2008) Net carbon dioxide losses of northern ecosystems in response to autumn warming. Nature 451:49–52

Raich JW, Schlesinger WH (1992) The global carbon dioxide flux in soil respiration and its relationship to vegetation and climate. Tellus 44B:81–99

Reichstein M, Beer C (2008) Soil respiration across scales: the importance of a model-data integration framework for data interpretation. J Plant Nutr Soil Sci 171:344–354

Reichstein M, Ciais P, Papale D, Valentini R, Running S, Viovy N, Cramer W, Granier A, Ogée J, Allard V, Aubinet M, Bernhofer C, Buchmann N, Carrara A, Grünwald T, Heimann M, Heinesch B, Knohl A, Kutsch W, Loustau D, Manca G, Matteucci G, Miglietta F, Ourcival JM, Pilegaard K, Pumpanen J, Rambal S, Schaphoff S, Seufert G, Soussana J-F, Sanz M-J, Vesala T, Zhao M (2007) Reduction of ecosystem productivity and respiration during the European summer 2003 climate anomaly: a joint flux tower, remote sensing and modelling analysis. Glob Change Biol 13:634–651

Stull RB (1988) An introduction to boundary layer meteorology. Kluwer Academic, Norwell, MA

Thum T, Aalto T, Laurila T, Aurela M, Hatakka J, Lindroth A, Vesala T (2009) Spring initiation and autumn cessation of boreal coniferous forest CO_2 exchange assessed by meteorological and biological variables. Tellus 61B:701–717

Tucker CJ, Slayback DA, Pinzon JE, Los SO, Myneni RB, Taylor MG (2001) Higher northern latitude normalized difference vegetation index and growing season trends from 1982 to 1999. Int J Biometeorol 45:184–190

Varga-Haszonits Z (2004) Az éghajlati változékonyság és a természetes periódusok (Climatic variability and natural periods). "AGRO-21" Füzetek 37:23–32

Vermeulen A (ed) (2007) CHIOTTO Final report. ECN-E--07-052. Available at http://www.ecn.nl/docs/library/report/2007/e07052.pdf

Zimov SA, Davidov SP, Zimova GM, Davidova AI, Chapin FSI, Chapin MC, Reynolds JF (1999) Contribution of disturbance to increasing seasonal amplitude of atmospheric CO_2. Science 284:1973–1976

Part II
Measurements and Estimations of Biosphere–Atmosphere Exchange of Greenhouse Gases

Chapter 5
Methodologies*

Csilla Farkas, Giorgio Alberti, János Balogh, Zoltán Barcza, Márta Birkás, Szilárd Czóbel, Kenneth J Davis, Ernő Führer, Györgyi Gelybó, Balázs Grosz, Natascha Kljun, Sándor Koós, Attila Machon, Hrvoje Marjanović, Zoltán Nagy, Alessandro Peressotti, Krisztina Pintér, Eszter Tóth, and László Horváth

Abstract Measurement of biosphere–atmosphere exchange of various greenhouse gases requires different techniques. In case of carbon dioxide, the net ecosystem exchange (NEE) is usually measured by the eddy covariance method. In the lack of these measurements in forests, the carbon dioxide uptake can be estimated by detecting changes in sequestrated carbon stocks or by using tree growth (dendrometric) measurements. The soil CO_2, CH_4, and N_2O efflux/exchange rates can be determined using in situ chamber techniques, or laboratory incubation measurements. Static and dynamic, manual and automatic chamber methods, as well as photo-acoustic, gas chromatography, and infrared detections can be used for this purpose. This chapter gives a general overview of the approaches applied in studies presented in this book for evaluating the greenhouse gas exchange between the biosphere and atmosphere.

Keywords Eddy covariance • Static and dynamic chambers • Gas chromatography • Infrared detection • Photo-acoustic detection • Laboratory incubation

---

*Citation: Farkas, Cs., Alberti, G., Balogh, J., Barcza, Z., Birkás, M., Czóbel, Sz., Davis, K. J., Führer, E., Gelybó, Gy., Grosz, B., Kljun, N., Koós, S., Machon, A., Marjanović, H., Nagy Z., Peresotti, A., Pintér, K., Tóth, E., Horváth, L., 2010: Measurements and estimations of biosphere–atmosphere exchange of greenhouse gases – Methodologies. In: Atmospheric Greenhouse Gases: The Hungarian Perspective (Ed.: Haszpra, L.), pp. 65–90.

C. Farkas
Bioforsk, Norwegian Institute for Agricultural and Environmental Research, Frederik A. Dahlsvei 20, Ås 1432, Norway

G. Alberti and A. Peressotti
Department of Agricultural and Environmental Sciences, University of Udine, I-33100 Udine via delle Scienze 208, Italy

J. Balogh, Z. Nagy and K. Pintér
Plant Ecology Research Group of Hungarian Academy of Sciences, H-2103 Gödöllő, Páter K. u. 1., Hungary
and
Department of Botany and Plant Physiology, Szent István University, H-2103 Gödöllő, Páter K. u. 1., Hungary

L. Haszpra (ed.), *Atmospheric Greenhouse Gases: The Hungarian Perspective*,
DOI 10.1007/978-90-481-9950-1_5,

Z. Barcza and Gy. Gelybó
Department of Meteorology, Eötvös Loránd University,
H-1117 Budapest, Pázmány P. s. 1/A, Hungary

M. Birkás
Institute of Crop Production, Szent István University, H-2103 Gödöllő, Páter K. u. 1., Hungary

Sz. Czóbel
Institute of Botany and Ecophysiology, Szent István University, H-2103 Gödöllő,
Páter K. u. 1., Hungary

K.J. Davis
Department of Meteorology, The Pennsylvania State University,
University Park, PA 16802, USA

E. Führer
Department of Ecology and Silviculture, Hungarian Forest Research Institute,
H-9400 Sopron, Paprét 17., Hungary

B. Grosz
Department of General and Inorganic Chemistry, Eötvös Loránd University,
H-1117 Budapest, Pázmány P. s. 1/A, Hungary

N. Kljun
Department of Geography, Swansea University, Singleton Park, Swansea SA2 8PP,
United Kingdom

S. Koós and E. Tóth
Research Institute for Soil Science and Agricultural Chemistry of the Hungarian Academy of
Sciences, H-1022 Budapest, Herman O. út 15., Hungary

A. Machon, Z. Nagy, and K. Pintér
Department of Botany and Plant Physiology, Szent István University, H-2103 Gödöllő,
Páter K. u. 1., Hungary

A. Machon
Center for Environmental Science, Eötvös Loránd University, H-1117 Budapest,
Pázmány P. s. 1/A, Hungary

A. Machon and L. Horváth (✉)
Hungarian Meteorological Service, H-1675 Budapest, P.O. Box 39, Hungary
e-mail: horvath.l@met.hu

H. Marjanović
Croatian Forest Research Institute, Cvjetno naselje 41, HR-10450, Jastrebarsko, Croatia

5.1 Introduction

Biosphere may act as either a source or a sink for atmospheric greenhouse gases. There are three greenhouse gases, namely carbon dioxide (CO_2), methane (CH_4), and nitrous oxide (N_2O), having measurable fluxes between the atmosphere and the biosphere. While fluxes of CO_2 and CH_4 are bidirectional (so the biosphere can be a source and a sink at the same time), N_2O is generally emitted by the biosphere.

Plants take up CO_2 molecules during photosynthesis and CO_2 is released by respiration processes. In the turbulent near-surface layer, turbulent motions transport molecules to and from the surface. In near-canopy and in-canopy (so-called quasilaminar) layers, mostly the molecular diffusion dominates, which is reported to be the primary mechanism for transporting CO_2 between the biosphere and the atmosphere (Freijer and Leffelaar 1996).

Soils are both sources and sink for carbon and are major contributors to the partitioning of greenhouse gas fluxes (Pumpanen et al. 2003). Carbon dioxide is released from the soil through soil respiration, which includes three biological processes, namely microbial respiration, root respiration, and faunal respiration primarily at the soil surface or within a thin upper layer where the bulk of plant residue is concentrated (Jorgensen and Wells 1973; de Jong et al. 1974; Edward 1975) and one nonbiological process, i.e., chemical oxidation, which could be pronounced at higher temperatures (Bunt and Rovira 1954). The contribution of each of these sources to the total soil CO_2 efflux is poorly understood and is probably extremely variable from site to site (Boone et al. 1998). Many studies have shown that soil CO_2 originates primarily from microbial oxidation of organic matter and root respiration (Macfayden 1963, 1970; de Jong et al. 1974; Edward 1975).

Depending on soil properties and environmental conditions, soils can act as a source or sink for CH_4. Well-aerated and dry soils can act as effective sinks for both atmospheric CH_4 and CH_4 produced in deeper soil layers. In the soil, CH_4 is predominantly used by methanotrophic bacteria (Prather et al. 1995; Steinkamp et al. 2001) as a source of carbon in a process called methane oxidation. On the other hand, CH_4 is produced in anaerobic condition in soils saturated with water. Wetland soils and shallow lakes are also important sources of CH_4 (Khalil and Shearer 1993; Le Mer and Roger 2001). Besides soil water content, soil temperature and nitrogen concentration can also be crucial factors in determining whether a particular soil will act as a sink or source for CH_4.

Soils have been classified by Intergovernmental Panel on Climate Change (IPCC) as major source of N_2O to the atmosphere (Houghton et al. 1992). Nitrous oxide is produced in the soil by nitrification (nitrifier nitrification) and dominantly by denitrification processes. Three gaseous intermediate products of these processes, namely the nitric oxide (NO), the nitrous oxide (N_2O), and the molecular nitrogen (N_2), are able to escape from the soil. The share of the emitted gases depends strongly on soil water status. In dry soils – in aerobic conditions – the production of NO dominates, while at medium and high soil water contents, N_2O is produced. N_2O remains in solution close to the saturation leading to N_2 production. The other main controlling factors determining N_2O emission from soils are soil temperature and mineral nitrogen concentration. Variations in these parameters along with changes in water-filled soil pore space can lead to substantial differences in N_2O emissions even from the same soils during the year (Buchkina et al. 2010).

Various techniques have been developed for measuring the greenhouse gas fluxes between the biosphere and the atmosphere. This chapter aims to overview the most widely used methods and to give detailed description of those applied in Hungary.

The net flux of CO_2, consisting of the canopy flux and the soil emission term, can be measured above the canopy using the eddy covariance (EC) method. The EC method is a routine micrometeorological method, which is the most frequently used and the most suitable approach for monitoring the net CO_2 flux at larger scales (Barcza et al. 2009).

For a better understanding of the processes standing behind the formation of the carbon balance in the soil-plant-atmosphere system, specific elements of this balance, like soil CO_2 efflux, should be identified. Methods used for evaluating soil CO_2 emission and intensity of soil respiration can be divided into direct and indirect approaches.

Indirect methods, based on determining losses of organic material from soil, allow coarse assessment of soil CO_2 emission over long periods only (Lopes de Gerenu et al. 2005). Nevertheless, changes in agricultural practices can result in changes in both the pool size and turnover rates of soil organic matter. Soil tillage produces a soil microenvironment favorable for accelerated microbial decomposition of plant and animal residues (Rastogi et al. 2002). Hence, the concept lying behind the indirect methods of quantifying soil CO_2 efflux is not suitable for analyzing the effects of soil management induced accelerated changes in soil respiration.

The most widely used diffuse transport and soil cover (chamber) methods belong to the direct methods of measuring soil CO_2 emission. The diffusion method (de Jong et al. 1974; Rolston 1978) is based on calculating soil respiration intensity from changes in air CO_2 concentration and alteration of CO_2 diffusion coefficient. Concentration measurements can be made directly in the soil (Kasparov et al. 1986; Larionova et al. 1988) or above the vegetation cover – aerodynamic approach (Lemon 1967; Kimball and Lemon 1970; Nakayama and Kimball 1988). However, in a detailed comparative study of different methods for determining soil CO_2 efflux, the diffusion method consistently yielded lower CO_2 emission estimates than the EC and the chamber methods (Parkin et al. 2006).

The chamber method has been used extensively for measuring gas exchange between soil surfaces and the atmosphere (Delle Vedove et al. 2007), and has the advantages of low cost and ease of use. Besides measuring soil respiration, it can also be applied for determining fluxes of CH_4 and N_2O. The major disadvantage of the chamber method is the poor spatial representativeness due to the small sampling area covered in each measurement.

Transparent chambers are used above short canopy, daytime, for measurements of the net flux including the effect of photosynthesis and soil exhalation.

Chamber methods can be used for both field and laboratory measurements. Incubation of soil samples in laboratory – under controlled temperature and soil water content – makes possible to study the dynamics of emission processes in a wide range of environmental variables, and can contribute to better understanding of processes, standing behind gas exchange between soil and atmosphere.

While the indirect method based on evaluating changes in organic carbon content in the studied object did not find wide application in soil respiration studies, it is broadly used for determining carbon balance of forest ecosystems. In Hungary, where no direct flux measurements in forests are available, we can estimate the magnitude of stored carbon by survey of carbon stocks as well as the dynamics of carbon sequestration by the observation of the above-ground and below-ground organic matter production in the forest ecosystems.

In this chapter, the EC and chamber methods applied for monitoring biosphere–atmosphere GHG exchange over grasslands, forests, and arable lands are described together with the carbon stock survey and annual organic matter production estimation methods for forest ecosystems. The measurements presented in the following chapters (Chapters 6–8) are based on these methods.

5.2 Eddy Covariance Technique for the Measurement of the Net Carbon Dioxide Flux Between the Atmosphere and the Biosphere

The EC method provides information about the net CO_2 flux at larger scales. It involves the parallel measurement of fluctuations of CO_2 concentration, as well as that of vertical component of wind velocity. Sensors are settled above the canopy where turbulent eddies rule the vertical exchange of CO_2 between the biosphere and the atmosphere. The height of the measurements is generally a few meters above vegetation up to around 100 m (e.g., in Hegyhátsál it is 82 m) for regional flux assessments. One of the most important criteria for micrometeorological flux measurements is the "good fetch," i.e., the need of surface homogeneity in the influence area (footprint) of the measurements. In this chapter, we describe the theoretical background of the EC approach and the calculation methods used for deriving the different fluxes.

5.2.1 *Theory Behind the Eddy Covariance Technique*

The surface flux of any passive scalar like CO_2 mixing ratio can be inferred from the vertical flux of the quantity measured at a given height and the change of scalar concentration profile below the measurement plane. Based on the mass conservation equation and the continuity equation, assuming zero divergence of the horizontal eddy flux and zero horizontal advection, Lee (1998) derives the following equation:

$$F_C = \left(\overline{w'c'}\right)_m + \int_0^{z_m} \frac{\partial \overline{c}}{\partial t} dz + \overline{w}_m \left(\overline{c}_m - \langle \overline{c} \rangle\right) = F_{Ctb} + F_{Cst} + F_{Cmf} \tag{5.1}$$

where F_C denotes net ecosystem exchange (NEE), $\left(\overline{w'c'}\right)_m$ is the turbulent flux of scalar c measured directly by the EC system (primes are departures from the mean, overbar is time averaging), z_m is the measurement height, $\overline{c}_m$ is the time-averaged scalar concentration at z_m, $\overline{w}_m$ is the mean vertical velocity at z_m, and $\langle \overline{c} \rangle$ is the average scalar concentration between the ground and the measurement height.

Note that Eq. (5.1) merely consists of elements that can be estimated using a single-point measurement with a supplementary estimate of the concentration profile, and all the other terms are neglected. For example, the above model assumes zero horizontal advection, which is not necessarily satisfied (Lee 1998; Paw U et al. 2000; Yi et al. 2000; Finnigan et al. 2003). Still, horizontal advection is very hard to measure accurately. However, neglecting horizontal advection while including vertical advection might introduce bias in the results (Aubinet et al. 2003; Feigenwinter et al. 2004).

The three terms on the right-hand side of Eq. (5.1) are the raw turbulent flux (F_{Ctb}), the storage flux (F_{Cst}), and the mass flow term (F_{Cmf}). Estimation of F_{Cmf} is problematic as currently it is not possible to accurately measure the mean vertical velocity due to limitations of the sonic anemometers (Yi et al. 2000). Consequently, in the majority of cases F_C is calculated only from the first two terms. Given their importance in the determination of NEE, and because tall tower NEE determination methodology is still under development, the two terms are discussed separately in the following subsections.

5.2.2 *Raw Turbulent Flux (F_{ctb})*

Several efforts can be made to avoid turbulent signal loss in the calculation of F_{Ctb}. The averaging time (usually 30 min for short towers, and 1 h for tall towers) and trend removal technique (linear trend removal) can be chosen by spectral and sensitivity analyses of the measured data, respectively (Barcza 2001; Berger et al. 2001). In case of closed-path infrared CO_2/H_2O analyzers, lag time caused by the air inlet tube can be determined for proper calculation of the fluxes. Since the flow rate in the inlet tube used for the closed-path CO_2/H_2O analyzers is usually fairly stable in time, a single lag time can be determined from the CO_2 mixing ratio and vertical wind speed time series, and stored for each day. During the flux calculations, the actual lag time can be chosen to correspond to the maximum covariance inside a narrow window around the predetermined lag time.

Prior to data processing, spectral analysis has to be performed to validate the response of the instruments, to inspect the "healthiness" of the power spectra and the cospectra (Kaimal et al. 1972), and thus the measured flux itself.

Depending on the type of the anemometer used, the angle of the attack adjustment has to be applied to the raw sonic anemometer data to ensure unbiased horizontal and vertical wind speed data (van der Molen et al. 2004). Three-dimensional wind vector rotation can be applied to align the coordinate system with the prevailing streamlines (Lee 1998). There are different methods to remove spikes from the time series. The simplest method is to remove data values outside $\pm 4\sigma$. A more sophisticated way using moving windows is described in Vickers and Mahrt (1997). Raw turbulent fluxes can be calculated from the covariances of the fluctuating time series, taking into account the delay time of the signals when necessary.

As part of postprocessing, spectral corrections have to be applied to account for flux loss (Berger et al. 2001); however, the terms of the corrections differ for open- and closed-path CO_2/H_2O analyzers. In case of closed-path systems, causes for the flux loss are signal damping inside the tube, limited time response, sensor separation, and sensor line averaging. For open-path systems, flux losses can appear due to the effects of sensor line averaging, sensor separation, and the limited response time of the anemometer and the open-path infrared gas analyzer (IRGA) (Moore 1986; Leuning and Moncrieff 1990; Lenschow and Raupach 1991; Massman 1991; Berger et al. 2001). Using the above methods, good agreement can be found between the theoretical and the corrected actual cospectra.

When using an open-path IRGA, a further correction has to be applied to correct the effect of density fluctuations (Webb et al. 1980).

Postprocessing can also be performed using commercial software like the EdiRe Data software (EC system developed at the University of Edinburgh; see Moncrieff et al. 1997).

A stationarity test has to be also performed for each averaging period (Foken and Wichura 1996), and the result has to be used during data integration.

5.2.3 The Storage Flux (F_{cst})

Determination of NEE requires accurate rate of change of CO_2 storage calculations for tall tower based measurements (Fan et al. 1990; Grace et al. 1995, 1996; Davis et al. 2003) based on the measured concentration profiles below the EC level. This can be calculated using a finite difference approach of term 2 on the right-hand side of Eq. (5.1):

$$\int_0^{z_m} \frac{\partial \overline{c}}{\partial t} dz = \frac{d}{dt} \int_0^{z_m} \overline{c} dz \approx \frac{\Delta\left(\sum c_z \Delta z\right)}{\Delta t} \tag{5.2}$$

For our case, the applied vertical integration step (Δz) is 1 m, and the averaging time (Δt) is 1 h. Usually, the concentration profiles (c_z) are calculated using simple linear interpolation between levels, where the concentration below the lowest level is considered to be constant. Storage calculation can also be performed using an appropriate similarity function and the measured concentration data (for details on the Hungarian measurements, see Haszpra et al. 2001, 2005).

For short tower sites, storage flux can be estimated by assuming constant CO_2 mixing ratio below the instrumentation level for each half hourly measurement period considering only the temporal change in the mixing ratio.

5.2.4 Low Turbulent Conditions During Nighttime

Significant nighttime flux loss can occur at EC sites due to advection (e.g., horizontal drainage flows) during calm conditions that may bias the year-round integrals of NEE (Aubinet et al. 2000). The logic of the so-called u_* (friction velocity) correction is to replace the fluxes measured during calm periods by values derived from windy periods, based on the temperature dependence of NEE (Massman and Lee 2002). However, this has to be done carefully if CO_2 respired and then accumulated below the level of the measurements system is not removed by drainage flows caused by the uneven terrain at the measurement, because it can be flushed out from the vegetation on the next morning, and captured by the EC system. Site-specific analysis has to be performed to decide on the necessity of the u_* correction and to estimate a threshold for the correction (Massman and Lee 2002; Lee et al. 2004).

5.2.5 Gap-Filling and Partitioning of NEE into Components

Multiple gap-filling methods were developed to gain defensible annual sums of NEE (Falge et al. 2001) in spite of the occasional data gaps, which inevitably occur during long-term monitoring. Methods used for gap-filling are generally also used for the partitioning of NEE into gross primary production (GPP) and total ecosystem respiration (R_{eco}), so they are discussed together.

Gap-filling and flux partitioning employ the measured NEE and the environmental variables measured at the site (air or soil temperature, photosynthetically active photon flux density (PPFD)). The standard method for flux partitioning of the measured NEE, i.e., to determine its main components GPP and R_{eco}, is based on nighttime NEE data (which equals total ecosystem respiration during nighttime) and extrapolates respiration to the day-time period (Reichstein et al. 2005). However, this method can lead to misinterpretations when applying to tall tower data. For an EC system located, e.g., at 100 m height, the mismatch between daytime and nighttime source areas can cause inconsistencies when extrapolating nighttime respiration to daytime. The relationships determined using nighttime fluxes might not be applicable to daytime data since the nighttime source area can be much larger than the daytime one, thus different regions are "seen" through the course of the day (Horst and Weil 1992; Wang et al. 2006). Moreover, prevailing wind might be different during night and day (Barcza et al. 2003); thus anthropogenic emission might also have different effects on the nighttime and daytime fluxes. This effect can especially be pronounced over a heterogeneous landscape.

The application of different gap-filling approaches simultaneously can help to overcome this difficulty. Light response curve based models that require daytime NEE data for both gap-filling and flux partitioning are appropriate methods to be used with tall tower data. By applying more than one method, we might get multiple estimates for GPP and R_{eco}, and also we will get an estimate of the uncertainty of the flux partitioning, which is useful for parameter estimation in ecosystem modeling (Hollinger and Richardson 2005).

We applied three methods discussed in Stoy et al. (2006), namely the Short-Term Exponential (STE), the Rectangular Hyperbolic (RH), and the Non-Rectangular Hyperbolic (NRH) methods. Note that the implementation of these methods is not completely analogous to those in Stoy et al. (2006), though the basic concepts are the same.

In the STE gap-filling and flux-partitioning method, missing measurement intervals of NEE are filled with the empirical light-NEE (for daytime gaps) and exponential nighttime temperature-NEE functions (for nighttime gaps; Lloyd and Taylor 1994) determined by nonlinear regression. The NEE-PPFD relationship was approximated by an RH function (Haszpra et al. 2005) according to the following formula:

$$NEE = \frac{a\text{PPFD}}{(b + \text{PPFD})} + R_d \tag{5.3}$$

where R_d, the intercept parameter, can be interpreted as daytime average rate of respiration, a is the mean apparent ecosystem quantum yield (the initial slope of the light response curve), and b is the maximum rate of photosynthesis. The nonlinear regression can be applied with time windows of variable length following the method of Reichstein et al. (2005). In the case of missing environmental data, the mean diurnal variation (MDV) method is applied (Falge et al. 2001). Usually, the monthly MDV of NEE is determined as the sum of the MDV of the CO_2 flux and that of the storage flux at tall tower sites (Haszpra et al. 2005).

Gross primary production and total ecosystem respiration can be determined using the temperature-dependent respiration function (estimated exclusively from nighttime NEE) of the soil/vegetation system described above. In the STE method, the fitted Lloyd and Taylor functions are used to estimate daytime ecosystem respiration, and then GPP is calculated as the signed difference between NEE and the total daily R_{eco} (Reichstein et al. 2005).

Gap-filling of NEE as well as flux partitioning can also be carried out using two alternative light response curve methods, where the procedure purely relies on daytime NEE data to ensure consistency of the source areas. RH and NRH (Gilmanov et al. 2003) light response curve methods can be chosen to describe PPFD-daytime NEE relationships. The two algorithms differ only in the regression curves, other details of the algorithms are the same.

In case of the RH model, daytime NEE and PPFD data can be used to determine daytime average R_{eco} (R_d in Eq. (5.3)). In the nonrectangular model, the fitted equation is the following:

$$\text{NEE} = \frac{a\text{PPFD} + b - \sqrt{(a\text{PPFD} + b)^2 - 4abc\text{PPFD}}}{2c} - R_d \tag{5.4}$$

where R_d, a, and b have the same meaning as for the RH model, while c is the curvature parameter. It has been shown that in most cases, the above model provides a better fit than the RH method does (for references, see Gilmanov et al. 2003). Note that the RH and NRH methods are basically different from the STE method because daytime respiration is not derived from the nighttime NEE but from day-time data.

To capture short-term variations in NEE, light-response curves can be fitted to NEE and PPFD data from a varying-size sliding window (maximum 10 days). Light-response curves and daytime R_{eco} results are accepted only if there are enough data for regression, i.e., the maximal 10 days long period provide at least 15 measurements. To obtain physiologically correct results, we set up acceptable ranges (maximum and/or minimum) in case of curve-fitting parameters (based on Stoy et al. 2006). Those regressions that result in unrealistic values (outside of their predefined ranges) for any of the parameters, i.e., lead to unrealistic light-response curves, are not accepted. If the regression process is successful, and none of the above criteria is violated, we accept the result as physiologically realistic. Accepted regression parameters are stored for NEE gap-filling.

The intercept of the light-response curves is considered as average respiration of the same time interval that the corresponding NEE data cover. The resulting daytime average rate of respiration data set is examined regarding its temperature dependency (Lloyd and Taylor 1994). The temperature value assigned to each daytime respiration value is the average of temperatures corresponding to the NEE data used in Eq. (5.3) or (5.4). Exponential curve fitting is performed for each period when temperature range exceeded 15°C and at least 10 daytime average rates of respiration data are available. This temperature range ensures that the regression will have a valid result, although does not provide information about the short-term changes in temperature dependency. R_{eco} time series are estimated based on this relationship and gap-filled temperature data.

Gap-filling of day-time NEE data series is based on the light-response curves described above. The gaps are filled using means of light-response curves derived using varying window lengths (maximum of 10 days; ±5 days) around the gap. If no curves are available in this time frame, gap-filling is performed in a second step using wider interval, and the resulting NEE data are flagged as less reliable. This situation often occurs in case of longer measurement gaps.

Nighttime hourly NEE data are assumed to be equal to respiration since according to the convention GPP is zero at night. Estimated R_{eco} values are compared with the measured NEE data and adjusted to the measured value if it is greater than NEE. Note that there are other methods for gap-filling and flux partitioning for tall tower based NEE data (Ricciuto et al. 2008).

5.2.6 Footprint Modeling

The source areas of the measured vertical fluxes can be estimated with state-of-the-art footprint models (Schmid 1994; Kljun et al. 2002; Soegaard et al. 2003;

Sogachev and Lloyd 2004; Wang et al. 2006). Kljun et al. (2004) presented a simple parameterization based on the LPDM-B Lagrangian stochastic footprint model. The underlying model LPDM-B is applicable for well-mixed conditions during convective, neutral, or stable stratification, inside or outside the surface layer. This latter feature makes the parameterized model appropriate for our purposes.

The LPDM-B parameterization estimates the crosswind integrated footprint function, and also the distance of the maximum source weight location (x_{max}) of the footprint function from the tower. Following Kljun et al. (2004), the maximum source weight location can be derived as

$$x_{max} \approx A_x \left(B - \ln z_0\right) z_m \left(\frac{\sigma_w}{u_*}\right)^{-\alpha}, \tag{5.5}$$

where z_m is the measurement height, σ_w is the standard deviation of the vertical wind speed fluctuation, u_* is the friction velocity, and A_x, B, and α are model parameters set to 2.59, 3.42, and 0.8, respectively (see Kljun et al. 2004 for details). The applicability of Eq. (5.5) is restricted to dynamically homogeneous terrain with the following conditions: $-200 \le z_m/L \le 1$; $u_* \ge 0.2$ m s^{-1}; $z_m > 1$ m; $z_m < h$ (L is the Obukhov length and h is the height of the planetary boundary layer). For the estimation of the hourly source regions of the Hungarian measurement sites, we used the roughness length classification scheme of the European Wind Atlas (Troen and Petersen 1989).

5.3 Dark Chamber Methods for Measuring Carbon Dioxide, Nitrous Oxide, and Methane Soil Flux

The chamber method commonly incorporates three measurement schemes: the open chamber method, the closed chamber static method, and the closed chamber dynamic method. During the measurement, chambers (dark or transparent, static or dynamic) are settled on the soil. In case of dark static chambers, changes in gas concentration as a consequence of soil emission or uptake are determined either by in situ measurements or by periodic sampling into vials after the closure of the chamber and by subsequent (e.g., gas chromatography) analysis. Flux is directly derived from the measured concentration change and the chamber size.

Depending on the presence or absence of air circulation, chamber methods have been categorized as either static or dynamic. CO_2, CH_4, and N_2O are nonreactive gases, so there is no substantial loss between the sampling and analysis. Therefore, the static chamber method is adequate for all these gases. The disadvantage of the static chamber method is the uncertain "effective" chamber volume, because the volume of the air-filled soil pores is not taken into account. Other disadvantage of using static chambers is the saturation effect phenomenon. Thus, while concentration increases in the chamber, emission and deposition processes tend to reach a dynamic equilibrium resulting in a saturation curve as the function of time, instead of linear concentration–time relationship. The deviation from linearity starts from in order of minutes – 10 min varying by components and chamber's dimensions.

The use of dynamic dark chambers for soil flux measurements – where air in the chamber is continuously leached by outside air – eliminates the disadvantage of the static chambers but needs in situ measurement of concentrations in the air entering and leaving the chamber. This method can be applied in case of intensive surface exchange of gases (typically for CO_2) to detect systematic differences between concentrations, and highly depends on the precision of the instruments.

Further, we introduce the dark chamber measurements, commonly used for measuring soil fluxes in Hungary.

5.3.1 Manual Static Chamber with Gas Chromatography Detection

Carbon dioxide emission measurements are carried out on experimental sites with different soil management practices, aiming to find coherences between soil properties and soil flux at various levels of soil disturbance. In situ measurements are completed and harmonized with laboratory incubation experiments, developed at the Research Institute for Soil Science and Agricultural Chemistry (see below). The methodology of in situ measurements is adopted from the Laboratory of Soil Science of the Kyoto University, Japan. In 2005, the air sampling method and the chamber type were further developed.

From the year of 2005, homemade chambers have been used for in situ emission measurements (Fig. 5.1). The chamber has two parts: a frame (40 cm × 50 cm × 12 cm) made of steel that can be inserted in the soil to a maximum depth of 10 cm. It has a slight rim filled with water for the insulation, to prevent the ventilation between inside and outside air. The other part of the chamber is the cover (40 cm × 50 cm × 15 cm) made of steel sheet. It fits into the rim on the frame. For the measurements, the frame is placed in the soil to a depth of 7 cm, leaving 20 cm high cover above the soil. Air samples can be taken through a hole on the chamber top closed with septum. Air samples are taken from the headspace of the chamber at the beginning of the measurements and one and a half hour later. Samples are injected into evacuated vials (Exetainer tube, Labco Limited, UK) and are analyzed using gas chromatograph as soon as possible. Control measurements justified the linearity of concentration changes in time during the 30- and 60-min closures (Fig. 5.2).

Parallel with sampling, soil water content and air temperature were also measured. Additionally, disturbed soil samples were also collected and analyzed for soil physical, chemical, and biological properties.

Fig. 5.1 Closed chamber for in situ emission measurements

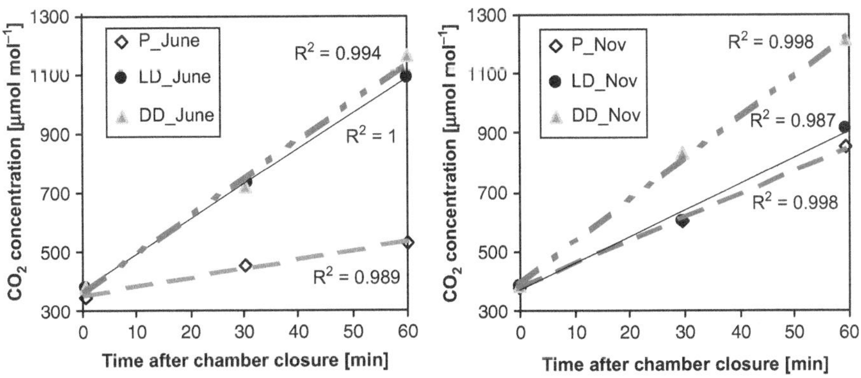

Fig. 5.2 Example linearity tests for CO_2 emission measurements performed at different times (in June and November 2005) in three different soil tillage practices. P, LD, and DD stand for ploughing, loosening combined with disking and direct drilling, respectively

5.3.2 *Automatic Static Chamber with Photo-Acoustic Detection*

During experiments, dark static chambers are settled on the bare soil, without rim at the start of the measurements. The volume of the chambers is 3854 cm^3, with height of 16.3 cm.

On arable lands, measurements were taken in series, five times directly after tillage and at the end of the first and the second days. Checks over longer periods are recommended in particular in the case of deeper soil disturbance. The INNOVA 1312 photo-acoustic multigas monitor was used for the measurements (Reth et al. 2005), together with the TESTO 535 monitor in three repetitions. The instrument measures CO_2 concentration in µmol mol^{-1} unit (ppm). Closed measuring chambers were placed on the soil at locations, representing the characteristic soil surface of the studied tillage systems. CO_2 concentrations were measured in the chambers at $t = 0$, 30, 45, 60, and 90 min, as well as 24 and 48 h after starting the experiment. After each measurement, the chambers were ventilated for 1 min by removing the top.

Soil CO_2 efflux (F) was calculated from the increase in CO_2 concentration over time according to Widén and Lindroth (2003) and Tóth et al. (2005):

$$F = \frac{1}{A} * \frac{pV}{RT} * \frac{\Delta C}{\Delta t} * M, \quad (5.6)$$

where F – CO_2 efflux (g m^{-2} day^{-1}), A – the area of the measuring chamber (m^2), V – the volume of the measuring chamber (m^3), p – the air pressure = 10132.5 N m^{-2}, R – gas constant = 8.314 J mol^{-1} K^{-1}, T – air temperature (K), ΔC – increase of mixing ratio of CO_2 in chamber during the closure (mol mol^{-1}), Δt = length of measurements (days), M – molar mass of CO_2 = 44.01 g mol^{-1}.

5.3.3 Dynamic Chamber with Infrared Detection

Soil respiration (R_s) measurements with dynamic chambers employ a LICOR-6400 Infra Red Gas Analyzer (IRGA) connected to a 6400-09 type soil chamber (Li-Cor Inc., NE, USA). During the measurements, the sharpened chamber base is inserted into the soil to ~3–5 mm depths for avoiding deeper disturbance of the soil and cut of near-surface roots (Wang et al. 2005).

Measurements were made in 3–5 spatial replications by placing the chambers on the soil surface for 1–3 min, depending on the respiration activity. Diurnal changes of soil respiration were also followed in every 2–3 h in whole day long measurement campaigns.

Soil water content and soil temperature as main drivers of soil respiration were measured at the same time (and the same place) as soil respiration by a TDR reflectometer (ML2, Delta-T Devices Co., Cambridge, UK) at 0–6 cm depth and by a digital soil thermometer at 5 cm depth.

Soil respiration data sets were fitted against soil temperature using the Lloyd–Taylor model (Lloyd and Taylor 1994):

$$R_s = R_{10} e^{\left(E_0 \left(\frac{1}{56.02} - \frac{1}{T - 227.13}\right)\right)}, \quad (5.7)$$

where R_{10} is the respiration rate at 10°C, T is the soil temperature at 5 cm (K), and E_0 is a parameter related to the activation energy (K).

This model was modified after Byrne et al. (2005) incorporating soil water content (SWC) (Eq. 5.8):

$$R_s = R_{10} e^{\left(E_0 \left(\frac{1}{56.02} - \frac{1}{Ts - 227.13}\right)\right)\left(-0.5 \ln\left(\frac{SWC}{SWC_{opt}}\right)^2\right)} \quad (5.8)$$

Using this modification, the values of R_{10}, E_0, and the optimal soil water content for soil respiration (SWC_{opt}, vol%, in Eq. (5.8)) were fitted simultaneously to the measured data.

Models based on Eqs. (5.7) and (5.8) use soil temperature and soil water content as driving variables for simulating daily sums of soil respiration.

Curve- and surface-fitting procedures were performed, and fitting statistics calculated by using Sigmaplot 8.0 (SPSS Inc).

5.3.4 Manual Chamber Method with Gas Chromatography Detection for Nitrous Oxide and Methane Soil Flux

Nitrous oxide and CH_4 fluxes were determined using dark, static (closed) chambers according to Christensen et al. (1990). Samplings, measurements, and calculations are described in detail in Horváth et al. (2006, 2008a, b, 2010) and Machon et al. (2010).

Two types of chambers were applied depending on the location: small ($A = 80$ cm^2, $V = 400$ cm^3, $h = 5$ cm) and large ($A = 2{,}500$ cm^2, $V = 12{,}500$ cm^3, $h = 5$ cm) chambers. Rims of chambers were pushed in the soil to 4 cm depth and were left permanently there to avoid the sudden emission peaks after the installation. The rims were installed so that no enclosure of tall plants could occur. The short grass within the rim area was not clipped before samplings. The rims were covered by the chamber body only during the 30 min of incubation, allowing normal light and precipitation exposure at all other times. Although there were not visible differences between soil conditions inside and outside the rims, some disturbances in soil moisture cannot be ruled out.

Air samples were taken 0, 10, 20, and 30 min after closure by a syringe. A total of 6–10 cm^{3}of samples was injected into evacuated vials of 5.6–10 cm^3 volume, respectively. Concentration changes of N_2O between 2002 and 2005 were measured by a gas chromatograph combined with mass-spectrometer (HP 5890 II.-HP 5972, equipped by HP-PlotQ column). From 2006, the concentration of N_2O and CH_4 were simultaneously measured by a gas chromatograph with GS-CARBONPLOT column, detected by electron capture detector (GC-ECD) and flame ionization detector (GC-FID), respectively. The NOAA, Environmental Monitoring and Diagnostics Laboratory, the Scott and the Messer companies provided the calibration and the standard gases.

Soil fluxes were calculated as

$$F_{N2O} = (\Delta C * 2 * A_N * V_{\mathrm{ch}} * 60 * f) / \left(V_{\mathrm{m}} * A_{\mathrm{ch}} * t\right) = 3.5 * \Delta C / t * f \quad (5.9)$$

and

$$F_{\mathrm{CH4}} = (\Delta C * M_{\mathrm{CH4}} * V_{\mathrm{ch}} * 60 * f) / \left(V_{\mathrm{m}} * A_{\mathrm{ch}} * t\right) = 2 * \Delta C / t * f, \quad (5.10)$$

where F is the flux (μg N m^{-2} h^{-1} or μg CH_4 m^{-2} h^{-1}), respectively, $\Delta C = C_{30} - C_0$ is the difference in mixing ratios (ppb) in chambers at the end and the start of samplings: $t = 30$ and $t = 0$ min, A_N is the atomic weight of N, M_{CH4} is the molecular weight of CH_4, V_{ch} the volume of chambers (m^3), 60 is the time conversion factor (min h^{-1}), f is the factor taking into account the residual pressure in the evacuated tubes (varies between 1.090 and 1.233), V_m is the molar volume – 24 l at laboratory temperature ($t = 20°C$) during measurements, A_{ch} is the surface of soil covered by chambers, t is the sampling time (min).

Precision and detection limit of concentration measurements were estimated as follows. According to statistical analysis, the nonsystematic bulk error (CV – coefficient of variation) of sampling and analysis estimated using parallel samplings was always below 10% for both components. CV was determined using at least 10 samples at the start of samplings ($t = 0$). Since samples were taken immediately after closure, $t = 0$ concentrations are close to the relatively constant atmospheric background values (320 ppb and 1.9 ppm for N_2O and CH_4, respectively). According to precision tests, coefficient of variation ranges from 4% to 6%. The detection limit was determined taking into account a minimum of 10% changes in concentration during sampling (30 min) from the initial background values. According to this criterion, the calculated detection limits for N_2O and CH_4 fluxes are 1.3 μg N m^{-2} h^{-1} and 10 μg CH_4 m^{-2} h^{-1}.

There are some evidences on the effect of nonlinear accumulation rate in the chamber on the calculated emission (e.g., Stolk et al. 2009). Sampling duration was optimized to reach as high accumulation rate as possible and to eliminate the effect of nonlinearity. Several pilot measurements justified the linearity of concentration

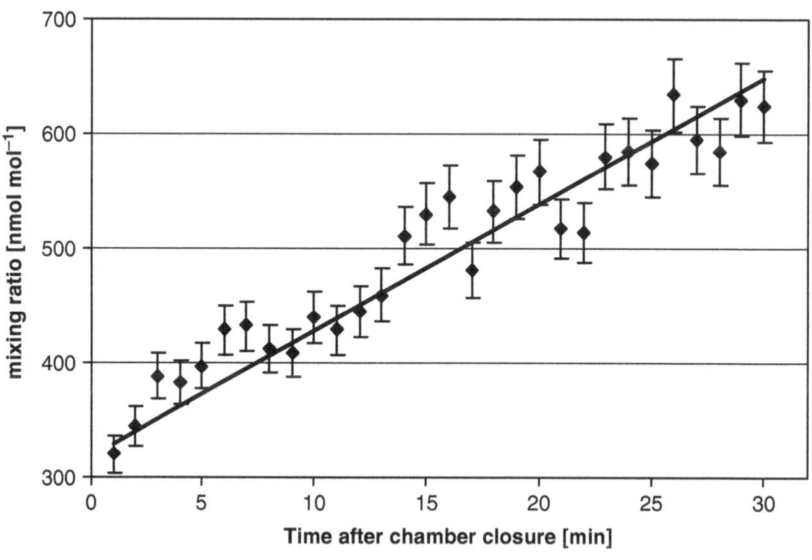

Fig. 5.3 Example linearity test for N_2O emission measurements. Error bars indicate the precision of the measurements

changes in the used static chambers in the first 30 min of the sampling (e.g., see Fig. 5.3). According to frequent tests, the error was always below 10%.

There is evidence for a systematic underestimation of the soil gas flux using closed chambers. For CO_2 soil efflux, Rayment (2000) pointed out that the bias was caused by the fact that the effective volume of the chamber was larger than the physical chamber volume, since it also included the volume of air-filled spaces in the soil. Underestimation of fluxes increased with the decrease of chamber's height. In our case ($h = 5$ cm), using Rayment's estimates for an equivalent depth of air in soil, N_2O soil fluxes are underestimated by 24%. Conen and Smith (2000) explained and modeled this phenomenon. They predict N_2O concentration increase in the soil air immediately after closure of the chamber, as part of produced gas is stored in the upper part of soil profile. Christensen et al. (1996) compared 10 different closed, dynamic chambers and micrometeorological methods and found good agreement between the results; however, using small chambers the magnitude of the underestimation was 20% against the standard method.

Gap-filling: as observations vary from weekly to monthly samplings, and the distribution of samplings is not even during the year, a simple gap-filling method was applied to calculate the yearly mean fluxes. For stations with continuous sampling (few samplings in each month), monthly means were calculated from the individual measurements. Annual N_2O and CH_4 emissions were calculated by summing up the monthly mean fluxes for the year. For stations with incomplete sampling during the year, where samples were taken solely in vegetation period, the ratio of fluxes in the examined and the missed period was determined by means of the flux ratio of stations with complete sampling protocol. This ratio was used as a gap-filling correction factor for the calculation of the average yearly flux sums. As a consequence of the discontinuous sampling, emission peaks that might have occurred – for example, after melting or high rainfall preceded by a drought period – could not be detected. However, because modeled and measured yearly fluxes for a site with the most frequent measurements (Bugac) are in good agreement (Nagy et al. 2007; Grosz et al. 2008), emission peaks probably play less important role in yearly account of fluxes.

5.3.5 Laboratory Incubation Method

This newly developed method was used for measuring soil CO_2 respiration from large undisturbed soil cores under controlled laboratory conditions to examine the effect of soil structure and pore-size distribution on soil biological processes and CO_2 emission.

Structureless undisturbed soil samples are frequently used for laboratory analyses to study structure-dependent soil properties and processes like CO_2 emission. This approach eliminates the effect of soil structure and pore-size distribution on ratio between solid, liquid, and air phases of soil, and their strong influence on soil biological processes and CO_2 emission. Measurements are usually carried out at similar soil water contents, which are calculated on mass base. It makes difficult to compare the results, obtained for soils belonging to different textural classes, because the same soil water content in, e.g., sandy and clay soils reflects totally different energetic status of water stored in the soil sample. The new measurement technique makes possible to account for the effect of soil structural status and soil water potential on soil CO_2 emission, and to calculate water balance elements on volume base.

The method was applied for studying soil microbial respiration of soil samples, taken from different soil tillage systems. As soil respiration concerns topsoil where soil microbial activity is the most intensive (Abgeko and Kita 2007), soil samples for the laboratory analyses were collected from the upper 10 cm layer of the soil. Twenty-three samples per treatment were collected with no disturbance in PVC tubes (20 cm long; 10.5 cm i.d.); 20 for laboratory CO_2 emission and 3 for determination of bulk density. The tubes were carefully inserted into the soil to the depth of 10 cm, and then they were taken out manually with the soil inside. The tubes were closed from the bottom during transportation. In addition, three undisturbed soil samples of 100 cm^3 volume were also taken from each of the examined treatments and bulk density, and soil water retention characteristics – water contents, corresponding to water potential of −1, −2.5, −10, −32, −100, −200 and of −500, −2,500, −15,850 hPa – were determined according to Várallyay (1973). Disturbed soil samples were collected and analyzed for soil physical, chemical, and biological properties such as texture, NH_4, NO_3, total N contents (Bremner 1965), AL-P_2O_5, AL-K_2O contents (Sarkadi et al. 1965), humus content, $CaCO_3$ content, organic carbon content, pH, substrate-induced respiration (SIR), and microbial biomass carbon content. Water-extractable organic carbon (WEOC) and water-extractable organic nitrogen (WEON) (0.01 M $CaCl_2$; 1:10 soil to solution) were measured by Apollo 9000 Combustion TOC Analyzer (Teledyne Tekmar Co., Mason, Ohio, USA).

The large undisturbed soil columns were placed in a climatic room and watered from the top to obtain four different soil water potential values inside the columns in five replicates. The amount of water added to the samples was determined from (i) the initial soil water contents and (ii) the water retention curves, as described further. Three large undisturbed columns per treatment were destroyed before starting the experiment so that their bulk densities and water contents could be measured. The oven-drying method was used for this purpose. Soil water contents, corresponding to the four desirable soil water potentials (−200, −500, −1,585, and −2,510 cm) were derived from the soil water retention curves of the small, 100 cm^3 volume soil cores. Further, the aimed soil water potential values in the 20 PVC tubes were set up on the basis of these data, using average values for calculations. Depending on the weight of the large soil columns, different amount of water was added to each of them to reach the wanted desired soil moisture status. These columns remained undisturbed during the whole measurement procedure. Thus, their initial soil water content and bulk density values could be only estimated at the first stage. Both values for each soil column were determined after the experiment ended and pF values of samples were revised on the basis of the real data.

Soil CO_2 emission measurements were carried out weekly in a climatic room at constant air temperature, humidity, and light conditions in five replicates.

After four measurement dates, samples were watered back to their initial mass. Because the primary objectives were to identify soil tillage-induced differences in soil respiration, no mineral or organic fertilizers were added to soil.

The air temperature in the climatic room was set to a constant value of 21°C. The mass of each soil column was determined at each air sampling time to calculate the volumetric soil water content and pF values after the experiment. The above described experimental setup enabled us to evaluate soil CO_2 emission on soil samples having undisturbed structure at permanent circumstances and at given volume-based soil water content values.

For the incubation, the top of the PVC tubes was also airtight closed with caps. Beforehand, the volume of the headspace was calculated for each column as a difference between the tube and soil sample volumes. Air sampling was performed from the headspace through a septum inserted into the hole of caps once a week during 3 months. At each sampling time, double samplings were carried out: one after closing the tubes and the other one 3 h later. At the first measurement event, additional samples were taken after 1 and 6 h of incubation in order to evaluate the effect of incubation time on measurement results. Air samples were collected into evacuated vials (Exetainer tube, Labco Limited, UK) and then the CO_2 concentration was analyzed by gas chromatograph as soon as possible.

Soil CO_2 fluxes were calculated from detected changes in CO_2 concentration during the incubation time using equation reported by Tóth et al. (2005).

5.4 Carbon Stock Measurements in Forest Ecosystems

Measurements of net carbon flows for forests generally lend themselves to the stock-change method, which estimates the amount of sequestered carbon as the net change in carbon stocks over time (Pearson et al. 2007).

The carbon stocks of trees are estimated most accurately and precisely by direct methods, e.g., through a field inventory, where all the trees in the sample plots above a minimum diameter are measured. The spatial representativeness of the results, obtained by the direct methods, however, highly depends on the selection of the number and location of the sampling plots. In many cases, when direct measurement of NPP (total net primary production) is not possible, particularly due to difficulties in measuring below-ground biomass growth and uncertainties related to below-ground biomass turnover, NPP has to be estimated indirectly.

One method for estimating NPP involves the use of EC flux measurement combined with respiration measurement in the forest stand. However, estimates of NPP from EC and respiration measurements are related to problems of carbon flux partitioning, particularly fluxes related to heterotrophic and autotrophic respiration (Hanson et al. 2000).

Alternative approach to obtain an independent estimate of yearly NPP of a forest stand involves classical forestry techniques of tree growth measurement. Biomass and carbon stock are estimated using appropriate allometric equations applied to the tree measurements. Tree biomass is often estimated from equations that relate biomass to tree diameter at breast height (dbh) or to dbh and tree height as independent variables. For estimating root biomass, the use of root/shoot ratios is a common method. Research in that field has recently intensified (Cairns et al. 1997; Levy et al. 2004; Balboa-Murias et al. 2006).

In this chapter, we introduce the direct method for measuring gross primary production that can be used for estimating the carbon exchange between forest ecosystems and atmosphere. The indirect approach, based on tree growth measurements using dendrometer bands has many site- and stand-specific aspects; therefore, it is described in detail in Chapter 7.

5.4.1 Determination of Organic Matter Quantity

Considering long-term production, the mass of organic material in the forest ecosystems consists of above- and below-ground components, which can be ranked into four groups. The above-ground organic material is located in foliage (leaves, etc.) (1st group), in stemwood and branches (2nd group), while the below-ground organic material is stored in stump and root swelling (root-wood) (3rd group) and in the root system (4th group).

Changes in all the four components can be monitored in surveying plots and measured as described in Sections 5.4.2–5.4.5. In the surveying plots, trees are marked by size-groups, considering the yield and the structural conditions of the stand. Each group contains two trees with average diameter and height, one tree above and another one below the average size. At the beginning of the so-called sustenance phase (in August) when organic material production ends, the selected trees are felled and on-site measurements are carried out by compartments (Führer and Jagodics 2009).

5.4.2 Foliage

The foliage of the felled trees is fully picked off and weighted in green condition. Foliage samples are taken afterwards for determining their dry weight and for chemical analyses in the laboratory. Assuming that the laboratory samples represent the foliage, quantitative determination of the complete foliage dry mass of the trees (leaves, crop, etc.) can be carried out. The foliage-mass of the sample trees is further converted to 1 ha by size-groups.

5.4.3 Determination of the Weight of the Stemwood and Branches

Determination of the weight of the stem (wood and bark) and of the larger branches is carried out for the usual diameter groups (6, 8, 10, 12 cm) by cutting the stems and branches downwards from the top (Fig. 5.4). The stem part, which is larger than

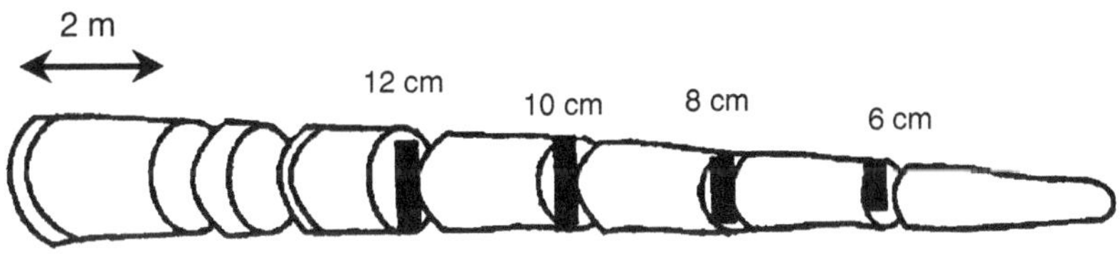

Fig. 5.4 Determination of the organic material of the stem, downwards from the top

12 cm in diameter, is cut into 2-m long pieces. The weight of all pieces is measured, and then average samples in the form of sample disks are taken from the branches (living and dead) in order to determine the absolute dry mass and to provide samples for further chemical analysis.

5.4.4 Weight Determination of the Root System

The estimation of the weight of roots requires complete excavation of the root system, which is impossible in practice. Therefore, the weight of the root section remaining in the soil can be estimated from subsamples (Van Laar and Akça 1997).

In the investigated stands, after felling, 1 m wide and 1.5 m deep (10–20 cm deeper than the growing depth) trenches touching the stumps were dug among three neighboring trees with an average trunk distance.

From the wall of the trench, 1–1 dm^3 of soil monoliths were taken with a monolith sampling shovel by the method of Járó (1995), starting horizontally from one tree to the second and third one. Monolith sampling was carried out vertically to 1.5 m, till roots were found. After sampling the whole profile-wall, the process was repeated in the following layer.

A map was drawn from the profile-wall with the numbers of monoliths. From the continuously numbered soil monoliths, the roots were collected in dry condition and categorized by their thickness. The absolute dry weights of the separated roots were determined, average samples were collected by diameter-groups for analyzing their chemical composition in the laboratory.

5.4.5 Determination of the Weight of Stump and Root Swelling Mass

After removing the soil monoliths and lifting of the stump and root swelling, all these materials were cleaned from the soil and measured in green condition. Average samples were taken to determine the absolute dry weight and the chemical composition.

5.4.6 Determination of Carbon Concentration

Average samples, taken from the tree components (leaves, branches, bark, stem, etc.) were combusted at 1,000°C in the laboratory with a C-N-S element analyzer (Eurovector EA 300). By use of the concentrations in the absolute dry samples, the data were converted into total weight of the tree components calculated for 1 ha.

References

Abgeko JF, Kita M (2007) A qualitative experiment to analyze microbial activity in topsoil using paper and handmade reflection photometer. J Chem Educ 84:1689–1690

Aubinet M, Grelle A, Ibrom A, Rannik U, Moncrieff J, Foken T, Kowalski AS, Martin PH, Berbigier P, Bernhofer C, Clement R, Elbers J, Granier A, Grunwald T, Morgenstern K, Pilegaard K, Rebmann C, Snijders W, Valentini R, Vesala T (2000) Estimates of the annual net carbon and water exchange of forests: the EUROFLUX methodology. Adv Ecol Res 30:113–175

Aubinet M, Heinesch B, Yernaux M (2003) Horizontal and vertical CO_2 advection in a sloping forest. Bound Layer Meteorol 108:397–417

Balboa-Murias M, Rojo A, Alvarez J et al (2006) Carbon and nutrient stocks in mature Quercus robur L stands in NW Spain. Ann For Sci 63:557–565

Barcza Z (2001) Long term atmosphere/biosphere exchange of CO_2 in Hungary. Ph.D. Thesis, Eötvös Loránd University, Department of Meteorology, Budapest. Also available online at http://nimbus.elte.hu/~bzoli/thesis/

Barcza Z, Haszpra L, Kondo H et al (2003) Carbon exchange of grass in Hungary. Tellus 55B:187–196

Barcza Z, Kern A, Haszpra L, Kljun N (2009) Spatial representativeness of tall tower eddy covariance measurements using remote sensing and footprint analysis. Agric For Meteorol 149:795–807. doi:10.1016/j.agrformet.2008.10.021

Berger BW, Davis KJ, Bakwin PS, Yi C, Zhao C (2001) Long-term carbon dioxide fluxes from a very tall tower in a northern forest: flux measurement methodology. J Atmos Oceanic Technol 18:529–542

Boone RD, Nadelhoffer KJ, Canary JD, Kaye JP (1998) Roots exert a strong influence on the temperature sensitivity of soil respiration. Nature 396:570–572

Bremner JM (1965) Total nitrogen. In: Black CA et al (eds) Methods of soil analysis Part 2 Chemical and microbiological properties. American Society of Agronomy, Madison, WI

Buchkina NP, Balashov EV, Rizhiya EY, Smith KA (2010) Nitrous oxide emissions from a light-textured arable soil of North-Western Russia: effects of crops, fertilizers, manures and climate parameters. Nutr Cycl Agroecosyst. Available at http://www.citeulike.org/journal/springerlink-100322

Bunt JS, Rovira AD (1954) Oxygen uptake and carbon dioxide evolution of heat sterilized soil. Nature 173:1242

Byrne AK, Kiely G, Leahy P (2005) CO_2 fluxes in adjacent new and permanent temperate grasslands. Agric For Meteorol 135:82–92

Cairns MA, Brown S, Helmer EH et al (1997) Root biomass allocation in the world's upland forests. Oecologia 111:1–11

Christensen S, Simkins S, Tiedje JM (1990) Spatial variation in denitrification: dependency of activity centers on the soil environment. Soil Sci Soc Am J 54:1608–1613

Christensen S, Ambus P, Arah JRM et al (1996) Nitrous oxide emission from an agricultural field: comparison between measurements by flux chamber and micro-meteorological techniques. Atmos Environ 30:4183–4190

Conen F, Smith A (2000) An explanation of linear increases in gas concentration under closed chambers used to measure gas exchange between soil and the atmosphere. Eur J Soc Sci 51:111–117

Davis KJ, Bakwin PS, Yi C et al (2003) The annual cycles of CO_2 and H_2O exchange over a northern mixed forest as observed from a very tall tower. Glob Change Biol 9:1278–1293

de Jong E, Schappeart HJV, Macdonald KB (1974) Carbon dioxide evolution from virgin and cultivated soil as affected by management practices and climate. Can J Soil Sci 54:299–307

Delle Vedove G, Alberti G, Peressotti A et al (2007) Automated monitoring of soil respiration: an improved automatic chamber system. Ital J Agron 4:377–382

Edward NT (1975) Effects of temperature and moisture on carbon dioxide evolution in a mixed deciduous forest floor. Proc Soil Sci Soc Am J 39:361–365

Falge E, Baldocchi D, Olson R et al (2001) Gap-filling strategies for defensible annual sums of net ecosystem exchange. Agric For Meteorol 107:43–69

Fan SM, Wofsy SC, Bakwin PS et al (1990) Atmosphere-biosphere exchange of CO_2 and O_3 in the central Amazon Forest. J Geophys Res 95D:16851–16864

Feigenwinter C, Bernhofer C, Vogt R (2004) The influence of advection on the short term CO_2-budget in and above a forest canopy. Bound Layer Meteorol 113:201–224

Finnigan JJ, Clement R, Malhi Y et al (2003) A re-evaluation of long-term flux measurement techniques part I: averaging and coordinate rotation. Bound Layer Meteorol 107:1–48

Foken T, Wichura B (1996) Tools for quality assessment of surface-based flux measurements. Agric For Meteorol 78:83–105

Freijer JI, Leffelaar PA (1996) Adapted Fick's law applied to soil respiration. Water Resour Res 32:791–800

Führer E, Jagodics A (2009) Carbon stocks in the stands of climate indicator tree species. "KLÍMA-21" Füzetek 57:43–55 (in Hungarian with English summary)

Gilmanov TG, Verma SB, Sims PL et al (2003) Gross primary production and light response parameters of four Southern Plains ecosystems estimated using long-term CO_2-flux tower measurements. Glob Biogeochem Cycles 17:1071

Grace J, Lloyd J, McIntyre J et al (1995) Fluxes of carbon dioxide and water vapour over an undisturbed tropical forest in south-west Amazonia. Glob Change Biol 1:1–12

Grace J, Malhi Y, Lloyd J et al (1996) The use of eddy covariance to infer the net carbon dioxide uptake of Brazilian rain forest. Glob Change Biol 2:209–217

Grosz B, Horváth L, Machon A (2008) Modelling soil fluxes of nitrogen and carbon gases above a semi arid grassland in Hungary. Cereal Res Com (suppl) 36:1523–1526

Hanson PJ, Edwards NT, Garten CT et al (2000) Separating root and soil microbial contributions to soil respiration: a review of method and observations. Biogeochemistry 48:115–146

Haszpra L, Barcza Z, Bakwin PS et al (2001) Measuring system for the long-term monitoring of biosphere/atmosphere exchange of carbon dioxide. J Geophys Res 106D:3057–3070

Haszpra L, Barcza Z, Davis KJ et al (2005) Long-term tall tower carbon dioxide flux monitoring over an area of mixed vegetation. Agric For Meteorol 132:58–77. doi:101016/jagrformet200507002

Hollinger DY, Richardson AD (2005) Uncertainty in eddy covariance measurements and its application to physiological models. Tree Physiol 25:873–885

Horst TW, Weil JC (1992) Footprint estimation for scalar flux measurements in the atmospheric surface layer. Bound Layer Meteorol 59:279–296

Horváth L, Führer E, Lajtha K (2006) Nitric oxide and nitrous oxide emission from Hungarian forest soils; linked with atmospheric N-deposition. Atmos Environ 40:7786–7795

Horváth L, Grosz B, Czóbel S et al (2008a) Measurement of methane and nitrous oxide fluxes in Bodrogköz Hungary; preliminary results. Acta Biol Szeged 52:119–122

Horváth L, Grosz B, Machon A et al (2008b) Influence of soil type on N_2O and CH_4 soil fluxes in Hungarian grasslands. Commun Ecol 9(suppl):75–80

Horváth L, Grosz B, MachonA et al (2010) Estimation of nitrous oxide emission from Hungarian semi-arid sandy and loess grasslands; effect of soil parameters grazing irrigation and application of fertilizer. Agric Ecosys Environ (accepted)

Houghton JT, Callander BA, Varney SK (eds) (1992) Climate change, 1992. The supplementary report on the IPCC scientific assessment. Supplementary report. Cambridge University Press, Cambridge, 200 pp

Járó Z (1995) Yearly organic matter production of the most important semi-natural-, second-growth and man-made forests. Report of OTKA-1385 research project (in Hungarian)

Jorgensen JR, Wells CG (1973) The relationship of respiration in organic and mineral soil layers to soil chemical properties. Plant Soil 9:373–387

Kaimal JC, Wyngaard JC, Izumi Y et al (1972) Spectral characteristics of surface-layer turbulence. Q J R Meteorol Soc 98:563–589

Kasparov SV, Minko OI, Amosova Ya M, Perova NE (1986) Approaches for studying soil gas profile functionality. Pochvovedenie 10:127–130 (in Russian)

Khalil MAK, Shearer MJ (1993) Sources of methane: an overview. In: Khalil MAK (ed) Atmospheric methane: sources sinks and role in global change. Springer, Berlin, pp 180–198

Kimball BA, Lemon ER (1970) Spectra of air pressure fluctuations at the soil surface. J Geophys Res 75:6771–6777

Kljun N, Rotach MW, Schmid HP (2002) A three-dimensional backward lagrangian footprint model for a wide range of boundary-layer stratifications. Bound Layer Meteorol 103:205–226

Kljun N, Calanca P, Rotach MW et al (2004) A simple parameterisation for flux footprint predictions. Bound Layer Meteorol 112:503–523

Larionova AA, Rozonova LN, Samoylov TI (1988) Gas exchange dynamics in a grey forest soil profile. Pochvovedenie 11:68–74 (in Russian)

Le Mer J, Roger P (2001) Production oxidation emission and consumption of methane by soils: a review. Eur J Soil Biol 37:25–50

Lee X (1998) On micrometeorological observations of surface-air exchange over tall vegetation. Agric For Meteorol 91:39–49

Lee X, Massman W, Law B (eds) (2004) Handbook of micrometeorology: a guide for surface flux measurement and analysis. Springer, Dordrecht, the Netherlands, p 250

Lemon E (1967) Aerodynamic studies of CO_2 exchange between the atmosphere and the plant. Harvesting the sun 1967:263–290

Lenschow DH, Raupach MR (1991) The attenuation of fluctuations in scalar concentrations through sampling tubes. J Geophys Res 96D:15259–15268

Leuning R, Moncrieff J (1990) Eddy-covariance CO_2 flux measurements using open- and closed-path CO_2 analysers: corrections for analyser water vapour sensitivity and damping of fluctuations in air sampling tubes. Bound Layer Meteorol 53:63–76

Levy PE, Hale SE, Nicoll BC (2004) Biomass expansion factors and root: shoot ratios for coniferous tree species in Great Britain. Forestry 77:421–430

Lloyd J, Taylor JA (1994) On the temperature dependence of soil respiration. Funct Ecol 8:315–323

Lopes de Gerenu VO, Kurganova IN, Zamolodchikova DG, Kudeyarov VN (2005) Methods for quantifying soil carbon dioxide efflux. In: Eskov AI et al (eds) Methods for analysign soil organic material. VNIPTIOU, Russian Academy of Sciences, Russian, pp 408–423

Macfayden A (1963) The contribution of the microfauna to total soil metabolism. In: Doeksen J, Van der Drift J (eds) Soil organisms. North Holland, Amsterdam, 346 pp

Macfayden A (1970) Soil metabolism in relation to ecosystem energy flow. In: Phillipson J (ed) Methods of study in soil ecology. IBP/UNESCO Symp, Paris, pp 167–172

Machon A, Horváth L, Weidinger T et al (2010) Estimation of net nitrogen flux between the atmosphere and a semi-natural grassland ecosystem in Hungary. Eur J Soil Sci DOI: 10.1111/j.1365-2389.2010.01264.x

Massman WJ (1991) The attenuation of concentration fluctuations in turbulent flow through a tube. J Geophys Res 96D:15269–15273

Massman WJ, Lee X (2002) Eddy covariance flux corrections and uncertainties in long-term studies of carbon and energy exchanges. Agric For Meteorol 113:121–144

Moncrieff JB, Massheder JM, de Bruin H, Elbers J, Friborg T, Heusinkveld B, Kabat P, Scott S, Sogaard H, Verhoef A (1997) A system to measure surface fluxes of momentum, sensible heat, water vapour and carbon dioxide. J Hydrol 188–189:589–611

Moore CJ (1986) Frequency response corrections for eddy correlation systems. Bound Layer Meteorol 37:17–35

Nagy Z, Pintér K, Czóbel S, Balogh J, Horváth L, Fóti S, Barcza Z, Weidinger T, Csintalan Z, Dinh NQ, Grosz B, Tuba Z (2007) The carbon budget of semi-arid grassland in a wet and a dry year in Hungary. Agric Ecosyst Environ 121:21–29. doi:10.1016/j.agee.2006.12.003

Nakayama FS, Kimball BA (1988) Soil carbon dioxide distribution and flux within the open – top chamber. Agron J 80:394–398

Parkin TB, De Sutter TM, Prueger JH, KasparTC, Sauer TJ (2006) Field comparison of methods for measuring soil CO_2 flux. In: ASA-CSSA-SSSA annual meeting abstracts. 12–16 Nov 2006, Indianapolis, IN (CD-ROM)

Paw UKT, Baldocchi DD, Meyers TP et al (2000) Correction of eddy-covariance measurements incorporating both advective effects and density fluxes. Bound Layer Meteorol 97:487–511

Pearson TRH, Brown SL, Birdsay RA (2007) Measurement guidelines for the sequestration of forest carbon. USDA Forest Service, Northern Research Station, General Technical Report NRS No. 18

Prather M, Drewent D, Enhalt P et al (1995) Other trace gases and atmospheric chemistry. In: Houghton J et al (eds) Climate change 1994. Cambridge University Press, Cambridge, pp 77–126

Pumpanen J, Ilvesniemi H, Hari P (2003) A process-based model for predicting soil carbon dioxide efflux and concentration. Soil Sci Soc Am J 67:402–413

Rastogi M, Singh S, Pathak H (2002) Emission of carbon dioxide from soil. Curr Sci 82:510–517

Rayment MB (2000) Closed chamber systems underestimate soil CO_2 efflux. Eur J Soil Sci 51:107–110

Reichstein M, Falge E, Baldocchi D et al (2005) On the separation of net ecosystem exchange into assimilation and ecosystem respiration: review and improved algorithm. Glob Change Biol 11:1424–1439

Reth S, Göckede M, Falge E (2005) CO_2 Efflux from agricultural soils in Eastern Germany – comparison of a closed chamber system with Eddy Covariance measurements. Theor Appl Climatol 80:105–120

Ricciuto DM, Butler MP, Davis KJ, Cook BD, Bakwin PS, Andrews AE, Teclaw RM (2008) Causes of interannual variability in ecosystem-atmosphere CO_2 exchange in a northern Wisconsin forest using a Bayesian synthesis inversion. Agric For Meteorol 148:309–327. doi:10.1016/j.agrformet.2007.08.007

Rolston DD (1978) Application of gaseous-diffusion theory to measurement of denitrification. In: Nielsen DR, McDonald JG (eds) Nitrogen in the environment 1: 309–336

Sarkadi J, Krámer M, Thamm B (1965) Determination of P-content of calcium and ammonium lactate soil extracts with the ascorbic acid-tin chloride method without heating. Agrokémia és Talajtan 14:75–86 (in Hungarian)

Schmid HP (1994) Source areas for scalars and scalar fluxes. Bound Layer Meteorol 67:293–318

Soegaard H, Jensen NO, Boegh E et al (2003) Carbon dioxide exchange over agricultural landscape using eddy correlation and footprint modeling. Agric For Meteorol 114:153–173

Sogachev A, Lloyd JJ (2004) Using a one-and-a-half order closure model of the atmospheric boundary layer for surface flux footprint estimation. Bound Layer Meteorol 112:467–502

Steinkamp R, Butterbach-Bahl K, Papen H (2001) Methane oxidation by soils of an N limited and N fertilized spruce forest in the Black Forest Germany. Soil Biol Biochem 33:145–153

Stolk PC, Jacobs CMJ, Moors EJ et al (2009) Significant non-linearity in nitrous oxide chamber data and its effect on calculated annual emissions. Biogeosci Discuss 6:115–141

Stoy PC, Katul GG, Siqueira MBS et al (2006) An evaluation of models for partitioning eddy covariance-measured net ecosystem exchange into photosynthesis and respiration. Agric For Meteorol 141:2–18

Tóth T, Fórizs I, Kuti L et al (2005) Data on the elements of carbon cycle in a Solonetz and Solonchak soil. Cereal Res Commun 33:133–136

Troen I, Petersen EL (1989) European Wind Atlas Published for the Commission of the European Communities Directorate General for Science Research and Development (Brussels Belgium), Risø National Laboratory Roskilde, Denmark, 656 pp

van der Molen MK, Gash JHC, Elbers JA (2004) Sonic anemometer (co)sine response and flux measurement II. The effect of introducing an angle of attack dependent calibration. Agric For Meteorol 122:95–109

van Laar A, Akça A (1997) Forest mensuration. Cuvillier, Germany, 419 pp

Várallyay Gy (1973) A new apparatus for the determination of soil moisture potential in the low suction range. Agrokémia és Talajtan 22:1–22 (in Hungarian)

Vickers D, Mahrt L (1997) Quality control and flux sampling problems for tower and aircraft data. J Atmos Ocean Technol 14:512–526

Wang WJ, Zu YG, Wang HM et al (2005) Effect of collar insertion on soil respiration in a larch forest measured with a LI-6400 soil CO_2 flux system. J For Res 10:57–60

Wang W, Davis KJ, Ricciuto DM, Butler MP (2006) An approximate footprint model for flux measurements in the convective boundary layer. J Atmos Oceanic Technol 23:1384–1394

Webb EK, Pearman GI, Leuning R (1980) Corrections of flux measurements for density effects due to heat and water vapour transfer. Q J R Meteorol Soc 106:85–100

Widén B, Lindroth A (2003) A calibration system for soil carbon dioxide – efflux measurement chambers: description and application. Soil Sci Soc Am J 67:327–334

Yi C, Davis KJ, Bakwin PS, Berger BW, Marr LC (2000) Influence of advection on measurements of the net ecosystem-atmosphere exchange of CO_2 from a very tall tower. J Geophys Res 105:9991–9999

Chapter 6
Grasslands*

Zoltán Nagy, Zoltán Barcza, László Horváth, János Balogh, Andrea Hagyó, Noémi Káposztás, Balázs Grosz, Attila Machon, and Krisztina Pintér

Abstract In this chapter, exchange dynamics of greenhouse gases over Hungarian grassland ecosystems are analyzed. Carbon dioxide (CO_2) exchange was measured by eddy covariance technique at three sites (Bugac, Mátra, and Hegyhátsál). Methane (CH_4) and nitrous oxide (N_2O) fluxes were occasionally measured by static chamber method partly at the same grassland sites and at a wetland site. Dry grasslands (Bugac and Mátra) were net sources of CO_2 in extreme drought years on annual time scale,

*Citation: Nagy, Z., Barcza, Z., Horváth, L., Balogh, J., Hagyó, A., Káposztás, N., Grosz, B., Machon, A., Pintér, K., 2010: Measurements and estimations of biosphere--atmosphere exchange of greenhouse gases – Grasslands. In: Atmospheric Greenhouse Gases: The Hungarian Perspective (Ed.: Haszpra, L.), pp. 91–119.

Z. Nagy (✉), J. Balogh and K. Pintér
Plant Ecology Research Group of Hungarian Academy of Sciences, H-2103 Gödöllő, Páter K. u. 1., Hungary
and
Department of Botany and Plant Physiology, Szent István University, H-2103 Gödöllő, Páter K. u. 1., Hungary
e-mail: nagy.zoltan@mkk.szie.hu

Z. Barcza and N. Káposztás
Department of Meteorology, Eötvös Loránd University, H-1117 Budapest, Pázmány P. s. 1/A, Hungary

L. Horváth and A. Machon
Hungarian Meteorological Service, H-1675 Budapest, P.O. Box 39, Hungary

A. Machon
Center for Environmental Science, Eötvös Loránd University, H-1117 Budapest, Pázmány P. s. 1/A, Hungary
and
Department of Botany and Plant Physiology, Szent István University, H-2103 Gödöllő, Páter K. u. 1., Hungary

A. Hagyó
Research Institute for Soil Science and Agricultural Chemistry of the Hungarian Academy of Sciences, H-1022 Budapest, Herman O. út 15., Hungary

B. Grosz
Department of General and Inorganic Chemistry, Eötvös Loránd University, H-1117 Budapest, Pázmány P. s. 1/A, Hungary

L. Haszpra (ed.), *Atmospheric Greenhouse Gases: The Hungarian Perspective*,
DOI 10.1007/978-90-481-9950-1_6, © Springer Science+Business Media B.V. 2011

while in other years they acted as sinks. The relatively humid Hegyhátsál was a net sink of CO_2 on annual time scale during the measurement period. The different soil types (light sandy soil, heavy clay soil, and loamy soil) and the variable amount of annual precipitation sums (ranging from 551 to 747 mm) provided a unique opportunity to analyze the response of CO_2 dynamics of grasslands to the soil type and effect of water stress. In case of the sandy grassland (Bugac), the exclusive role of the annual precipitation sum in the determination of annual NEE is strongly coupled to the soil type. Although the sandy grassland expressed adaptation to drought, the CO_2 sink--source precipitation threshold was within the standard deviation (112 mm) of the annual precipitation, showing the high risk of desertification in this ecosystem. Grassland ecosystem on heavy clay soil was more vulnerable to drought stress than the grassland on sandy soil due to the worse water management properties of the clay soil. At the grassland on loamy soil in the more rainy western part of the country, gross primary production was occasionally limited by high soil water contents. This grassland has good adaptation capabilities to uneven precipitation distribution owing to the high silt fraction of the soil resulting in high water storage capacity. CH_4 flux above grasslands was within the range of –54 to 58 mg CH_4 m^{-2} $year^{-1}$ (negative flux means uptake by the soil). Wetland soils are generally CH_4 emitters. Annual mean of N_2O soil emission varied between 0.005 and 0.17 g N_2O m^{-2} $year^{-1}$ with 6-year averages of 0.08 and 0.03 g N_2O m^{-2} $year^{-1}$ for sandy/loess and clay soils, respectively. Mean soil flux of N_2O over wetlands was similar to N_2O emission over dry grasslands.

Keywords Carbon balance • Eddy covariance, Grasslands • Greenhouse gas balance • Interannual variation • Net biome production • Soil type

6.1 Introduction

Grasslands' share in the terrestrial ecosystems, their characteristics of being prone to seasonal droughts, and their importance in global food webs and ruminant nutrition are facts to be considered when looking at the possible threatens grassland ecosystems may face under global climate change. Grasslands are also important elements of the global carbon cycle because of their large area and potentially significant sink or source capacities (Suyker and Verma 2001; Hunt et al. 2004; Novick et al. 2004; Xu and Baldocchi 2004). Measurements proved that significant carbon losses occur during droughts in water-limited grassland ecosystems (Novick et al. 2004; Nagy et al. 2007; Pereira et al. 2007; Reichstein et al. 2007). Interestingly, in a humid grasslands ecosystem, CO_2 uptake can be less in a wet year (due to inundation effects) than in a relatively dry one as it was found by Jaksic et al. (2006). Long-term CO_2 exchange measurements proved the strong interannual variability in the annual carbon balance of grassland ecosystems driven by the variation in the environmental factors (Xu and Baldocchi 2004; Jacobs et al. 2007; Nagy et al. 2007; Aires et al. 2008). Variation in the net ecosystem exchange (NEE) of grasslands is largely constrained by the amount of precipitation (Flanagan et al. 2002;

Pintér et al. 2008). Responses of NEE to environmental factors are also subject to change with the phenological stage (Yuste et al. 2004).

About 10% of the global carbon pool is in the soil of temperate grasslands (Eswaran et al. 1993; Schimel 1995; Batjes 1998). Their diversity makes them able to store more carbon than arable lands even if the soil and precipitation conditions are poorer (Frank 2004). Unmanaged grasslands' carbon storage potential is about half of that of forests, although grasslands "pump" more carbon into the soil than forests (~60 and ~20 g C m^{-2} $year^{-1}$, respectively, Schulze et al. 2009). Greenhouse gas (GHG) balance of grasslands is largely shifted to source direction by nitrous oxide (N_2O) and (to a lesser degree) by CH_4 fluxes. N_2O emissions are largely determined by fertilizer (or organic manure) applications, and so are related to the intensification of the management. Soil CH_4 fluxes of grasslands, while dependent on the water status, are usually very small and the CH_4 fluxes in grazed grassland ecosystems are mainly determined by the fermentation process in ruminants (Soussana et al. 2007).

Differences in response of grasslands to environmental drivers are important considering both the primary functions of grasslands in interaction with climate and grasslands' role in biodiversity conservation, soil conservation, carbon storage, or food production. Without information on the stability and resilience of our grassland ecosystems, we have no possibility to develop sound adaptation and mitigation strategies to follow under changing climate.

In this chapter, we summarize the Hungarian eddy covariance CO_2 exchange measurements and the chamber measurements of non-CO_2 GHG exchange between grasslands and the atmosphere. We have to face the issues posed by increasing level of droughts in the region in the near future (Bartholy and Mika 2005). Knowledge of interannual variation in net GHG exchange as related to climate variation and ecosystem level features (e.g.. stability and resilience) may therefore prove to be of high value, also in practical terms.

6.2 Site Characteristics, Instrumentation, and Climate

CO_2 exchange was measured at three grassland sites by eddy covariance technique (see Section 5.2 for details about the method). The first of these sites was established in the relatively wet western part of the country, near Hegyhátsál (Barcza et al. 2003). The second one is located in the dry central region of Hungary, in Kiskunság National Park, near Bugac (Nagy et al. 2007), and the third one is in the north eastern region in Mátra Mountains with mean annual precipitation sums between those of the other two sites (Pintér et al. 2008).

Static chamber method (see Section 5.3) was used to measure N_2O and CH_4 fluxes occasionally during measuring campaigns at two of the eddy sites (Bugac and Mátra), at a loess grassland site, and at a wetland site. Site characteristics are presented for all measurement sites in this section. As the CO_2 dynamics of the eddy covariance sites cannot be interpreted without detailed meteorological data, climate information will be given in more detail for Bugac, Mátra, and Hegyhátsál sites (Table 6.1).

Table 6.1 Climatic features at the Bugac, Mátra, and Hegyhátsál sites. Yearly average temperature (T_{year}), mean temperature in the growing season (T_{grow}), annual sum of precipitation (P_{year}), sum of precipitation in the growing season (P_{grow}), number of summer days (N_{sd}, daily maximum temperature above 25°C), number of heat days (N_{hd}, daily maximum temperature above 30°C), and yearly maximum of daily maximum temperatures ($T_{max, daily}$) are given for each study year. In case of Hegyhátsál, climate data are available only from the two closest meteorological observatories (Farkasfa, 46°55′N, 16°19′E, 312 m asl, and Rábagyarmat, 46°57′N, 16°25′E, 211 m asl). The long-term averages are measured at Farkasfa, while the annual data are taken from Rábagyarmat

	T_{year}(°C)	T_{grow}(°C)	P_{year}(mm)	P_{grow}(mm)	N_{sd}	N_{hd}	$T_{max, daily}$ (°C)
Bugac							
1995–2004 mean	10.4		562				
2003	9.8	15.3	456	371	120	59	38.4
2004	10	14.3	728	512	80	20	37
2005	9.2	14.1	587	457	79	18	35
2006	9.9	14.5	524	403	84	36	36.5
2007	11.1	15.4	461	307	101	54	41
2008	11	15	579	425	97	42	39.3
Mátra							
1991–2000 mean	10.2		622				
2004	10.1	14.6	635	462	53	10	33.5
2005	9.7	14.6	704	497	49	6	33.1
2006	10.5	15.3	650	577	46	12	33
2007	11.6	16.2	560	402	76	24	37.5
2008	11.4	15.3	610	439	63	9	33.9
Hegyhátsál							
1961–1990 mean	8.9	13.2	759	584			
1999	10.5	15.2	735	582	57	8	33
2000	11.6	15.7	606	441	72	22	38
2007	11.4	15.3	786	628	83	26	38.5
2008	11.4	15.1	661	543	81	17	33
2009	11	15.5	937	681	80	13	33

6.2.1 Semi-arid Grassland on Sandy Soil: Bugac

The eddy covariance measurements at Bugac (46.69°N, 19.60°E, 111 m asl) were started in July 2002, within the framework of the Greengrass project of the European Commission's 5th R&D Framework Programme, and have been running since then. The monitoring site is located in the Hungarian Great Plain, which is frequently droughted. The terrain is flat around the measurement site, and the surface is slightly undulating. The soil type is sandy chernozem (Calcic Chernozem according to the World Reference Base (WRB) classification; see http://www.fao.org/nr/land/soils/soil/en/) with high sand and low clay contents (Table 6.2). Total organic carbon content in the upper 10 cm soil layer is 5.2%. The characteristic plant species are *Festuca pseudovina* Hack. ex Wiesb., *Carex stenophylla* Wahlbg., and *Salvia pratensis* L. The grass is regularly burnt out during summer droughts, characteristic in the Danube–Tisza plain. The area belongs to Kiskunság National Park, and is grazed by a herd of an ancient cattle variety (Hungarian gray cattle) in years with adequate

Table 6.2 Soil properties at the three sites monitored continuously (eddy covariance sites)

Site name	Hegyhátsál	Bugac	Mátra
Sand fraction (%, 0–10 cm)	30	78	7
Silt fraction (%, 0–10 cm)	50	9	58
Clay fraction (%, 0–10 cm)	20	13	35
Organic carbon content(wt%, 0–10 cm)	1.84	5.2	1.5
Bulk density (g cm^{-3}, 0–10 cm)	1.33	1.25	1.3
Soil water content at field capacity (pF = 2.3)	50	25.7	44.4
Soil water content at wilting point (pF = 4.2)	15	7.6	34.6
Plant available water content (vol %)	35	18.1	9.8

precipitation supply (Nagy et al. 2007). The management is extensive (Soussana et al. 2007); in drought years during the grazing period, the grass is not enough to support animals even at a grazing pressure as low as 0.75 animal ha^{-1}. Usually, the grazing period starts at 1 May; it is interrupted at the end of June and restarts at the beginning of autumn, if there is sufficient regrowth during the regeneration period. The grassland is managed extensively; selective grazing and trampling is typical under the traditional management, as the herd (all age groups together) slowly moves during grazing, passing several kilometers a day on the 550 ha area of the grassland.

At Bugac, the eddy-covariance system measuring CO_2 fluxes consists of a CSAT3 sonic anemometer (Campbell Scientific Inc, Logan, Utah, USA) and a LI-COR 7500 (LI-COR Inc., Lincoln, Nebraska, USA) open path infrared gas analyzer (IRGA) operated at 10 Hz sampling frequency. The measuring height is 4 m. Ancillary measurements include the monitoring of soil temperature at 5 cm depth and soil water content (SWC). Soil water content was measured at 5 cm depth from July 1, 2002 to June 15, 2005, using a CS615 Water Content Reflectometer (Campbell Scientific Inc.). Since June 24, 2005, a CS616 Water Content Reflectometer (Campbell Scientific Inc.) has been used at 3 cm depth. Site instrumentation is described in more details in Nagy et al. (2007) and Pintér et al. (2008).

At Bugac, N_2O and CH_4 fluxes were measured in weekly intervals in freeze-free periods between 2002 and 2007. Samples were taken with dark static chambers and analyzed by gas chromatograpy as described in Section 5.3.

In the region of Bugac, the mean annual (10 years average, 1995–2004) temperature and sum of precipitation are 10.4°C and 562 mm, respectively. Annual mean temperatures ranged from 9.2°C to 11.1°C (2003–2008), year 2007 was the hottest (Table 6.1). Years with strongly differing course of annual temperature may be characterized by similar mean annual temperatures; therefore, important interannual differences can be masked. For example, the effect of the heat wave experienced in summer 2003 was balanced by the negative temperature anomalies measured in January and February. During the 6 measurement years, heat waves were experienced in 2 years (2003 and 2007). In these years, the annual precipitation sum at Bugac was less than the 10-year mean (1995–2004). The difference between the annual totals of precipitation in the driest (2003) and the wettest (2004) years was 270 mm. This variation exceeds the half of the annual average total precipitation. The ratio of heat days (when daily maximum temperature exceeds 30°C) to summer days (when daily maximum temperature exceeds 2°C) was 0.5 in heat wave years and about 0.25 in the other years. This strong interannual variability of weather at the sandy grassland site presents the principal constraint to the ecosystem's function.

6.2.2 Dry Grassland on Mountain Heavy Clay Soil: Mátra

The study site in the Mátra Mountains (Mátra, near Szurdokpüspöki, 47.85°N, 19.73°E, 300 m asl) was set up in June 2003, within the framework of the Carbomont project of the European Commission's 5th R&D Framework Programme. The soil type of the grassland is heavy clay, brown forest soil (WRB: Haplic vertisol) with 34.6% clay content in the upper 10 cm, resulting in disadvantageous soil water management in terms of water conductivity and narrow range of plant available soil water (Hagyó 2009). Total organic carbon content in the upper 10 cm soil layer is 1.5% (Table 6.2). The main species are *Festuca pseudovina* Hack. ex Wiesb, *Arrhenatherum elatius* L., *Poa pratensis* L., and *Plantago lanceolata* L.

The eddy covariance system for CO_2 flux measurements, mounted at 3 m above the ground, consists of a CSAT3 sonic anemometer (Campbell Scientific Inc, Logan, Utah, USA) and a LI-COR 7500 (LI-COR Inc., Lincoln, Nebraska, USA) open path IRGA operated at 10 Hz sampling frequency, and it is controlled by a CR5000 datalogger (Campbell Scientific Inc, Logan, Utah, USA). Soil temperature is measured at 5 cm depth, while SWC is monitored by means of a CS616 Water Content Reflectometer (Campbell Scientific Inc.) device in 8 cm depth. Further details on the measurement program and instrumentation are given in Nagy et al. (2007) and Pintér et al. (2008).

N_2O and CH_4 fluxes were measured in weekly intervals in freeze-free periods between 2004 and 2007. Samples were taken with dark static chambers and analyzed by gas chromatograpy as described in Section 5.3.

While not as dry and hot as the sandy grassland (Table 6.1), this site may also be considered as a regularly droughted one. Weather variability is slightly less than that at the sandy grassland as shown by the smaller difference between the annual precipitation sums in the driest and the wettest years. Heat waves and droughts were not as severe here as for the sandy grassland in terms of heat days to summer days ratio in heat wave years (ratio: 0.3) and normal years (ratio: 0.15), respectively.

6.2.3 Moderately Wet Grassland on Silt Soil: Hegyhátsál

The measurement site was established in western Hungary (Hegyhátsál, 46.95°N, 16.65°E, 248 m asl) over a managed, species-rich, seminatural grass field (hay meadow) surrounded by agricultural fields. The area of the site was used as arable land previously and was turned into a grassland around 1990. The dominant species of the grassland are *Arrhenatherum elatius*, *Taraxacum officinale*, *Poa pratensis*, *Agropyron repens*, *Anthoxanthum odoratum*, *Dactylis glomerata*, *Holcus lanatus*,

Briza media, and *Festuca pratensis.* The grass is mowed two times a year, and the mowed grass is taken away from the site and utilized as fodder. The soil type in the region is "*Lessivated brown forest soil*" (Alfisol, according to USDA system; Haplic Cambisol according to the WRB classification). The organic matter content of the upper 15 cm thick layer is 1.3–1.9%. The soil type is loam/silt loam (Table 6.2).

At the Hegyhátsál grassland site (co-located with the tall tower atmospheric and eddy covariance measurements, see Chapters 2 and 8), the monitoring system was originally based on a LI-COR Model LI-6262 closed path, fast response infrared gas analyzer (IRGA), and a three-dimensional, fast response sonic anemometer–thermometer (Kaijo-Denki, model DA-600) (1999–2000; see Barcza et al. 2003). Due to malfunction of the anemometer, the measurements were ceased in 2001. In 2007, the measurements were resumed when the ultrasonic anemometer was replaced with a new model (Solent Research R3-50, Gill Instruments Ltd., Lymington, United Kingdom). The sonic anemometer and the inlet tube of the IRGA are mounted on a mast at 3 m elevation above the grass-covered surface. The inlet tube is mounted at the elevation of the active center of the ultrasonic anemometer, 25 cm away from it horizontally. Raw voltage data generated by the fast response sensors were collected and digitized by means of a TEAC data logger between 1999 and 2000, while since 2007 the digital signals of the sensors have been recorded by a PC also at 5 Hz frequency. Soil water content measurements were started in September 2001, using a Campbell CS615 water content reflectometer (Campbell Scientific, Inc., Logan, Utah, USA). Soil moisture is measured as the mean SWC in the 15–30 cm soil layer. Detailed description of the ancillary measurements can be found in Haszpra et al. (2005).

The long-term average (1961–1990) annual precipitation in the region is around 750 mm, while the average temperature is 8.9°C. During the measurement period, mean annual temperatures were higher while annual precipitation amounts were generally lower than the long-term means (Table 6.1).

6.2.4 Semi-arid Grassland on Loess Soil: Isaszeg

Isaszeg study site is located at the edge of the Hungarian Great Plain (47.53°N, 19.32°E, 255 m asl), on a Chernozem-type sandy loess soil (cambisol, Horváth et al. 2008a, 2010a) with sand:silt:clay ratio of 46:36:18%. The climate is semi-arid, temperate continental, mean annual temperature is 10.7°C. Average annual total precipitation in the period of 1989–2006 was 579 mm. The vegetation studied was xeric, species-rich, tall loess grassland (*Salvio-Festucetum rupicolae*) climatic zonal grassland vegetation dominated by *Festuca rupicola, Salvia nemorosa,* and *Chrysopogon gryllus.*

Vegetation-covered loess monoliths were transplanted in spring 1998, to the Botanical Garden of the Szent István University, Gödöllő. N_2O and CH_4 fluxes were measured on the monoliths in weekly intervals in freeze-free periods between 2002 and 2007, as well as on the plots at the Isaszeg site.

6.2.5 Wetland on Acidic Soil: Óbodrog

The wetland site is situated on the northernmost edge of the Hungarian Great Plain, in Northeast Hungary, at the bank of Óbodrog oxbow (48.32°N, 21.57°E; 105 m asl, Horváth et al. 2008b). Mean annual precipitation in the region is 690 mm and mean annual temperature is 9.6°C. The acidic soil, which is poor in nitrogen and carbon stocks, is Fluvisol. The plant community dominated by *Glyceria maxima* accompanied by *Persicaria lapathifolia*, *Salvinia natans*, and *Utricularia vulgaris* favors the moving, oxygen-rich waters; its stands appear along the edges of lakes, oxbows, slowly flowing ditches. N_2O and CH_4 fluxes were measured in weekly intervals in freeze-free periods between 2006 and 2009.

6.3 Carbon Dioxide Exchange of Grassland Ecosystems

Data from the three Hungarian eddy covariance sites (Bugac, Mátra, and Hegyhátsál) located in different environments provide invaluable information about the CO_2 exchange dynamics of grasslands of the Carpathian Basin, which is a relatively small geographical region. As the climate and the soils of the sites are different (see Table 6.1 and 6.2), it is interesting to look at the similarities and discrepancies in the CO_2 exchange dynamics of the sites and their relation to the environmental factors. Air temperature, precipitation, soil temperature, and volumetric (%) soil water content data were included in the analysis. Start of the growing season, growing season length (GSL; days), and length of the intensive growth period (LIGP; days) were used to study the annual carbon exchange dynamics of the three sites.

GSL was estimated on the basis of the daily maximum half hourly CO_2 uptake (i.e., maximum of minimum of half hourly – NEE) using the method of Suni et al. (2003). According to the method, growing season starts when the daily maximum half hourly CO_2 uptake exceeds a predefined threshold first in the year, where the threshold is typically 20% of the maximum half hourly CO_2 uptake in the given year. End of the growing season is estimated in the same way. It ends when the daily maximum half hourly CO_2 uptake sinks below the threshold. In order to avoid the short-term variation and to get robust estimates for the beginning and end of the growing season, we smoothed the data series of the daily maximum half hourly fluxes. The method also registers the second active period of the grassland if growing season is split into two parts, but excludes the summer inactive period.

Identification of the start of the intensive growth period (IGP) was based on the slope of the cumulative daily NEE curve against time. Linear regression lines were fitted using 5-day moving time windows using 1-day time steps. IGP starts when the slope of the fitted line becomes negative first in a given year. IGP ends with the onset of source activity, i.e., when the slope becomes positive. LIGP is calculated from the beginning and end of the IGP. Using the above definition, beginning of IGP is not necessarily the same as the beginning of the growing season defined following the Suni et al. (2003) method.

In the present study, NEE is negative by definition if CO_2 is taken up from the atmosphere by the vegetation. Gross Primary Production (GPP) and total ecosystem respiration (R_{eco}) are positive by definition (NEE = R_{eco} – GPP). Net biome production (NBP, defined as –NEE minus horizontal carbon removal) is positive if the ecosystem gains carbon. Note that we ignore leaching of dissolved organic and inorganic carbon (DOC/DIC) and other carbon losses that cannot be quantified in the calculation of NBP.

6.3.1 Short-Term Variability of Carbon Dioxide Exchange

6.3.1.1 Bugac

The daytime minimum half hourly average NEE (corresponding to the maximum CO_2 uptake rate) is a good indicator of the productivity. Minimum of daytime average NEE during the study period (2003–2008) ranged between –0.5 mg CO_2 m^{-2} s^{-1} (in 2003) and –1.4 mg CO_2 m^{-2} s^{-1} (in 2004). At Bugac, the 5-percentile values of daytime minimum half hourly NEE for the period of the highest uptake (March–May) ranged between –0.49 mg CO_2 m^{-2} s^{-1} (2007) and –1.12 mg CO_2 m^{-2} s^{-1} (2004). Nighttime NEE (ecosystem respiration, R_{eco}) in the dormant periods typically ranged between 0.1 and 0.2 mg CO_2 m^{-2} s^{-1}. During the growing season, R_{eco} typically varied between 0.4 and 0.6 mg CO_2 m^{-2} s^{-1}.

The maximum daily net uptake rate ranged between 2.4 and 7.9 g C m^{-2} day^{-1}. Occurrence of positive daily NEE sums was regular during mid-summers. During the six measurement years (2003–2008), maximum daily ecosystem respiration was 8.4 g C m^{-2} day^{-1} and maximum daily GPP sum was 14 g C m^{-2} day^{-1}, respectively.

6.3.1.2 Mátra

Minimum of daytime half hourly NEE was around –0.8 to –0.9 mg CO_2 m^{-2} s^{-1} in normal years and –0.5 mg CO_2 m^{-2} s^{-1} during the heat wave year, 2007. The maximum uptake rates were smaller and the interannual variability of maximum uptake rates was considerably lower here than at Bugac. Five-percentile values of daytime minimum NEE for the March-May period ranged between –0.39 mg CO_2 m^{-2} s^{-1} (2007) and –0.86 mg CO_2 m^{-2} s^{-1} (2006).

Nighttime NEE (ecosystem respiration) during the winter dormant period varied between 0.03 and 0.1 mg CO_2 m^{-2} s^{-1} in all years. Nighttime R_{eco} varied between 0.4 and 0.6 mg CO_2 m^{-2} s^{-1} during the growing season (in 2008 it reached 0.8 mg CO_2 m^{-2} s^{-1} for a short period). These values are similar to those measured at Bugac.

The highest daily uptake rate (4.6 g C m^{-2} day^{-1}) was measured in 2008. Maximum uptake rate was smaller than at Bugac in spite of the more favorable precipitation conditions (Table 6.1). Maximum daily ecosystem respiration was 5.1 g C m^{-2} day^{-1} (2005). Maximum GPP was 7.2 g C m^{-2} day^{-1} (2005).

6.3.1.3 Hegyhátsál

At Hegyhátsál, minimum daytime half hourly mean NEE reached −1.2 to −1.4 mg CO_2 m^{-2} s^{-1} in all years. The latter value is the same as the measured minimum at Bugac. Five-percentile values of daytime minimum NEE for the period of the highest uptake (March–May) ranged between −0.83 mg CO_2 m^{-2} s^{-1} (2009) and −1.29 mg CO_2 m^{-2} s^{-1} (2007).

Total ecosystem respiration (nighttime NEE) in the dormant period was typically between 0.01 and 0.1 mg CO_2 m^{-2} s^{-1}. During the growing season, R_{eco} typically varied between 0.2 and 0.4 mg CO_2 m^{-2} s^{-1} but for short periods it was as high as 0.5–0.7 mg CO_2 m^{-2} s^{-1}.

Daily net CO_2 uptake showed its maximum in 2007 (8.7 g C m^{-2} day^{-1}). This value is the largest among those measured at the three eddy covariance sites. Maximum daily ecosystem respiration was 13 g C m^{-2} day^{-1} (also in 2007). It seems, the effects of the heat wave and drought in 2007 were not as severe here as at the other sites. Daily GPP showed maximum in 1999 (16.1 g C m^{-2} day^{-1}). Maximum daily R_{eco} and GPP is about twice as large at Hegyhátsál than at Mátra. While maximum uptake and maximum GPP were similar at Bugac and Hegyhátsál, maximum respiration was lower at Bugac as compared to the other site.

6.3.2 *Annual Cycles of NEE, GPP, and R_{eco}*

6.3.2.1 Bugac

Carbon dioxide source and sink periods are shown by the cumulative daily NEE curves (Fig. 6.1). During winter, the grassland was source of CO_2 with the exception of the beginning of 2004, when the cumulative curve was flat, showing that respiration and photosynthesis cancelled out on a daily scale. The intensive growth period (for definition, see the beginning of Section 6.3) started on 69–109th day of the year (DOY) during the measurement period discussed in this work (2003–2008). Length of the intensive growth period ranged between 37 (in 2005) and 125 days (in 2008) at this sandy grassland. In spring 2005, a transient heat spell caused an approximately 4 days long source period, which was enough to interrupt the intensive growth period according to the method applied. However, this heat spell was short, the vegetation recovered quickly, and continued to sequester carbon till the end of autumn. The carbon uptake during the longest intensive growth period was 251 g C m^{-2}, while it was 55 g C m^{-2} during the shortest one.

Important feature of the grassland's annual carbon cycle is the secondary growing period (or regeneration period) that usually lasts from late August until end of November in this region. Depending on the meteorological conditions, this recovery may also show remarkable interannual variation. The sandy grassland was a source of CO_2 under the heat wave in 2003 from July, and remained source until the next spring, while showed strong sink activity during the secondary growing period in

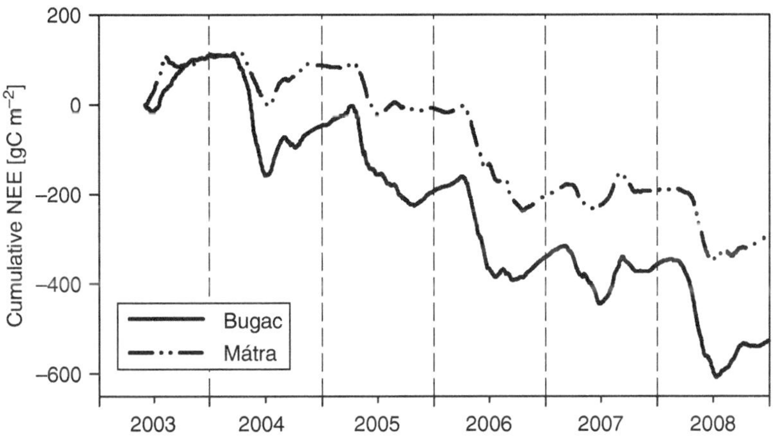

Fig. 6.1 Cumulative values of NEE at the Bugac (sandy grassland) and Mátra (heavy clay grassland) sites for the period 2003–2008

2004 after the late summer rains. In 2004, from the beginning of September till the first third of October, 41 g C m^{-2} was absorbed. In 2007, the annual cumulative sum of NEE was already positive (as calculated from the beginning of that year) at the end of summer, when secondary growth decreased the cumulative net emission by 31 g C m^{-2}. As the summers are expected to be drier, and the long-lasting freezes to occur later during the year in the future (Bartholy and Mika 2005), the autumn-winter regeneration may become critical in the annual carbon balance of the grassland.

6.3.2.2 Mátra

The intensive growth period started between 89th and 108th DOY in the study years. LIGP was a little shorter at this grassland than at Bugac, ranging from 36 (2007) to 72 days (2006) (Table 6.3). CO_2 uptake during the longest intensive growth period was 128 g C m^{-2}, while it was 38 gC m^{-2} during the shortest period of intensive growth.

Sink activity during the autumn significantly influenced the annual balance as it was the case in 2007 when the grassland was a net source of CO_2 already after the main growing season (with a cumulative value of 36 g C m^{-2}) and recovered to a weak sink during the autumn regeneration (Fig. 6.1).

GPP was usually the highest in May, with values typically close to 200 g C m^{-2} month^{-1} in normal years and below 100 g C m^{-2} month^{-1} in the drought year, 2007. These values are less than the monthly sums for Bugac and Hegyhátsál, and the differences cannot be explained by differences in the precipitation sum (Table 6.1.) as one may expect for grassland ecosystems in the same geographical region but with different precipitation supply.

Table 6.3 Annual sums of NEE, GPP, R_{eco}, and NBP for the measurement years at the three grassland sites with eddy covariance measurements for CO_2. Length of the intensive growth period (LIGP) and growing season length (GSL) are also given for all three sites. See text for details on the definition of the fluxes and the methods

	NEE (g C m^{-2} $year^{-1}$)	GPP (g C m^{-2} $year^{-1}$)	R_{eco} (g C m^{-2} $year^{-1}$)	NBP[a] (g C m^{-2} $year^{-1}$)	LIGP based on the cumulative NEE method	GSL based on the Suni et al. (2003) method
Bugac						
2003	106	601	707	−106	47	71
2004	−168	1,162	994	168	99	158
2005	−151	1,128	977	151	37	195
2006	−147	1,050	903	147	105	146
2007	−17	726	709	17	53	133
2008	−171	935	764	171	125	133
Mátra						
2004	−13	739	726	13	56	96
2005	−98	964	866	98	53	161
2006	−197	1066	869	197	72	186
2007	14	650	664	−14	36	139
2008	−102	932	830	102	71	180
Hegyhátsál						
1999	−183	1,786	1,603	–	63	211
2000	−226	1,689	1,463	3	60	251
2007	−227	1,523	1,295	–	51	229
2008	−165	1,681	1,516	−49	50	232
2009	−96	1,557	1,462	−17	31	211

[a] NBP for Bugac and Mátra is estimated as –NEE due to lack of information about lateral carbon flux. See text for details

6.3.2.3 Hegyhátsál

CO_2 uptake and source periods are shown by the cumulative NEE curves (Fig. 6.2). If we compare the cumulative curves with those of the two other sites (Fig. 6.1), it is clear that usually the growing season at Hegyhátsál is not interrupted by a mid-summer source period common at the other sites. Effect of mowing is visible in the cumulative NEE curves presented in Fig. 6.2 (mowing events are marked by arrows). It can be seen that the cumulative net CO_2 exchange was disturbed for a couple of days after mowing, and then the normal trend returned. In 2000, the grass was first mowed at the beginning of June. Unfortunately, it cannot be detected because the measurement was suspended during that time due to technical problems and the gap-filling procedure could not reproduce the human intervention. The second mowing happened in September; thus it did not affect the cumulative trend significantly.

The effect of mowing is very spectacular in both 2008 and 2009. Mowing took place in the middle of the growth periods in both years. After mowing, the disturbance

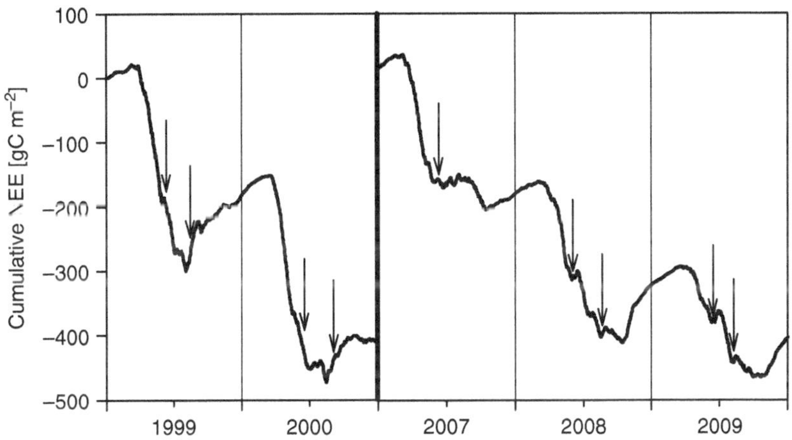

Fig. 6.2 Cumulative values of NEE at the Hegyhátsál grassland site for the periods 1999–2000 and 2007–2009, respectively. The arrows indicate the mowing events

caused sharp changes in the daily NEE sums. The grass turned into a net source from sink for a couple of days after mowing.

At Hegyhátsál, in 1999 and 2000, intensive CO_2 uptake started around 90th DOY (Fig. 6.2), while it started earlier in 2007 and 2008. Beginning of 2007 was unusually mild, which also initiated the early winter wheat growth in the region (Barcza et al. 2009).

In 1999 and 2000, during the April–June period, the grassland was generally a net sink of CO_2. In 2007 and 2008, during the same period, CO_2 uptake was strong in April and May, but decreased in the beginning of June (monthly NEE sum in June was around −70 g C m^{-2} in 1999 and 2000, while it was only −3 and −22 g C m^{-2} in 2007 and, 2008, respectively). This was most probably caused by the less favorable precipitation distribution and soil moisture content in 2007 and 2008.

6.3.3 *Annual Sums of NEE, GPP, and R_{eco}*

6.3.3.1 Bugac

The annual sums of NEE varied between −171 and 106 g C m^{-2} at the Bugac grassland site (Table 6.3). These values are within the range of grassland ecosystems as summarized by Novick et al. (2004). The 6-year mean NEE was −91 g C m^{-2} year^{-1}. Maximum annual GPP sum was 1,162 g C m^{-2} year^{-1} (2004), while the minimum was about the half of it in 2003 (601 g C m^{-2} year^{-1}) due to the continental scale drought (Reichstein et al. 2007). Maximum annual ecosystem respiration was 994 g C m^{-2} year^{-1} (2005), while the minimum was 707 g C m^{-2} year^{-1} in 2003.

Bugac is the most drought-prone grassland out of the three investigated ones, and is also the only one where the annual precipitation sum seems to govern the annual

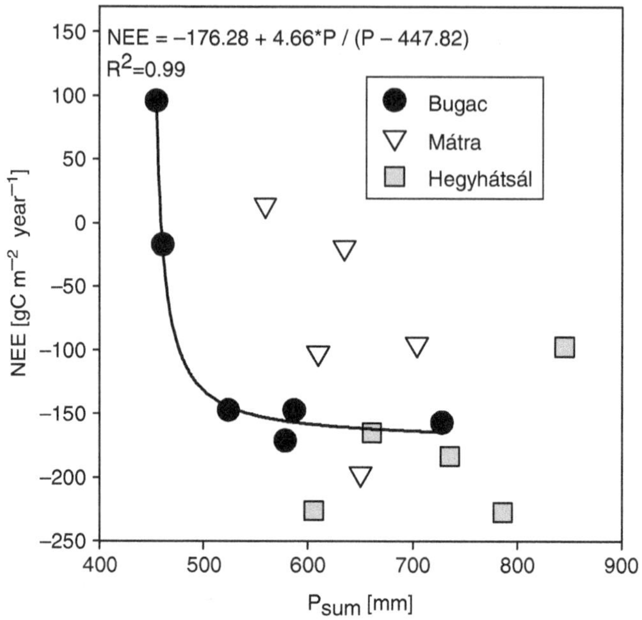

Fig. 6.3 Relation between the annual sum of precipitation and annual NEE at the three grassland sites in Hungary. The relationship is highly significant (R^2 = 0.99) for Bugac when fitting a Michaelis–Menten-type relation (NEE = −176.3 + 4.66* $P/(P$ − 447.82)), while it is not significant in statistical terms in case of the grassland on heavy clay (Mátra) or on loamy soil (Hegyhátsál) with higher annual precipitation sums

NEE exclusively. Using a Michaelis–Menten-type relation, variation in the annual precipitation explained 99% of the variance in the annual NEE sums over the 6 years considered in the present study (Fig. 6.3). Although the number of data points is low, the relation is statistically highly significant (P < 0.0001). The 10-year average annual precipitation sum is close to the threshold (~460 mm) where the annual carbon balance of the grassland may switch between sink and source.

In this grassland, the soil water content is available to plants at a much lower water content range than in the grassland on the heavy clay soil or on the silt soil. The steep relation between the average uptake rate during the intensive growth period and SWC (Fig. 6.4) well represents the main difference between the sites. This difference is also an important aspect when the interannual variation of NEE is characterized.

Annual GPP and R_{eco} sums are closely related to mean temperatures during the March–October period (R^2 = 0.87 and 0.96, respectively), and to annual precipitation (R^2 = 0.68 and 0.62, respectively). There is a similar relationship between the fluxes and precipitation sum during March–October (R^2 = 0.68 for GPP, R^2 = 0.71 for R_{eco}).

Five-percentile of daytime minimum NEE during March–May explains 2/3 of the variability of annual NEE (R^2 = 0.66). It means that NEE during spring has strong effect on the annual NEE sum, though, as it was demonstrated, the lack or presence of the secondary growing period also has strong impact on the annual NEE sum.

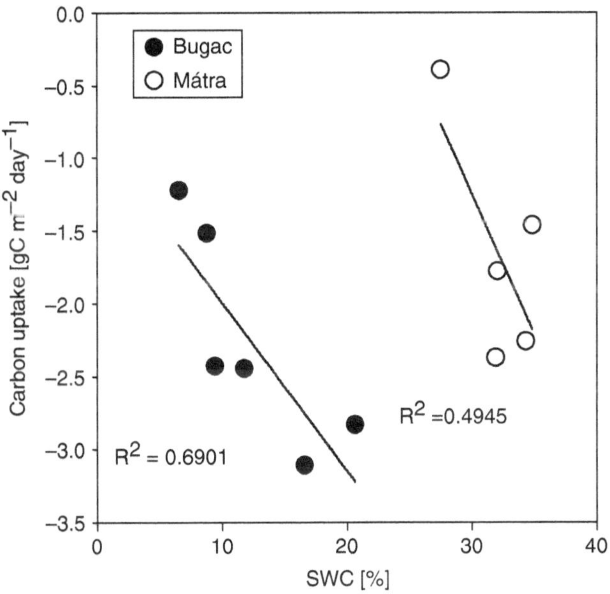

Fig. 6.4 Relationship between the average daily sink strength and the volumetric soil water content during the intensive growth period at the sandy grassland and at the grassland on heavy clay soil. Each data point stands for a particular year

Length of the intensive growth period and growing season length are supposed to be important in determining the sink strength in a given year. LIGP estimated with the cumulative NEE method showed weak covariance with the annual NEE ($R^2 = 0.36$). GSL estimated by the method of Suni et al. (2003) correlates better with the annual NEE ($R^2 = 0.66$).

6.3.3.2 Mátra

Annual sums of NEE varied between −197 and 14 g C m^{-2} (Table 6.3) at this grassland on heavy clay at the northern part of the country. The mean NEE is −79 g C m^{-2} year^{-1}, which means that this grassland was also a net sink from the point of view of the atmosphere on average.

The highest annual GPP was 1066 g C m^{-2} year^{-1} (2006), while the lowest was 650 g C m^{-2} year^{-1} (2007). Maximum annual ecosystem respiration was 869 g C m^{-2} year^{-1} (2006), while the minimum was 707 g C m^{-2} year^{-1} (2007).

Annual NEE sums differed strongly between years with the same amount of annual precipitation (~630–650 mm, years 2004 and 2006), or were similar when the precipitation amount was markedly different between the years (around −100 gC m^{-2} year^{-1} in 2005 and 2008). As it is shown in this study, the sandy grassland site receives less annual precipitation than the Mátra site, but the water use efficiency of the sandy grassland is better.

Annual precipitation sums did not correlate with annual NEE sums ($R^2 = 0.29$). Annual NEE was basically independent from the annual mean temperature.

Annual GPP had no correlation with the annual mean temperature ($R^2 = 0.1$), and the same holds true for R_{eco} ($R^2 = 0.21$). GPP and R_{eco} sums are not governed by the monthly mean temperatures during the March–October period ($R^2 = 0.15$ and 0.25, respectively), which is not in accordance with the experiences at Bugac. Annual GPP and R_{eco} sums have a stronger dependence on annual precipitation ($R^2 = 0.46$ and 0.59, respectively), and the fluxes also co-vary with the precipitation sum during the March–October period ($R^2 = 0.68$ for GPP and $R^2 = 0.56$ for R_{eco}).

Five-percentile daytime minimum NEE during March–May explains 89% of the variability in annual NEE. It means that at Mátra, NEE during springtime has a very strong and clear effect on the annual NEE sums, independently from the occurrence of the secondary growing period after the summer drought.

LIGP estimated with the cumulative NEE method has a strong covariance with annual NEE ($R^2 = 0.67$). GSL estimated by the Suni et al. (2003) method also correlates well with annual NEE ($R^2 = 0.64$).

6.3.3.3 Hegyhátsál

At Hegyhátsál, the grass field acted as a net sink of CO_2 in all years (Table 6.3). Sink activity was the strongest in 2000 and 2007 (~226 g C m^{-2} year^{-1}), and weakest in 2009 (−96 g C m^{-2} year^{-1}), respectively. This interannual variability may indicate a considerable dependence of NEE on the climatic conditions.

Total ecosystem respiration varied between 1,603 g C m^{-2} year^{-1} (1999) and 1,295 g C m^{-2} year^{-1} (2007). Year-round GPP varied between 1,786 g C m^{-2} year^{-1} (1999) and 1,524 g C m^{-2} year^{-1} (2007). It is worth noting that the magnitude of R_{eco} and GPP was the lowest in 2007, while their difference (NEE) was the highest in magnitude.

Annual NEE sums increased (which means less uptake) with increasing annual precipitation ($R^2 = 0.5$) though the relationship is not significant due to the low number of data points. The relatively high annual precipitation may suppress photosynthesis if the soil is saturated (see also Fig. 6.5). There is no correlation between the mean annual temperature and NEE ($R^2 = 0.16$). Mean temperature in the growing season (typically March–October) is uncorrelated with annual NEE. Annual GPP and R_{eco} did not correlate significantly with mean annual temperatures. Annual GPP and R_{eco} sums were also independent of the growing season's mean temperatures. Annual precipitation sum explained 35% of the variability of GPP (for R_{eco} R^2 is close to zero). Growing season precipitation has similar effect on the fluxes. Therefore, the production was apparently not limited by annual scale values of the driving variables (precipitation and temperature).

We can relate NEE measured during the beginning of the growing season to the annual NEE sums. It was found that, the 5-percentile daytime minimum NEE during March–May is a good predictor for annual NEE in this grassland. Eighty-five percent of the variability in annual NEE can be explained by the 5-percentile values. This suggests that the environmental conditions in springtime can have a profound effect

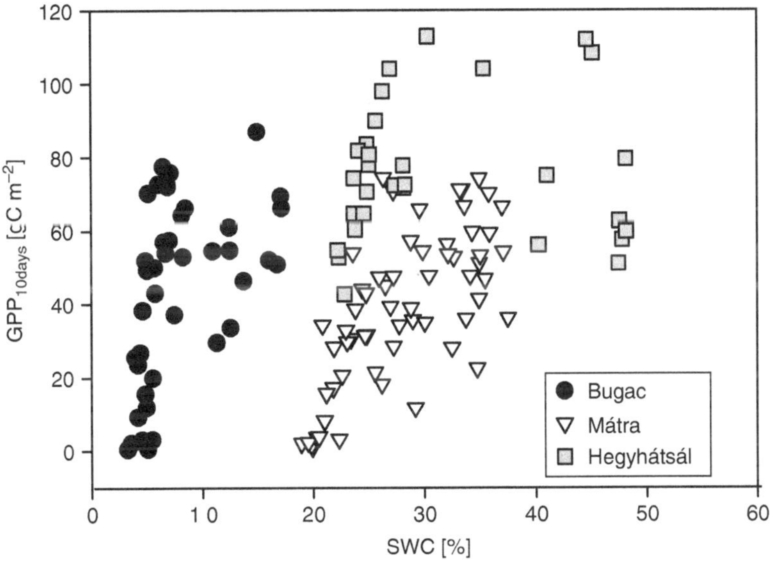

Fig. 6.5 Relationship between 10-day totals of GPP and 10-day average SWC at the eddy covariance sites between May and August from all measurement years

on the annual evolution of NEE both at Hegyhátsál and at Mátra (and also at Bugac, to a lesser extent).

It might be interesting to relate the measured data to the length of the growing season and the intensive growth period. Using LIGP derived from the cumulative NEE data has a fairly good correlation with annual NEE ($R^2 = 0.65$). GSL estimated with the Suni et al. (2003) method also correlates well with annual NEE ($R^2 = 0.45$).

6.3.4 Net Biome Production of the Grassland Sites

For the estimation of the full CO_2 balance of the grasslands, we have to take into account lateral carbon fluxes and other carbon fluxes not accounted for by the eddy covariance technique (see Chapter 13). In the present context, we neglect dissolved inorganic and organic carbon fluxes and consider only lateral carbon flux caused by management.

At Bugac, animals are most likely causing horizontal carbon flux due to grazing and excreta. As we do not have a method to estimate the net effect of grazing (due to the lack of control on grazing intensity, significance of selective grazing, and trampling) on the CO_2 balance, we assume that horizontal carbon replacement is zero. It means that we approximate NBP with -NEE (Soussana et al. 2007) (Table 6.3).

At Mátra, the grassland was unmanaged during part (2006–2008) of the study period, so as a first approximation we also assume that NBP equals -NEE (Table 6.3).

At Hegyhátsál, the grass was cut twice every year, and the mowed grass was taken away and utilized as fodder. This may affect NEE when comparing to the results from other sites, since the decomposition of part of the organic matter happened elsewhere (typically within 1 year; Ciais et al. 2007). Due to the removal of the harvested grass, local ecosystem CO_2 release via respiration became smaller, and the magnitude of year-round NEE became larger (i.e., more negative) than net biome production (NBP = −NEE − harvest, where harvest has positive sign).

To quantify NBP at Hegyhátsál, we need information about the biomass removed by mowing. This was possible for 2000, 2008, and 2009 when the harvested grass was accounted and the removed biomass could be estimated. Assuming that the water content of hay is 16% and carbon content of dry biomass is 43%, the removed carbon was 223 g m^{-2} in 2000, 214 g m^{-2} in 2008, and 113 g m^{-2} in 2009. Assuming that there was no other carbon loss from the ecosystem, NBP was close to zero in 2000 and negative in the other 2 years (Table 6.3). According to the results, the grass ecosystem was a net source of CO_2 (in terms of NBP) to the atmosphere in 2008 and 2009.

The results suggest that land use conversion from managed cropland to managed grassland does not necessarily imply of turning NBP into positive (i.e., increase in soil organic carbon content). This is a first estimate of NBP and more data and more precise estimate of harvested biomass are needed to reinforce these results.

6.3.5 NEE Variability of Grasslands as Affected by Soil Texture

The range of SWC differed markedly among the sites, ranging between 7.6% and 25.7% on the sandy soil at Bugac, 34.6% and 44.4% on the heavy clay at Mátra, and 22% and 48% at Hegyhátsál. Soil water management characteristics dependent on the soil physical structure (Table 6.2), like soil water potential at specific water content, differed accordingly between the sites. The soil water content range with water potentials between pF2.3 and pF4.2 is commonly referred to as plant available soil water and it was either measured (Mátra) (Hagyó 2009) or estimated from the sand, silt, and clay fractions (Bugac and Hegyhátsál) using the software by Fodor and Rajkai (2005). Plant available soil water content is 9.8% in the heavy clay soil (Mátra), 18.1% in the sandy soil (Bugac), and 35% in the loamy soil (Hegyhátsál) at field capacities (Table 6.2). The optimum water content for evaporation varied similarly to GPP (Fig. 6.5) as shown by the relationship between SWC and daily mean evapotranspiration (Fig. 6.6). While this suggests that production was frequently limited by available water at the stomatal level at Bugac and Mátra, it also shows that the primary functions of the grassland on loamy soil (Hegyhátsál site) are generally not limited by the available soil water content.

There is about 25% difference between the optimum domains for evapotranspiration among the sites that can be attributed to the different soil characteristics. However, this difference is not the only one: conductivity for water is also largely different among these soils with potentially serious consequences on precipitation runoff. Water content ranges, optimal for GPP, also differed strongly among the sites

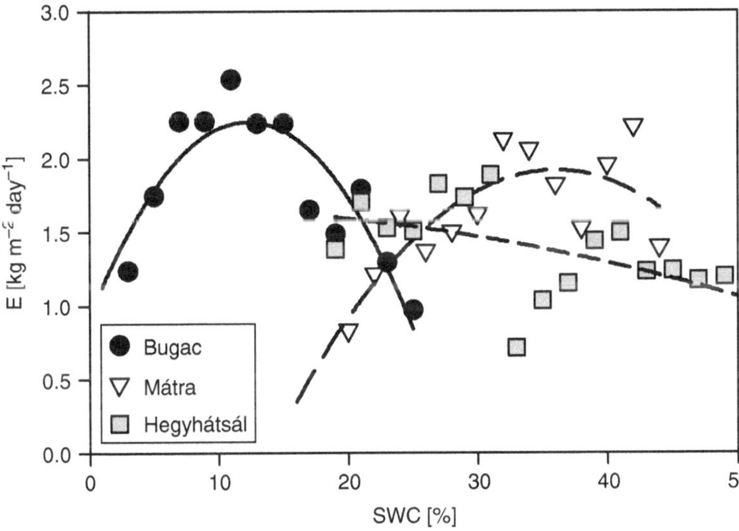

Fig. 6.6 Daily mean evapotranspiration rates (*E*) from eddy covariance measurements as averaged for 2% SWC bins at the three measuring sites between April and September

(Fig. 6.5), apparently explained by the differences in plant available soil water contents among the sites (Table 6.2). Negative effect of high SWC on GPP was shown only in the case of the Hegyhátsál grassland probably caused by waterlogging.

Interannual variation of carbon balance was primarily constrained by the annual sum of precipitation at the sandy grassland. Other factors (i.e., timing of precipitation events) might also play important roles in the case of the grassland on heavy clay (Fig. 6.3), while precipitation was not limiting the annual NEE at all in the case of the loamy grassland (Fig. 6.3). At the loamy soil grassland, water supply was much more abundant than at the other two sites (Table 6.1) and the water management properties of the soil are also the best considering the three grasslands (Table 6.2). The net result is that on annual scale the carbon balance is dependent upon the interplay between precipitation amount and soil water management properties. In some cases, these two seem to be well harmonized (Bugac) and the ecosystem is probably well adapted to the prevailing weather. In the other case (Mátra), the adaptation of ecosystem's function (as measured by annual carbon balance) is less effective. The decisive factor here is that due to soil water management features, this soil is not well suited to absorb water from rainstorms, but from more regular (and less intense) precipitation. Consequently, the soils' mechanical composition can explain the differences between the responses of annual NEE to annual precipitation, or in other words, the ecoystem scale adaptation.

The importance of using SWC in addition to soil temperature (T_s) when characterizing dependence of soil respiration (R_s, as the major component of R_{eco}) on environmental drivers is shown in Fig. 6.7 for Bugac. The variance explained by the fit was significantly increased by inclusion of SWC as an independent factor in addition to T_s (Fig. 6.7). The importance of considering this dependence is evident,

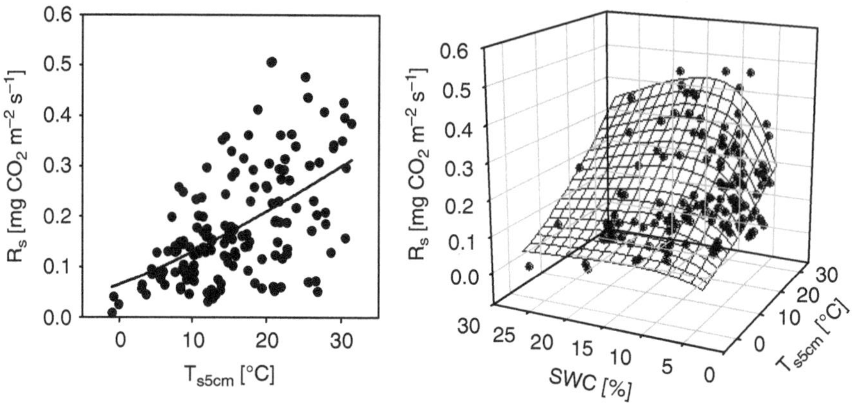

Fig. 6.7 Dependence of soil respiration on daily means of soil temperature (at 5 cm depth) and on daily means of soil temperature and soil water content (0–6 cm, integrative), as measured simultaneously at the Bugac site during 2005–2008. Data points are averages of 3–5 spatial replications of soil respiration measurements from field measurement campaigns during the growing season. The regressions and all the fitted coefficients are significant at the $P < 0.0001$ level. Goodness of fit and coefficients of the fitted functions (Eq. 5.7 on the left panel and Eq. 5.8 on the right panel) are the following. $R^2 = 0.31$, $n = 161$, $R_{10} = 0.13$, $E_0 = 176.66$ for the figure on the left; $R^2 = 0.47$, $n = 161$, $R_{10} = 0.15$, $E_0 = 215.22$, $SWC_{opt} = 14.19$ for the figure on the right

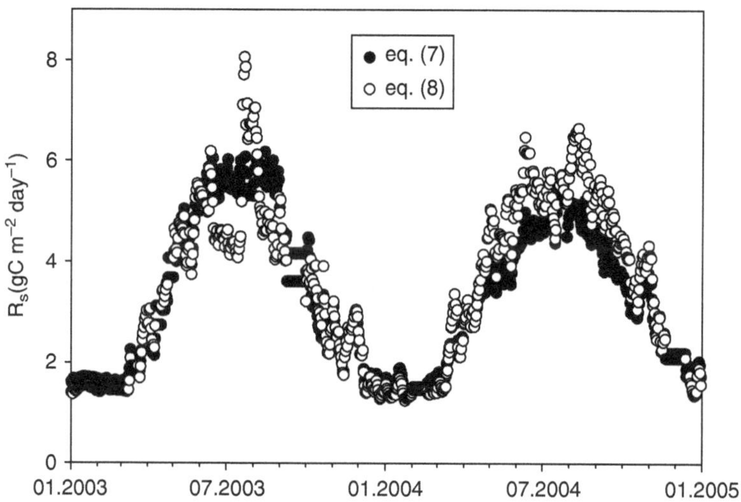

Fig. 6.8 Modeled daily soil respiration values (R_s) in 2003–2004 at Bugac, as based on daily means of temperature (Eq. 5.7, *filled circles*) and daily mean soil water contents (Eq. 5.8, *open circles*)

e.g., in modeling (or in gap-filling half hourly eddy covariance data) as shown in Fig. 6.8: the simulated soil respiration values differ markedly during the growing season showing the effect of the additional driving variable. Considering the temperature dependence of respiration alone is therefore not recommended probably in most of the modeling exercises. While it may be tempting in modeling studies to use temperature as a single driving variable, it is simply misleading when modeling one of the large contributing processes (R_{eco}) in drought-prone ecosystems. Further argument of using SWC as the second predictor is that SWC is routinely measured at most eddy covariance sites.

6.4 Nitrous Oxide and Methane Exchange of Grassland Ecosystems

6.4.1 Nitrous Oxide and Methane Soil Fluxes at the Dry Sandy Grasslands (Bugac), Heavy Clay Grassland (Mátra), and Loess Grassland (Isaszeg)

The measurement program is presented in Table 6.4 (Horváth et al. 2008a). The results of individual measurements of CH_4 soil flux can be seen in Fig. 6.9. CH_4 flux varies within a wide range of −80 to 40 μg CH_4 m^{-2} h^{-1} (negative sign means uptake by the soil). The magnitude and the direction (emission or uptake by soil) of CH_4 fluxes are strongly dependent on the soil moisture content and temperature (Pol-van Dasselaar et al. 1998). Uptake is increasing with temperature; optimum soil moisture content for uptake is between 20% and 35%. Maximum uptake can be observed at high soil temperature and medium soil moisture (20–35%) ranges. In dry soil (SWC < 5%), CH_4 uptake by the soil is inhibited by the water stress as exerted on methanotroph bacteria. In wet soil (SWC > 50%), CH_4 production dominates against CH_4 uptake. CH_4 fluxes of sandy and loess soil (Table 6.5) showed large variation, while mean fluxes (either sink and source) were small. CH_4 emission and uptake were balanced above these types of soils during the two

Table 6.4 Sampling protocols used at each site for measurement of soil N_2O and CH_4 fluxes (small chamber's base area: $A = 80$ cm^2, large chamber's base area: $A = 2{,}500$ cm^2)

Station	Period	Frequency	Chamber type	Repetitions	N_2O	CH_4
Bugac	August 2002–December 2004	~Bi-weekly	Small	10	Yes	No
Bugac	August 2006–November 2007	Weekly	Large	8	Yes	Yes
Isaszeg	August 2002–June 2004	~Bi-weekly	Small	10	Yes	No
Gödöllő	August 2002–December 2004	~Bi-weekly	Small	18	Yes	No
Gödöllő	August 2006–December 2007	Monthly	Small	18	Yes	Yes
Mátra	2004–2005, growing season	Monthly	Small	10	Yes	No
Mátra	2006–2007, growing season	Monthly	Small	10	Yes	Yes

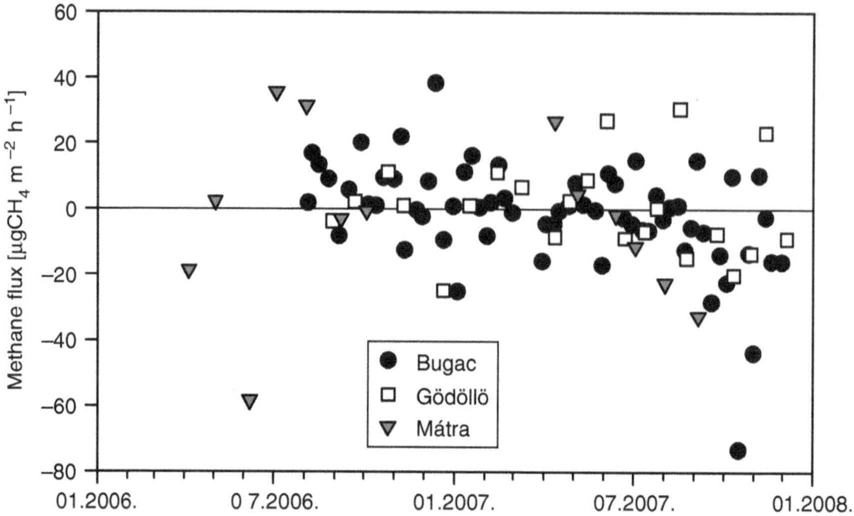

Fig. 6.9 Measured CH_4 fluxes above different Hungarian grassland soils between 2006 and 2007 (Horváth et al. 2008a)

Table 6.5 Annual sums of soil CH_4 and N_2O fluxes as calculated from periodic measurements (Horváth et al. 2008a). SD means standard deviation

	Bugac	Isaszeg/Gödöllő	Mátra
CH_4 flux (mg CH_4 m^{-2} $year^{-1}$)			
2006	58	–25	–14
2007	–49	17	–54
Mean	4.5 (SD: 76)	–4 (SD: 30)	–34 (SD: 28)
N_2O flux (g N_2O m^{-2} $year^{-1}$)			
2002	0.118	0.174/0.115	–
2003	0.126	0.105/0.145	–
2004	0.116	0.118	0.019
2005	–	–	0.042
2006	0.027	0.005	0.020
2007	0.022	0.028	0.055
Mean	0.082 (SD: 0.123)	0.085 (SD: 0.171)	0.035 (SD: 0.033)

years of observation (2006–2007). For the grassland on clay soil, net CH_4 uptake was observed in both years, but again, the large variation indicated sources and sinks to be nearly balanced in this case, as well.

Nitrous oxide fluxes varied within the range of –31.4 and 125.7 µg N_2O m^{-2} h^{-1} (Fig. 6.10) showing high standard deviations with low means for different years and ecosystems (Table 6.5). At Bugac and Isaszeg, the N_2O emission of the

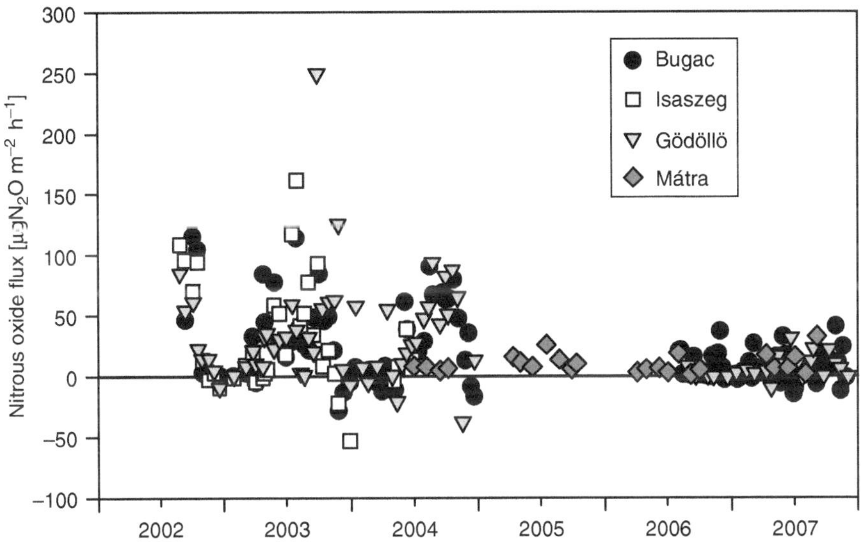

Fig. 6.10 Nitrous oxide fluxes from grassland soils in Hungary as measured between 2002 and 2007 (Horváth et al. 2008a)

soil has decreased by a factor of 6 from the period of 2002–2003 (mean 0.12–0.13 g N_2O m^{-2} $year^{-1}$) to the years 2006–2007 (mean 0.02 g N_2O m^{-2} $year^{-1}$), suggesting that substantial changes in soil characteristics might have occurred during the last few years (Machon et al. 2010). Interannual changes in fluxes were probably caused by different rates of denitrification–nitrification processes in relation to change of soil physical, chemical, and biological characteristics as experienced by Flechard et al. (2007) for 10 different European grassland sites (including intensively managed grasslands) for the period 2002–2004 with average annual N_2O flux of 0.15 g N_2O m^{-2} $year^{-1}$ (standard deviation (SD): 0.2 g N_2O m^{-2} $year^{-1}$), a value very close to the one mentioned above. Our data suggest that sandy and loess soil grasslands emit similar amounts of N_2O (0.08 g N_2O m^{-2} $year^{-1}$), while in the case of clay soil in the mountainous region, lower emissions was measured (0.03 g N_2O m^{-2} $year^{-1}$). The difference cannot be explained by the different rates of N-fixation, because the share of leguminous species at the different sites was approximately the same (15–18%).

Atmospheric deposition and biological N-fixation are the primary N-sources for nonfertilized grasslands. The rate of wet and dry N-fluxes (calculated in N_2O equivalents here) were determined for grasslands in the Hungarian Great Plain by Kugler et al. (2008) as −0.74 ± 0.04 (wet) and −1.43 ± 0.09 g N_2O m^{-2} $year^{-1}$ (dry), respectively. Considering these figures, 0–8% of the deposited N-compounds are converted to N_2O in the soil mostly by denitrification, and released subsequently into the atmosphere.

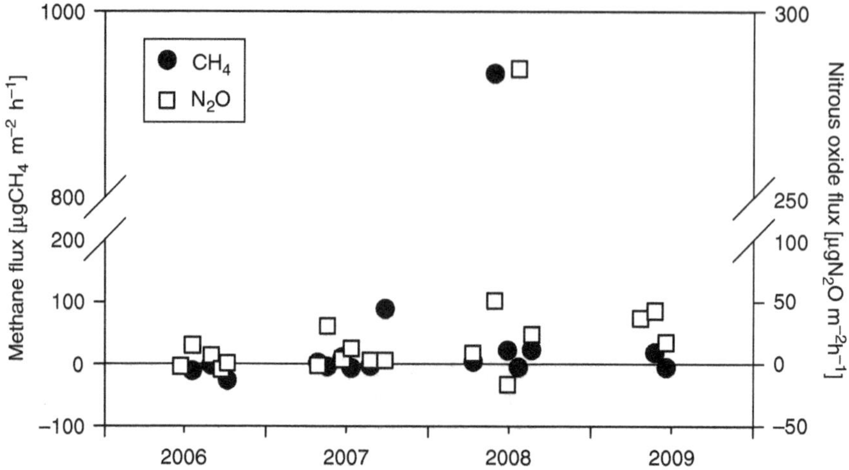

Fig. 6.11 Measured *CH_4* and N_2O fluxes from wetland soils between 2006 and 2009 (Horváth et al. 2010b)

6.4.2 *Nitrous Oxide and Methane Soil Fluxes at Óbodrog Wetland*

Pilot measurements (19 sampling events) were conducted between 2006 and 2009 at the wetland as described in Horváth et al. (2010b) using 5-10 parallel small ($A = 80$ cm^2) chambers. Average *CH_4* flux of the soil was +17.5 µg CH_4 m^{-2} h^{-1} (neglecting the outlier value in Fig. 6.11). In contrast, the mean soil CH_4 flux from another four nonwetland sites at the same region (Bodrogköz) was −6.6 µg CH_4 m^{-2} h^{-1} (deposition). Results of these pilot measurements are in accordance with the theory, namely that dry soils are sinks for CH_4, while in wetlands, anaerobic decomposition of soil organic matters results in net CH_4 emission in these regions. Mean soil flux of N_2O (neglecting the outlier value in Fig. 6.11) was 6.44 µg N_2O m^{-2} h^{-1}. It is in the same order of magnitude as the average value measured in Bodrogköz region above nonwetland soils (11.47 µg N_2O m^{-2} h^{-1}). High soil wetness does not seem to favor the production of N_2O. Preliminary measurements suggest that wetlands may have an important role in biosphere–atmosphere CH_4 exchange in Hungary.

6.5 Net GHG Balance of Grassland Ecosystems

To estimate the full GHG balance of the grasslands, we have to take into account the effect of non-CO_2 GHGs. It was possible in case of the Bugac and Mátra sites, where both CO_2 and non-CO_2 GHG exchange was measured during the same period.

To quantify the net greenhouse gas balance (NGB; see Schulze et al. 2009) of Bugac and Mátra grassland sites, N_2O and CH_4 fluxes were converted to CO_2

equivalents using the global warming potential (GWP) conversion factors according to IPCC (1996) taking GWP for N_2O as 310, while that for CH_4 as 21 on 100-year time horizon. NGB is positive if the biosphere is a net sink of GHGs.

At Bugac, NGB is estimated for the period of 2003–2007. NBP was 75.4 g C m^{-2} $year^{-1}$ for this period. The CO_2-equivalent CH_4 flux was estimated to be 0.03 g C m^{-2} $year^{-1}$ (emission) based on data from 2006 to 2007 (Table 6.5). The temporal mismatch between the NBP and the CH_4 flux estimate may cause uncertainties though the contribution of CH_4 is clearly negligible when calculating NGB. The CO_2-equivalent N_2O flux was 6.1 g C m^{-2} $year^{-1}$ (emission) for the 2003–2007 period. NGB for Bugac was 69.3 g C m^{-2} $year^{-1}$ (expressed in CO_2--C equivalent unit), which is still close to NBP.

At Mátra, NGB was estimated for the 2004–2007 period. Mean NBP was very close to the Bugac data (74 g C m^{-2} $year^{-1}$) for the 4-year period. CO_2-equivalent CH_4 flux was –0.2 g C m^{-2} $year^{-1}$ (uptake) for the 2006–2007 period (Table 6.5). Similarly to Bugac, this flux is negligible even if it was estimated on the basis of data only from 2 years. CO_2-equivalent N_2O flux was 2.9 g C m^{-2} $year^{-1}$ (emission). NGB was 71.3 g C m^{-2} $year^{-1}$ (expressed in CO_2–C equivalent unit), which is also close to NBP.

Though the time series of the annual data is still short, we can conclude that the two grasslands where we could estimate the full GHG balance are net sinks of GHGs from the point of view of the atmosphere on average. The third grassland where carbon balance was estimated (Hegyhátsál) is already a source to the atmosphere without considering non-CO_2 GHGs (NBP is negative: see Table 6.3). As N_2O flux is expected to shift NGB to even higher emission, and the contribution of CH_4 is expected to be low, the Hegyhátsál grassland is most likely a net source of GHGs to the atmosphere.

6.6 Conclusions

Both length of growing season (Baldocchi 2008) and maximum uptake capacity (Lagergren et al. 2008) have been noted as candidate factors determining the annual total NEE of grasslands, but the relative weight of these characteristics probably also depends on the geographical latitude. At high latitudes, maximum CO_2 assimilation rate can become important during the short periods of abundant light (and adequate water supply) conditions, while at longer growing season lengths the duration (with generally high CO_2 assimilation rate) becomes more decisive. The background probably includes that, at sites of long growing seasons, the relative share of leaf development and senescence periods may be much smaller than during the short growing seasons at higher latitudes. In dry environments, both the season length and the maximum uptake capacity are linked, both being largely affected by the amount of water available at the start of the growing season.

The presented results show that largely different precipitation sums are needed to reach the field capacities on the light sand, heavy clay, and the loamy grassland soils. The amount of plant available water at field capacity also largely differs among the three soil types. The annual NEE and the uptake capacity during the main vegetation period are limited by the available soil water in the sandy and heavy clay grassland ecosystems, while precipitation generally does not limit the production at the loamy grassland. The heavy clay grassland is more dependent on the frequency of precipitation events than the sandy grassland, since the plant available soil water content range is narrower; therefore, more frequent filling up is necessary to maintain favorable soil water conditions. Further, when dried out (upper soil layer), the conductivity of clay soil dramatically decreases (Hagyó 2009); therefore, the runoff fraction is larger than in the case of the sandy soil. The sandy soil, on the other hand, shows wider range of plant available soil water content, it has much higher conductivity for water, and therefore it is better suited for utilizing rainstorm precipitation. Water management characteristics of the silt soil (high capacity to store water in the plant available water potential range) is the best of the soil types considered and the annual precipitation sum is the highest in that region (Hegyhátsál). These two factors together mean that the Hegyhátsál grassland ecosystem probably will not have to face the danger of desertification even if annual precipitation will decrease in the future. Grassland ecosystems on sandy soils with high humus content seem to be better adapted to droughts than the ones on heavy clay soils, most susceptible to droughts. Grassland ecosystems on heavy clay soils are also less resilient than the ones on sandy soils because they are less able to utilize rainstorms interrupting droughts.

A consequence of the above characteristics was that sink activity of the sandy grassland was generally stronger than that of the grassland on heavy clay soil, in spite of the fact that the sandy grassland received less precipitation on annual basis (Table 6.1). Therefore, the main cause of discrepancy in sink activity between the grasslands is not the precipitation difference. Since frequency of both droughts and interrupting rainstorms are predicted to increase in the region (Bartholy and Mika 2005), the grassland ecosystem on the heavy clay soil is predicted to be more vulnerable to weather-induced carbon loss than the sandy grassland. In spite of the ecosystem scale, adaptation to drought by the sandy grassland ecosystem, the danger of turning into a source of CO_2 on annual basis is still high, since the average annual total precipitation is very close to the sink–source threshold precipitation limit. The wet grassland on the silty soil is potentially the least vulnerable to droughts primarily because of its favorable soil physical structure.

During a changing climate, the CO_2 dynamics of the grassland sites can be altered as it is foreshadowed by the NEE data from the recent drought years (2003 and 2007). At present, Bugac and Mátra sites are net sinks of CO_2. (The Hegyhátsál grassland is a net source to the atmosphere in spite of the favorable soil quality probably because of the regular mowing.) The positive NBP (at Bugac and Mátra) does not necessarily mean that these grasslands are actually mitigating the climate change. We have to take into account CH_4 and N_2O soil fluxes to estimate NGB for the ecosystems (Soussana et al. 2007; Schulze et al. 2009).

Net GHG balance was estimated for the Bugac and Mátra sites using the available eddy covariance and non-CO_2 GHG exchange measurements. The results indicate that Bugac and Mátra sites are GHG sinks, while the Hegyhátsál site is a net GHG source to the atmosphere. This latter can be associated with the management (regular mowing) at Hegyhátsál. As management practices can have a profound effect on the net GHG balance of grassland ecosystems (Soussana et al. 2007), it would be important to quantify NGB of other intensively and extensively managed grassland ecosystems in Hungary. Clearly, more research is needed to quantify the full GHG balance of grasslands on the country level also taking into account management practices and soil type.

Acknowledgments The present work was supported by the Carboeurope-IP (EU GOCE-CT2003-505572), NITROEUROPE-IP (017810-2) of the European Comission's 6th R&D Framework Programme, the Hungarian Ministry of Economy and Transport (GVOP-3.2.1.-2004-04-0107/3.0), and the Hungarian Scientific Research Fund (OTKA K75638). The authors highly appreciate the help and contribution of Giorgio Alberti, Alessandro Peressotti (University of Udine, Udine, Italy), István Tóth (technician at the Hegyhátsál site), Susumu Yamamoto, Hiroaki Kondo, and Nobuko Saigusa (National Institute of Advanced Industrial Science and Technology, Tsukuba, Japan).

References

Aires LMI, Pio CA, Pereira JS (2008) Carbon dioxide exchange above a Mediterranean C3/C4 grassland during two climatologically contrasting years. Glob Change Biol 14:539–555

Baldocchi D (2008) Breathing of the terrestrial biosphere: lessons learned from a global network of carbon dioxide flux measurement systems. Aust J Bot 56:1–26

Barcza Z, Haszpra L, Kondo H, Saigusa N, Yamamoto S, Bartholy J (2003) Carbon exchange of grass in Hungary. Tellus 55B:187–196

Barcza Z, Kern A, Haszpra L, Kljun N (2009) Spatial representativeness of tall tower eddy covariance measurements using remote sensing and footprint analysis. Agric For Meteorol 149:795–807

Bartholy J, Mika J (2005) Időjárás és éghajlat – cseppben a tenger? (Weather and climate – Sea in the drop?). Magy Tud 166:789–796

Batjes NH (1998) Mitigation of atmospheric CO_2 concentrations by increased carbon sequestration in the soil. Biol Fertil Soils 27:230–235

Ciais P, Bousquet P, Freibauer A, Naegler T (2007) Horizontal displacement of carbon associated with agriculture and its impacts on atmospheric CO_2. Glob Biogeochem Cycles 21:GB2014. doi:doi:10.1029/2006GB002741

Eswaran H, Vandenberg E, Reich P (1993) Organic-carbon in soils of the world. Soil Sci Soc Am J 57:192–194

Flanagan LB, Wever LA, Carlson PJ (2002) Seasonal and interannual variation in carbon dioxide exchange and carbon balance in a northern temperate grassland. Glob Change Biol 8:599–615

Flechard CR, Ambus P, Skiba U, Rees RM, Hensen A, van Amstel A, van den Pol-van Dasselaar A, Soussana J-F, Jones M, Clifton-Brown J, Raschi A, Horvath L, Neftel A, Jocher M, Ammann C, Leidfield J, Fuhrer J, Calanca PL, Thalman E, Pilegaard K, Di Marco C, Campbell C, Nemitz E, Hargreaves KJ, Levy P, Ball BC, Jones S, van de Bulk WCM, Groot T, Blom M, Domingues R, Kasper G, Allard V, Ceshia E, Cellier P, Laville P, Henault C, Bizouard F, Abdalla M, Williams M, Baronti S, Berretti F, Grosz B (2007) Effects of climate and management intensity on nitrous oxide emissions in grassland systems across Europe. Agric Ecosyst Environ 121:135–152

Fodor N, Rajkai K (2005) Számítógépes program a talajok fizikai és vízgazdálkodási jellemzőinek egyéb talajjellemzőkből történő számítására (A computer program for computation of physical and water management properties of soils from other soil characteristics), (TALAJTANonc 1.0). Agrokémia és Talajtan 54:25–40

Frank AB (2004) Six years of CO_2 flux measurements for a moderately grazed mixed-grass prairie. Environ Manage 33:426–431

Hagyó A (2009) A növényzet és földhasználat hatása barna erdőtalajok nedvesség-dinamikájára. Ph.D. értekezés, Gödöllő (Effect of vegetation and land use on water content dynamics of brown forest soils. Ph.D. thesis, Gödöllő) pp 136

Haszpra L, Barcza Z, Davis KJ, Tarczay K (2005) Long-term tall tower carbon dioxide flux monitoring over an area of mixed vegetation. Agric For Meteorol 132:58–77

Horváth L, Grosz B, Machon A, Balogh J, Pintér K, Czóbel S (2008a) Influence of soil type on N_2O and CH_4 soil fluxes in Hungarian grasslands. Commun Ecol 9(Suppl):75–80. doi:10.1556/Com.Ec.9.2008.S11

Horváth L, Grosz B, Czóbel Sz, Nagy Z, Péli E, Szerdahelyi T, Szirmai O, Tuba Z (2008b) Measurement of methane and nitrous oxide fluxes in Bodrogköz, Hungary; preliminary results. Acta Biologica Szegediensis 52:119–122

Horváth L, Czóbel Sz, Grosz B, Tuba Z (2010a) Bodrogközi vizes élőhelyek CH_4 és N_2O kibocsátása (2006–2009) (CH_4 and N_2O effluxes from wetlands in the Bodrogköz region). In memoriam Zoltán Tuba. Szent István University, Gödöllő

Horváth L, Grosz B, Machon A, Tuba Z, Nagy Z, Czóbel Sz, Balogh J, Péli E, Fóti Sz, Weidinger T, Pintér K, Führer E (2010b). Estimation of nitrous oxide emission from Hungarian semi-arid sandy and loess grasslands; effect of soil parameters, grazing, irrigation and use of fertilizer. Agric Ecosyst Environ (accepted)

Hunt JE, Kelliher FM, McSeveny TM, Ross DJ, Whitehead D (2004) Long-term carbon exchange in a sparse, seasonally dry tussock grassland. Glob Change Biol 10:1785–1800

IPCC (1996) Climate change 1995: The science of climate change. In: Houghton JT, Meira Filho LG, Callander BA, Harris N, Kattenberg A, Maskell K (eds) Contribution of Working Group I to the second assessment report of the Intergovernmental Panel on Climate Change. Cambridge University Press, Cambridge, UK and New York USA

Jacobs CMJ, Jacobs AFG, Bosveld FC, Hendriks DMD, Hensen A, Kroon PS, Moors EJ, Nol L, Schrier-Uijl A, Veenendaal EM (2007) Variability of annual CO_2 exchange from Dutch grasslands. Biogeosciences 4:803–816

Jaksic V, Kiely G, Albertson J, Oren R, Katul G, Leahy P, Byrne KA (2006) Net ecosystem exchange of grassland in contrasting wet and dry years. Agric For Meteorol 139:323–334

Kugler S, Horváth L, Machon A (2008) Estimation of nitrogen balance between the atmosphere and Lake Balaton and a semi natural grassland in Hungary. Environ Pollut 154:498–503

Lagergren F, Lindroth A, Dellwik E, Ibrom A, Lankreijer H, Launiainen S, Molder M, Kolari P, Pilegaard K, Vesala T (2008) Biophysical controls on CO_2 fluxes of three Northern forests based on long-term eddy covariance data. Tellus Ser B Chem Phys Meteorol 60:143–152

Machon A, Horváth L, Weidinger T, Grosz B, Pintér K, Tuba Z, Führer E (2010) Estimation of net nitrogen flux between the atmosphere and a semi-natural grassland ecosystem in Hungary. European Journal of Soil Science DOI: 10.1111/j.1365-2389.2010.01264.x

Nagy Z, Pintér K, Czóbel Sz, Balogh J, Horváth L, Fóti S, Barcza Z, Weidinger T, Csintalan Z, Dinh NQ, Grosz B, Tuba Z (2007) The carbon budget of a semiarid grassland in a wet and a dry year in Hungary. Agric Ecosyst Environ 121:21–29

Novick KA, Stoy PC, Katul GG, Ellsworth DS, Siqueira MBS, Juang J, Oren R (2004) Carbon dioxide and water vapor exchange in a warm temperate grassland. Oecologia 138:259–274

Pereira JS, Mateus JA, Aires LM, Pita G, Pio C, David JS, Andrade V, Banza J, David TS, Paco TA, Rodrigues A (2007) Net ecosystem carbon exchange in three contrasting Mediterranean ecosystems - the effect of drought. Biogeosciences 4:791–802

Pintér K, Nagy Z, Barcza Z, Balogh J, Czóbel Sz, Csintalan Z, Tuba Z (2008) Interannual variability of grasslands' carbon balance depends on soil type. Commun Ecol 9:43–48

Reichstein M, Ciais P, Papale D, Valentini R, Running S, Viovy N, Cramer W, Granier A, Ogee J, Allard V, Aubinet M, Bernhofer C, Buchmann N, Carrara A, Grunwald T, Heimann M, Heinesch B, Knohl A, Kutsch W, Loustau D, Manca G, Matteucci G, Miglietta F, Ourcival JM (2007) Reduction of ecosystem productivity and respiration during the European summer 2003 climate anomaly: a joint flux tower, remote sensing and modelling analysis. Glob Change Biol 13:634–651

Schimel DS (1995) Terrestrial Ecosystems and the carbon-cycle. Glob Change Biol 1:77–91
Schulze E-D, Luyssaert S, Ciais P, Freibauer A, Janssens IA, Soussana JF, Smith P, Grace J, Levin I, Thiruchittampalam B, Heimann M, Dolman AJ, Valentini R, Bousquet P, Peylin P, Peters W, Rödenbeck C, Etiope G, Vuichard N, Wattenbach M, Nabuurs GJ, Poussi Z, Nieschulze J, Gash JH, the CarboEurope Team, (2009) Importance of methane and nitrous oxide for Europe's terrestrial greenhouse-gas balance. Nat Geosci 2:842–850. doi:10.1038/ngeo686
Soussana JF, Allard V, Pilegaard K, Ambus P, Amman C, Campbell C, Ceschia E, Clifton-Brown J, Czobel Sz, Domingues R, Flechard C, Fuhrer J, Hensen A, Horvath L, Jones M, Kasper G, Martin C, Nagy Z, Neftel A, Raschi A, Baronti S, Rees RM, Skiba U, Stefani P, Manca G, Sutton M, Tuba Z (2007) An assessment of the greenhouse gas balance of nine European grassland sites. Agric Ecosyst Environ 121:121–134
Suni T, Berninger F, Vesala T, Markkanen T, Hari P, Mäkelä A, Ilvesniemi H, Hänninen H, Nikinmaa E, Huttula T, Laurila T, Aurela M, Grelle A, Lindroth A, Arneth A, Shibistova O, Lloyd J (2003) Air temperature triggers the recovery of evergreen boreal forest photosynthesis in spring. Glob Change Biol 9:1410–1426
Suyker AE, Verma SB (2001) Year-round observations of the net ecosystem exchange of carbon dioxide in a native tallgrass prairie. Glob Change Biol 7:279–289
van den Pol-van Dasselaar A, van Beusichem ML, Onema O (1998) Effects of soil moisture content and temperature on methane uptake by grasslands on sandy soils. Plant Soil 204:213–222
Xu LK, Baldocchi DD (2004) Seasonal variation in carbon dioxide exchange over a Mediterranean annual grassland in California. Agric For Meteorol 123:79–96
Yuste JC, Janssens IA, Carrara A, Ceulemans R (2004) Annual Q(10) of soil respiration reflects plant phenological patterns as well as temperature sensitivity. Glob Change Biol 10:161–169

Chapter 7
Forests*

Hrvoje Marjanović, Giorgio Alberti, János Balogh, Szilárd Czóbel, László Horváth, Anikó Jagodics, Zoltán Nagy, Maša Zorana Ostrogović, Alessandro Peressotti, and Ernö Führer

Abstract Fluxes of carbon dioxide, methane, nitrous oxide, and carbon stocks were measured at selected Hungarian forests and at a Croatian stand 120 km far from the Hungarian border. Annual carbon balance for Hungarian forests was also determined. Carbon stock of dendromass in selected Hungarian beech, hornbeam-pedunculate oak, and Turkey oak forests was between 191 and 292 t C ha^{-1} in 2003–2005. The total carbon stock (dendromass and soil) of Hungarian forests is estimated to be 377 MtC (1991–2000), and the annual carbon uptake of dendromass

---

*Citation: Marjanović, H., Alberti, G., Balogh, J., Czóbel, Sz., Horváth, L., Jagodics, A., Nagy, Z., Ostrogović, M. Z., Peressotti, A., Führer, E., 2010: Measurements and estimations of biosphere-atmosphere exchange of greenhouse gases – Forests. In: Atmospheric Greenhouse Gases: The Hungarian Perspective (Ed.: Haszpra, L.), pp. 121–156.

L. Horváth (✉)
Hungarian Meteorological Service, H-1675 Budapest, P. O. Box 39, Hungary
e-mail: horvath.l@met.hu

H. Marjanović and M.Z. Ostrogović
Department for Forest Management and Forestry Economics,
Croatian Forest Research Institute, HR-10450 Jastrebarsko, Cvjetno naselje 41, Croatia

G. Alberti and A. Peressotti
Department of Agricultural and Environmental Sciences,
University of Udine, I-33100 Udine via delle Scienze 208, Italy

J. Balogh and Z. Nagy
Plant Ecology Research Group of Hungarian Academy of Sciences, H-2103
Gödöllő, Páter K. u. 1., Hungary
and
Department of Botany and Plant Physiology, Szent István University, H-2103
Gödöllő, Páter K. u. 1., Hungary

Sz. Czóbel
Institute of Botany and Ecophysiology, Szent István University, H-2103
Gödöllő, Páter K. u. 1., Hungary

A. Jagodics and E. Führer
Department of Ecology and Silviculture, Hungarian Forest Research Institute, H-9400
Sopron, Paprét 17., Hungary

L. Haszpra (ed.), *Atmospheric Greenhouse Gases: The Hungarian Perspective*, 121
DOI 10.1007/978-90-481-9950-1_7, © Springer Science+Business Media B.V. 2011

is 6.9 MtC. According to the measurements and estimations, the gross primary production (GPP) in 2008–2009 at Jastrebarsko pendunculate oak forest was 1,428 and 1,633 g C m^{-2} $year^{-1}$, while ecosystem respiration of CO_2 was 1,044 and 1,049 gC m^{-2} $year^{-1}$, yielding a net ecosystem exchange (NEE) of −384 and −584 g C m^{-2} $year^{-1}$, respectively. Net primary production (NPP) of Hungarian forests (1991–2000) was 377 g C m^{-2} $year^{-1}$ (246 g C m^{-2} $year^{-1}$ excluding leaves) part of which is released back to the atmosphere by heterotrophic respiration and harvest.

Comparison of CO_2 sinks using forest inventory approach (and not considering carbon storage in durable wood products) of Hungary's forest (2.26 t CO_2 ha^{-1} $year^{-1}$) and pedunculate oak forests in Croatia (4.57 t CO_2 ha^{-1} $year^{-1}$) gives a good agreement considering differences between forests and management practice. However, flux measurements from Jastrebarsko indicate that forests are probably much stronger sink.

The share of the other greenhouse gases (CH_4 and N_2O) to the greenhouse balance between atmosphere and forest ecosystem (expressed in CO_2 equivalent) is negligible in lowland forests of Bodrogköz region (average uptake of CH_4 of 0.03 t CO_2-eq ha^{-1} $year^{-1}$ and emission of N_2O of 0.05 t CO_2-eq ha^{-1} $year^{-1}$). However, for the forests in Mátra mountain region, emissions of N_2O are not negligible with 0.61 and 0.83 t CO_2-eq ha^{-1} $year^{-1}$ for spruce plantation and sessile oak-hornbeam forest, respectively. Due to the lack of long-term observations, these results have to be regarded as first approximations.

Keywords Forest ecosystem • carbon sequestration • carbon dioxide • methane • nitrous oxide • net ecosystem exchange • soil respiration

7.1 Introduction

The forested area in Hungary, according to CORINE land cover 2000 database, amounts to 1.832 million ha (19.7% of country's territory), of which 86% are deciduous and 14% coniferous forests (ÁESZ 2002). In contrast to the relative low share of forested area, forest ecosystems play an important role in the atmospheric budget of greenhouse gases (GHGs).

Forests significantly influence the biosphere–atmosphere cycles of three GHGs, namely that of carbon dioxide, methane, and nitrous oxide. Among these, the most important is the exchange of CO_2, especially in the growing period when forests act as net sinks for atmospheric CO_2. On the other hand, carbon sequestration in dendromass is climate-dependent, i.e., expected climate change, according to a positive feedback mechanism, may decrease the carbon sequestration by forest ecosystems.

Due to the effect of the assumed climate change, warmer and drier climate conditions can be expected in Hungary (Bartholy 2006). The quantity of precipitation, but mainly its distribution within a year, could change. All these changes would have an effect on the productivity of forest sites, determining not only tree species composition of forests, but also affecting their organic matter production either directly (e.g., decrease a productivity of the existing species) or indirectly (e.g., replacement of the existing tree species with others that are less productive but better adapted to the changed conditions).

The Hungarian system for classification of forest sites (Járó 1968) is based on five dominant factors which are: climate, hydrologic conditions (water-losing, i.e., free draining, nonprecipitation water sources like ground-water, inundation waters), genetic soil type, physical make up, and rootable depth. The four types of climate (which are called forestry climate classes) have been named after natural forest communities (associations) characteristic of particular climate type. Different forestry climate classes (beech forest, hornbeam-oak forest, sessile and Turkey oak forest, and forest-steppe) used by foresters in Hungary have different productivities. While for the beech forests the average standing volume in the near optimum climate circumstances is 374 m^3 ha^{-1} (for the first yield class), the standing volume at the time of final cut is 548 m^3 ha^{-1}, and the average current annual increment is 8.8 m^3 ha^{-1} $year^{-1}$, in the less productive Turkey oak forestry climate class, corresponding values for the Turkey oak forests (also for the first yield class) are by 40% lower (ÁESZ 2002).

Change of the existing weather pattern would inevitably result in the changes and/or shifting of existing forest types. This would probably affect the quantity and quality of the organic matter production, and also raise costs related to forest management (Führer and Járó 1992). Furthermore, it is probable that forests with lesser potential productivity of organic matter would sequestrate lesser amount of atmospheric carbon dioxide.

The accurate and precise measurement of carbon stored and sequestered in forests is increasingly gaining global attention (Brown 2002). Eddy covariance is a sophisticated method that has been used in estimating carbon stored and released from forest ecosystems, but it has some important technical limitations (see Chapter 5.2). In addition, required technical expertise of the staff and financial requirements have limited the expansion of the method, and only a few eddy covariance stations operate in Central and South-eastern Europe. Because of the lack of direct flux measurements of carbon dioxide exchange between the atmosphere and forest ecosystems in Hungary, we aimed to estimate the magnitude of net CO_2 uptake and carbon sequestration in indirect way using a forest inventory approach. First step of the accounting was the survey of carbon stocks in forest ecosystems involving the dendromass and the soil reservoir as well. By estimating forest carbon stocks and rate of their change (growth, decomposition, harvest), we are able to infer the role of forest as a sink of carbon dioxide. Some authors believe that through activities of forest management, which would result in an increased sinking capacity of forest, the negative effects of anthropogenic greenhouse gases emissions could be mitigated (Führer et al. 1991; Führer and Járó 1992). However, others express reserves to this notion, emphasizing that the growing need for biomass for energy and changes in wood processing industry (which can successfully process also smaller size lumber by making plywood, etc.) push toward shorter rotations in forestry, which decreases average carbon stock (Luyssaert et al. 2010).

The importance of forests and forest management for the global climate was recognized in articles 3.3 and 3.4 of the Kyoto Protocol. In order to account the role of forests and forest management in the regional carbon budget of Hungary, it is necessary to answer the following questions: (i) how much carbon is accumulated in the forests in Hungary; (ii) how the carbon stock stored in forests is changing; (iii) what kind of silvicultural measures can elevate the carbon fixing capacity of the forests.

There are numerous papers dealing with the carbon sequestration capacity of forests, forest organic matter and carbon stock in different pools in forests ecosystems (Bornkamm and Bennert 1989; Dewar 1991; Ziegler 1991; Dewar and Cannel 1992; Weber and Burschel 1992; Burschel et al. 1993; Cannel and Milne 1995; Harrison et al. 1995; Böswald 1996).

A number of researches have been conducted in Hungary dealing with the effects of climate change on forests (Járó 1990; Führer and Járó 1992; Führer 1995; Mátyás 2005, 2010; Führer and Jagodics 2007; Somogyi 2008), and on carbon stocks and carbon sequestration of forests (Führer and Molnár 2003; Führer 2004; Führer and Mátyás 2005; Horváth 2006; Somogyi and Horváth 2006; Führer and Jagodics 2007; Führer and Jagodics 2009). Overview of some of the methods used in those researches for obtaining the carbon content and carbon concentration of the organic matter in the average tree and in different tree components, per hectare, is presented in Chapter 5.

The carbon accumulated in forests is stored partly in the dendromass and partly in the soil. While the quantity of organic material, accumulated in the living stock above the cutting surface, can be relatively precisely calculated from the forest management plans, the dendromass in the soil (stump, root system, etc.), as well as the humus topsoil and the carbon content of the organic material in mineral soil can be only roughly estimated.

Within the research project "Forest-Climate" (NKFP 3/B/0012/2002) of the Hungarian National Research and Development Program, average quantity of organic matter in Hungarian forests was studied. One of the aims of the project was to obtain estimates of the magnitude of carbon stocks in different tree components (foliage, branches, stem, root system) for the three main forest types, growing under optimal ecological conditions in Hungary, namely forests of: common beech, hornbeam-pedunculate oak, and Turkey oak.

Though there are not available direct measurements of net ecosystem exchange (NEE) of carbon dioxide at forest site in Hungary, at one of Croatian measurement stations (Jastrebarsko forest), approximately 120 km from the Hungarian border, detailed investigations have started in the year of 2007 including eddy covariance measurements. The elevation and climate conditions, which generally greatly influence the forest productivity, at Jastrebarsko site are close to the Hungarian conditions in the neighboring counties (annual mean air temperature, $t = 10.4°C$, mean yearly precipitation sum, $p = 900$ mm year^{-1}). Results from Croatian research station are described in this chapter in detail. Although result of direct CO_2 flux measurements from Jastrebarsko forest site are only preliminary at the moment, they are unique for this region and can be used as a first approximation in the estimate of the net ecosystem exchange and the net primary production (NPP) for Hungarian forests of the similar age and type (e.g., hornbeam-pedunculate oak).

Beside direct flux measurements in neighboring Croatia and local indirect estimation of carbon (dioxide) uptake and stocks, several researches involving direct soil flux measurements of carbon dioxide, methane, and nitrous oxide have been performed in Hungary (Horváth et al. 2006; Horváth et al. 2008). Although the number of soil CO_2 respiration, CH_4 uptake and N_2O emission measurements is limited, these measurements are important for approximation of the order of magnitude of soil fluxes, particularly those of non-CO_2 GHG.

7.2 Carbon Stocks and CO_2 Fluxes in Pedunculate Oak Forest at Jastrebarsko, Croatia

7.2.1 Overview

The ability to measure the carbon stored and sequestered in forests accurately and precisely is increasingly gaining global attention in recognition of the role forests have in the global carbon cycle, particularly with respect to mitigating carbon dioxide emissions (Brown 2002). Until the mid-twentieth century, most of European forests were heavily depleted of carbon in both soil and aboveground biomass in owing to harvesting and litter raking (Nabuurs et al. 2003). Since then, because of the improvement in silvicultural systems and enhanced fertility, a sharp growth and increase in carbon stocks have been registered across European forests (Ciais et al. 2008). Obviously, depending on forest type, stand development stage (Magnani et al. 2007), period of observation (Goulden et al. 1996), climate conditions (Valentini 2003), direct and indirect human activity, a forest stand can act as a net sink or a net source of greenhouse gases, primarily carbon dioxide.

The implementation of the Kyoto Protocol and rising concern for the well-being of natural forest ecosystems triggered a wave of new scientific research aiming to fill the knowledge gap in forest carbon balance and processes governing forest growth and carbon turnover. Thus, there is a growing need not only to study net ecosystem exchange (NEE) rates of the main vegetation types but also to understand the observed variability of NEE in different global vegetation types in order to quantify global carbon balance (Valentini et al. 2000).

Area of forests in Croatia is similar to the area of forests in Hungary (2.1 Mha – Croatia; 1.8 Mha – Hungary). It is estimated that forests in Croatia will remove between 39.5 and 47.7 Mt CO_2 in the period 2008–2012 (Marjanović et al. 2007). Parts of these forests are located in the plains and dominated by pedunculate oak. They appear to be the most productive forest ecosystems in Croatia and represent an important economic resource for the local community and the state. However, alterations in temperature and precipitations pattern, as manifestations of climate change, are to be expected in this area (Bates et al. 2008). These changes, combined with local anthropogenic impacts (habitat fragmentation, melioration, agriculture, etc.), are likely to cause negative effects on forests. It is believed that some of those negative effects already reflect through irregular and/or reduced acorn masts and increased tree dieback (Vukelić and Baričević 2000).

At present, most long-term eddy flux studies have focused on various temperate coniferous and broadleaved forests, tropical and boreal forests, but no measurements have been performed in the area of Southeastern Europe in highly productive lowland oak forests.

In the present chapter, we investigate the carbon cycle of a 35 years pedunculate oak (*Quercus robur* L.) stand in Jastrebarsko forest near Zagreb performed by continuous monitoring of CO_2 fluxes at ecosystem scale by the eddy covariance method, continuous soil respiration measurements, and periodic growth analysis using dendrometer bands. Moreover, we compare NPP measurements derived from eddy covariance and continuous soil respiration with NPP measurements derived from dendrometer bands.

7.2.2 *Material and Methods*

7.2.2.1 Site Description

Eddy covariance tower for CO_2 flux measurement was installed (45.62°N, 15.69°E) in a mixed stand of pedunculate oak (*Quercus robur* L.) and became operational in September 2007. The forest compartment where the tower is located is part of the state owned Forest Management Unit "*Jastrebarski lugovi*" (Eng. Jastrebarsko forest), which is part of the 13,600 ha large forest complex of *Pokupski* basin (i.e., the river *Kupa* basin). The terrain is mainly flat with altitudes ranging from 106 m above sea level at the central part of the basin up to 120 m and 130 m in the south-western and northern parts, respectively. At the center of the basin fishing ponds lay, made after logging 650 ha of forest at the beginning of twentieth century (Rauš 1996).

Soil is mainly gleysol with low vertical water conductivity (Mayer 1996). A dense network of channels has been laid during the last century in order to modify the hydrological conditions of the area. During winter and early spring, parts of the forest are flooded with stagnating water or partly waterlogged due to poor vertical water conductibility of soil, while during summer the soil dries out. In the absence of summer precipitations, trees, and oak in particular, strongly depend on groundwater.

The climate of the area is maritime temperate (Köppen classification). Average temperature for Jastrebarsko is 10.4°C (average over the period 1981–2007) with mean monthly temperatures of −0.2°C and 20.7°C in January and July, respectively. Fog is frequent in autumn, while occurrence of late frost in May is possible. Average yearly precipitation is around 900 mm $year^{-1}$, out of which around 500 mm falls during the vegetation period (April-September).

As stated above, water and microrelief play a key role in the formation of forest communities in the whole *Pokupski* basin. According to Rauš (1996), six different forest communities can be identified at the site: in four of them, the dominant tree species is pedunculate oak (*Quercus robur* L.), while in other two the dominant species are narrow-leafed ash (*Fraxinus angustifolia* L.) and black alder (*Alnus glutinosa* Gaernt.). Stands around the flux tower are dominated by pedunculate oak with dyer's greenweed (*Genista tinctoria* L. ssp. *elata*). Other tree species are hornbeam (*Carpinus betulus* L.), black alder, and narrow-leafed ash, while European white elm (*Ulmus laevis* Pall.), wild pear (*Pyrus pyraster* (L.) Burgsd.), poplars (Populus sp.), and common hazel (*Corylus avellana* L.) appear sporadically. Bushes are hawthorn (*Crataegus* sp.), common dogwood (*Cornus sanguinea* L.), breaking buckthorn (*Rhamnus frangula* L.), and blackthorn (*Prunus spinosa* L.). Tree age at the flux site was between 35 and 40 years in the year 2009. Forest stands around the eddy covariance tower are the result of regeneration cuts of old pedunculate oak stands in the early 1970s.

7.2.2.2 Carbon Stocks and Net Primary Production Measurements

During March–April 2007, and February 2008, a total of 65 permanent circular plots with a radius between 8 and 10 m were set around tower at the nodes of a 100×100 m grid for the assessment of carbon stocks of the forest. Plots that were set up in 2007 were remeasured in February 2008, when the remaining plots were established. At each plot the following measurements were performed: diameter at breast height (dbh; 1.3 m above ground) of standing trees (live and dead) greater than 2 cm with calipers at 1 mm precision; position of each tree (dbh > 5 cm); height of trees with dbh > 5 cm using Vertex III hypsometer (Haglof Instruments). In November 2008, tree heights were measured again on all trees in plots in order to estimate tree height increment in 2008. Volume of stem with branches (thicker than 3cm on thinner end) was calculated using Schumacher – Hall function (Schumacher and Hall 1933) with local, species-specific parameters (Špiranec 1975; Cestar and Kovačić 1982; Cestar and Kovačić 1984).

After preliminary footprint analysis of eddy covariance flux measurements, 24 plots having high probability to be in the footprint were selected for installation of manual dendrometer bands to estimate radial growth and aboveground net primary production (ANPP). Dendrometric values for the forest, based on data from the 24 chosen plots, are given in Table 7.1. A total of 643 aluminum dendrometers, custom-made at the laboratory using the method described by Keeland and Young (2006), were installed on all trees with dbh > 7.5 cm. Cumulative stem circumference increment was measured with precision of 0.01 mm every week during growing season 2008 and 2009 using small calipers with electronic display. Weekly measured total increment in circumference was corrected for stem curvature, converted into dbh increment and added to the initial value of dbh that was measured at the day of dendrometer installation. Error of the measurement was estimated to be around 1% of the value of corresponding dbh increment (Marjanović 2009).

Volume increment of stem and branches was converted into dry mass using species specific wood basic densities reported by IPCC (2003, 2006). Assuming that biomass of thin branches (<3 cm) and twigs is 5% of total aboveground biomass (Balboa-Murias et al. 2006), total aboveground tree biomass increment was then calculated. Below ground NPP (BNPP) was computed using a root to shoot ratio of 0.257 (Cairns et al. 1997).

For the assessment of leaf and litter production (LNPP) a total of 12 baskets (45 cm in diameter) were placed at centers of randomly selected plots with dendrometer bands. Two of the baskets sustained damage, and data from them were discarded.

Total NPP was calculated as the sum of ANPP, BNPP, and LNPP. Dry biomass was converted to carbon using a conversion factor of 0.50 (IPCC 2003, 2006).

Table 7.1 Basic dendrometric values of the forest in the footprint of the Jastrebarsko eddy covariance flux tower, shown as: live tree stock, and standing snags. Results shown here are obtained using data from all trees thicker than 5 cm at breast height (1.3 m above ground)

	Live tree stock (spring 2008, age 35 years)							Snags		
	N	G	d_g	$\langle h \rangle$	d_{g100}	h_{100}	V	N	G	V^*
Tree species	(n ha^{-1})	(m^2 ha^{-1})	(cm)	(m)	(cm)	(m)	(m^3 ha^{-1})	(n ha^{-1})	(m^2 ha^{-1})	(m^3 ha^{-1})
Quercus robur L.	576	12.55	11.8	16.7	-	-	119.6	232	1.51	<10.3
Carpinus betulus L.	414	3.37	7.2	12.8	-	-	26.2	26	0.21	<1.7
Alnus glutinosa Gearnt.	382	6.54	10.5	16.1	-	-	52.8	42	0.43	<3.2
Fraxinus angustifolia L.	154	2.63	10.5	14.1	-	-	19.2	12	0.12	<0.8
Other tree sp.	34	0.33	7.9	10.9	-	-	2.5	6	0.14	<1.1
Total	1560	25.42	10.2	15.1	25.5	20.1	220.3	317	2.40	<17.1

N–number of stems per ha; G–basal area per ha; d_g–quadratic mean diameter; $\langle h \rangle$–average height; d_{g100}–quadratic mean diameter of 100 thickest trees per ha; h_{100}–top height; V–volume of stems and branches (>3 cm on thinner end) per ha; V^*–volume of standing snags per ha (V^*–is calculated using height curves of living trees, therefore, values shown here are most likely overestimated and should be considered a rough estimate of upper limit)

7.2.2.3 Meteorological Measurements

A weather station was placed at the eddy site to measure soil temperature at three depths using thermocouples (5, 15, and 25 cm from the top), soil water content (0–20 cm) using two time-domain reflectometers (CS616, Campbell Scientific), incoming short-wave radiation (CMP3, Kipp and Zonen), incoming and outgoing photosynthetic photon flux density (PPFD) (LI-190SL quantum sensor, Li-Cor), net radiation (NR-LITE, Kipp and Zonen), air temperature and humidity (HMP45AC, Vaisala), soil heat flux (5 and 15 cm) using four soil heat flux plates (HFT3, REBS), and total rainfall (52,202 tipping bucket rain gauge, R.M. Young). Corrected surface values of soil heat flux were computed according to Ochsner et al. (2007) and Cava et al. (2008). All variables were measured at 0.1 Hz using a CR1000 data logger (Campbell Sci. Inc. Lincoln Nebraska, USA) and then averaged half-hourly.

7.2.2.4 Eddy Covariance Measurements

The eddy-covariance flux tower was installed in September 2007, at the site to assess mass, momentum, and energy ecosystem exchanges. The measurement height was 23 m above soil (3–5 m above canopy at the time of installation), and the eddy covariance system was made up of a sonic anemometer (81,000 V, Young, USA) and an open path infrared gas analyzer (LI-7500, Li-Cor, USA). The LI-7500 was pointed toward the North by an angle of 20° to minimize solar radiation influence and to facilitate the shedding of water droplcts from the sensor lenses after rain events. Data from the sonic anemometer and the open path IRGA were recorded at a frequency of 20 Hz by a hand computer (Matese et al. 2008). Ecosystem fluxes of CO_2, momentum, sensible (H), and latent heat (LE) were averaged on a half-hourly base. The applied methodology was based on the EuroFlux protocol (Aubinet et al. 2000) with a further adjustment such as the Webb Pearman Leuning correction (WPL; Webb et al. 1980). All the postprocessing elaborations and frequency response corrections have been performed using EdiRe Data software (University of Edinburgh), and quality assessment and quality check analysis (QA and QC) have been applied according to Foken and Wichura (1996). Open path IRGA provides inadequate and erroneous data during rainy or foggy conditions, or when water condensation occurs on the instrument optical lens, especially in autumn. Typically, the malfunctioning of infrared gas analyzer (IRGA), in such conditions, causes the occurrence of spikes, and in this case, a spike analysis algorithm is applied to accept or discard data before the QA/QC analysis. A gap-filling procedure was applied to obtain daily fluxes (Reichstein et al. 2005) when the QA/QC criteria were not satisfied and when a lack of turbulent transport was evident from the data.

CO_2 concentrations at different heights within the canopy (1, 2, 4, 8, 16, and 24 m from the soil surface) were measured every 6 min using a valve system and pumping down the air to an infrared gas analyzer (IRGA, SBA-4, PP-System). Storage term within the canopy and real NEE were then computed according to Aubinet et al. (2000).

7.2.2.5 Soil Respiration Measurements and Partitioning

Continuous soil respiration measurements were performed every 4 h using an automated soil respiration monitoring system. A detailed description of the system is reported in Delle Vedove et al. (2007). Briefly, the system is as a closed dynamic system according to Livingston and Hutchinson (1995), and can operate up to six automated soil respiration chambers. In our case, a system with two chambers was placed at the eddy site in order to measure soil CO_2 efflux. Each chamber consists of a steel collar (16 cm of diameter and 8 cm height) and a DC motor closing a steel lid. Lid is positioned to end up on the northern side of the collar when the chamber is open to avoid shadowing of the chamber. During the operation, air is circulated between the soil chamber and the infrared gas analyzer (IRGA, SBA-4, PP-System), at a constant flow rate (0.5 L min^{-1}). Air humidity, pressure, and temperature are measured using the additional sensors, provided by the PP-System and are acquired by parsing the digital output of the analyzer. The sequential sampling of the chambers is electronically controlled by a data logger (CR1000) and a 16-channel controller (SDM-CD16AC, Campbell Scientific) through the activation of paired solenoid valves connected to the inlet and outlet of each chamber. The system uses the rate of increase of CO_2 within the chamber to estimate, by an empirical diffusion model, the gas efflux (μmol m^{-2} s^{-1}).

Flux gap filling and flux partitioning was performed using the model proposed by Reichstein et al. (2003). Tang and Baldocchi (2005) reported that share of root respiration in oak–grass savannah ecosystem ranges from 39% to 41% of total soil respiration. Since variability in the root respiration can be in the range of 10–90% of total soil respiration (Hanson et al. 2000), we assumed that heterotrophic respiration R_h in our forest ecosystem amounts to 50% of total soil respiration.

7.2.3 Results and Discussion

During the measurement period, the air temperature ranged between the absolute minimum of −17.2°C and the absolute maximum of 34.4°C, while daily average ranged between −11.4°C and 26.1°C (Fig. 7.1a). Rain was well distributed during the study period, and total rainfall was 855 mm and 939 mm in 2008 and 2009, respectively. From December up to the end of March, the majority of forest was flooded because of the topography (lowland forest) and because of the high precipitations occurred in that period. In fact, soil water content (SWC) was around 58% v/v, but, with the beginning of budding and leaf formation, SWC experienced a significant decrease. Acceleration of the downtrend in SWC corresponded with the end of leaf formation in oak at the beginning on May and the downward trend in SWC continued throughout the whole vegetation period with intermittent spikes resulting from rain episodes (Fig. 7.1a). In any case, soil water content was never below 25% v/v.

A good agreement between energy fluxes measured at the eddy station and energy balance, calculated using net radiation and soil heat fluxes at the weather stations, was found (R^2=0.88, n=9340, P<0.001). In fact, the energy balance closure (0.64) is

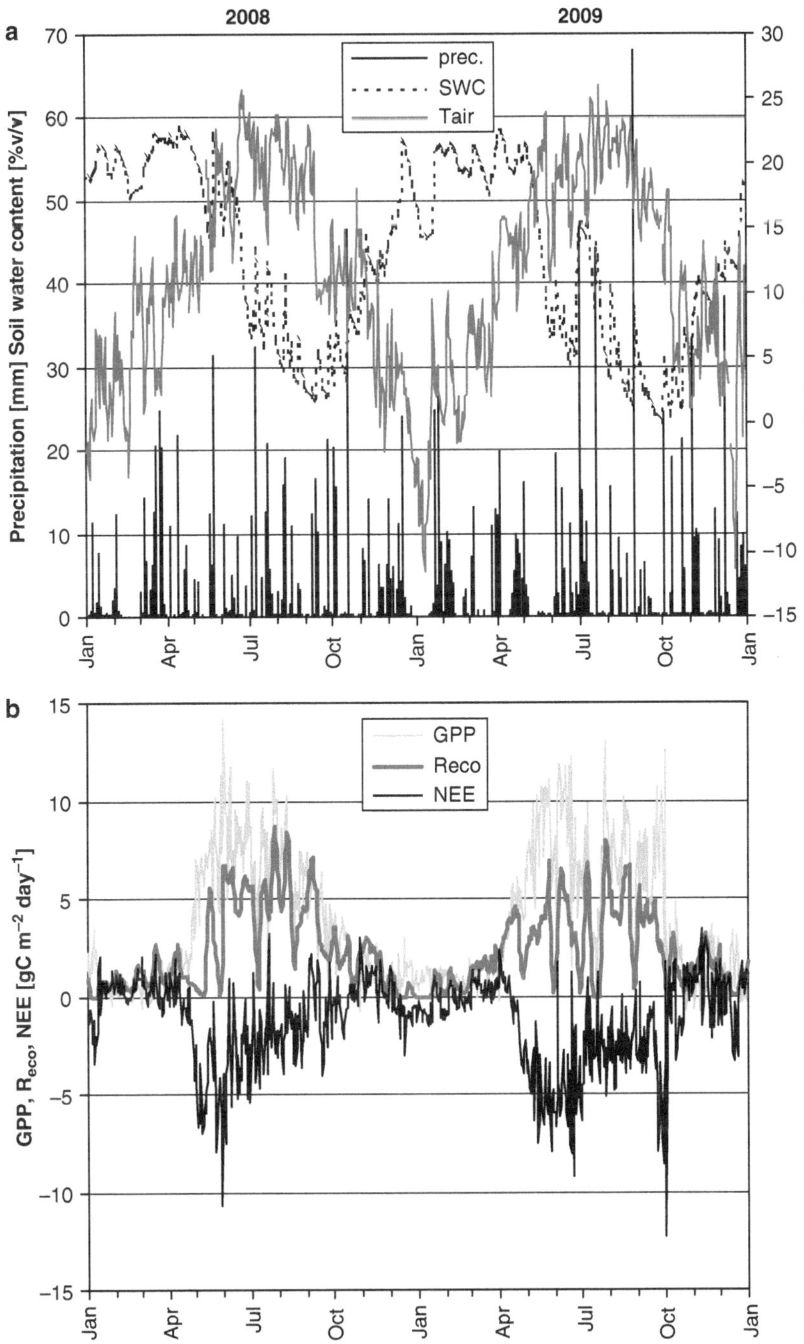

Fig. 7.1 **Panel A**: daily average soil water content with daily precipitation (*left y-axes*) and average air temperature (*right y-axes*). **Panel B**: daily gross primary production (GPP), ecosystem respiration (Reco), and net ecosystem exchange (NEE) at the study site in Jastrebarsko forest

within the range found by Twine et al. (2000) and Wilson et al. (2002) who examined data from FLUXNET sites throughout the United States and Europe.

Daily net ecosystem exchange values relative to the whole study period (January 2008–December 2009) were computed by using about 58% of measured eddy flux data and 42% of gap filled data (Fig. 7.1b). The discarded data were mainly due to lack of turbulent transport and water condensation on the IRGA, especially during autumn and winter nights. Daily NEE showed an annual course with CO_2 release during winter and net CO_2 uptake during summer. A maximum rate of carbon net uptake of −10.7 and −12.2 g C m^{-2} day^{-1} was reached in 2008 and 2009, respectively. NEE values were around zero during wintertime when soil water content was around 55% v/v, and no leaves were present. The forest acted as active CO_2 sink in both years: the cumulative NEE was −385 and −584 g C m^{-2} year^{-1} in 2008 and 2009, respectively, total evapotranspiration calculated from latent heat flux measurements (LE) was 468 and 475 mm H_2O, gross primary production (GPP) was 1,428 and 1,633 g C m^{-2} year^{-1}, meaning a water use efficiency (WUE) of 11.2 and 12.6 mg CO_2 per g H_2O (Table 7.2). NEE yearly data are close to the average for Europe reported by Luyssaert et al. (2010).

Daily soil respiration fluxes measured by chambers responded to soil temperature (Tsoil) according to an exponential function (SR=0.41 exp$^{0.14\ \mathrm{Tsoil}}$; R^2=0.88; n=561). As mentioned before, a gap filling procedure was applied according to the daily time step model proposed by Reichstein et al. (2003). A good agreement between measured and modeled fluxes was found (slope=0.87; R^2=0.87; n=561). Total soil respiration was 877 and 882 g C m^{-2} year^{-1} in 2008 and 2009, respectively, and a strong seasonal pattern was detected: daily values ranged from a minimum of 0.35 g C m^{-2} day^{-1} during winter time to a maximum of 5.83 g C m^{-2} day^{-1} at the end of June (Fig. 7.2). Q10 is 1.2, which is well within the range reported by Reich and Schlesinger (1992).

Average leaf litter production (LNPP) was 199 g C m^{-2} year^{-1}and 209 g C m^{-2} year^{-1}, respectively. LNPP data have been combined with weekly dendrometer band data in a first attempt to obtain an independent estimate of NPP to be compared with NPP derived by eddy NEE and heterotrophic respiration (Rh) (Fig. 7.3). A very good agreement has been found at the beginning of the growing season for both of the considered years. However, NPP derived from eddy and SR data is higher from the end of July. This difference may be due to a shift in carbon allocation (from stem and leaves to roots) or to a shift in Rh:SR ratio with a decrease of Rh due to higher temperature and less soil water content availability. However, further researches are needed to better estimate Rh such as the measurement of soil CO_2 efflux in root exclusion subplots.

Table 7.2 Components of carbon fluxes obtained from eddy covariance, soil respiration measurements and tree growth measurements

	NEE	GPP	R_{eco}	R_s	R_h	NPP[a]	NPP[b]	NPP[c]
Year	(g C m^{-2} year^{-1})							
2008	−384	1428	1044	877	438	818	578	777
2009	−584	1633	1049	882	441	1025	637	845

GPP gross primary production; NEE net ecosystem exchange; R_{eco}–total ecosystem respiration; R_s–total soil respiration; R_h–heterotrophic respiration; NPP net primary production from: [a]eddy covariance measurements; [b]tree growth measurements not including leaf production; [c]including leaf production

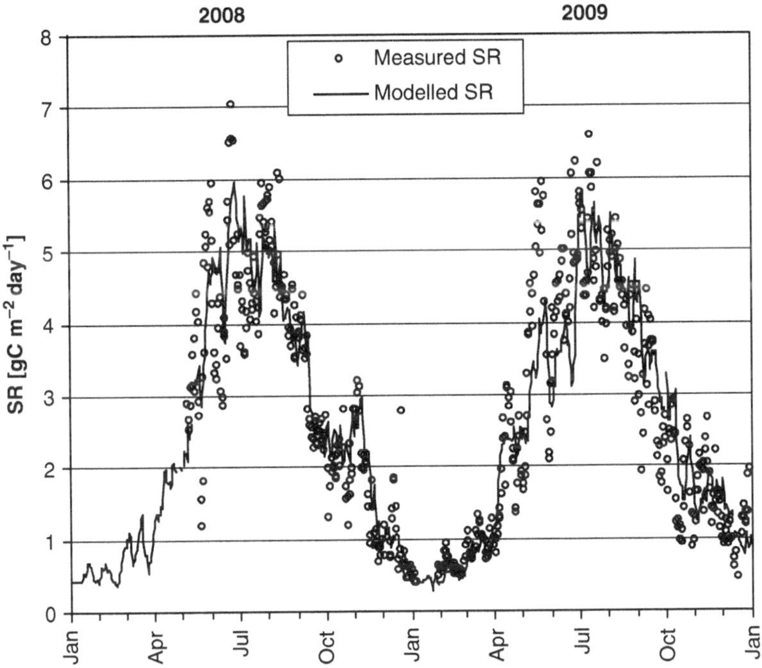

Fig. 7.2 Measured (*open symbols*) and modeled (*continuous line*) daily soil respiration using the daily time step model proposed by Reichstein et al. (2003). A good agreement between measured and modeled fluxes was found (slope = 0.87; R^2 = 0.87; n = 561)

It is important to compare data of carbon uptake from flux measurements with data that could be obtained using Croatian national forest management plan (ŠGOP 2006). According to ŠGOP (2006), area covered with forests is 2.403 Mha but 0.513 Mha is covered with degraded forest, mainly in the coastal and hinterland areas. Thus, area covered with nondegraded forests is 1.890 Mha, which is almost identical to the forest area of Hungary (1.832 Mha). Stands classified as pedunculate oak stands cover 0.216 Mha (11.4% of nondegraded, or 9.0% of total). Using estimated increment data for pedunculate oak forests (ŠGOP 2006) and values for root:shoot, basic wood density and conversion factor of 0.257, 0.55 and 0.5, respectively (see Section 7.2.2.2) we obtained an estimate for the average NPP of 301 g C m^{-2} year^{-1}. For stands younger than 20 years, there is no data on increment in ŠGOP (2006); therefore, we approximated it with the average increment of the remaining older stands. It is important to stress here that this estimate does not include carbon stored in thin trees (dbh < 10 cm), thin branches (<7 cm), and leaves, nor possible change in carbon stock of forest soils.

To assess the role of pedunculate oak forest in the carbon budget, it is necessary to include carbon removals (harvest, forest fires, etc.). Using the same approach as for the increment, the average carbon release through harvest, calculated based on National forest management plan (ŠGOP 2006), in pedunculate oak stands is estimated to be 176 g C m^{-2} year^{-1}. Difference between the increment and the harvest

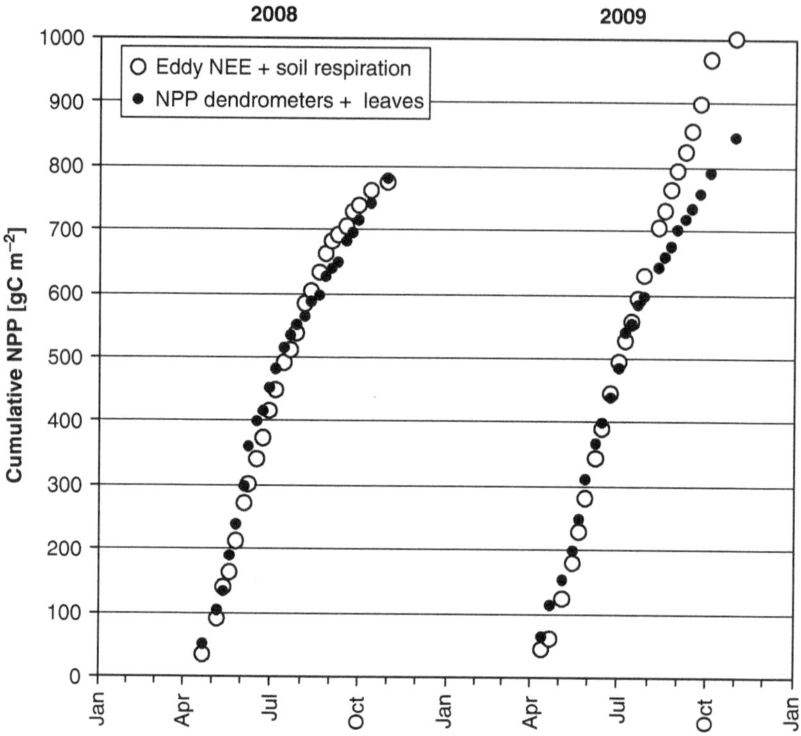

Fig. 7.3 Comparison of cumulative NPP estimates for a 35-year-old stand of pedunculate oak in *Jastrebarski lugovi* forest using NEE from eddy covariance measurements together with soil respiration measurements (*open circles*), and dendrometer bands with simple model for estimating carbon storage in tree's reserves for the production of leaf biomass in the following year (*solid circles*)

(125 g C m^{-2} $year^{-1}$) remains in forests and gradually increases the existing stock. Forest fires in pedunculate oak forests have been rare, and so far did not contribute significantly to the carbon emission.

It would be wrong to conclude that pedunculate oak forests in Croatia are net carbon sink of 125 g C m^{-2} $year^{-1}$ on average. The true sink could be greater or smaller, depending mainly on the possible change in the soil carbon pool. In obtaining the value of 125 g C m^2 $year^1$ we did not take into account the role of heterotrophic respiration, production of leaves and fine branches, litter, or the fact that significant part of oak wood is used for production of durable wood products. So far, estimates of carbon stored in durable products have not been performed in Croatia. But, if we assume that the share of felled wood used for durable wood products in Croatia equals to that in Hungary (37.9%; see Section 7.4.2, Table 7.7), we obtain an estimate that approximately 67 g C m^{-2} $year^{-1}$ is fixed in durable wood products. In that case, carbon sink would be significantly greater, namely 192 g C m^{-2} $year^{-1}$. However, whether durable wood products should be considered as real carbon storage is a matter of debate. Looking at a time-scale of several centuries it is likely that most of the carbon in the present "durable" wood products will be released.

Taking into account that pedunculate oak forests cover 0.216 Mha in Croatia, we can conclude that pedunculate oak forests are net carbon sink, after harvest removals and not taking into account soil carbon changes, of approximately 0.269 MtC year^{-1} (i.e., 0.413 MtC year^{-1} if we would take into account the carbon stored in durable forest products).

It is difficult to compare the value for carbon sink in pedunculate oak forests obtained at national scale using forest management plan (ŠGOP) with the value from eddy covariance flux measurements at local scale. However, we can compare NPP estimates. NPP estimate using data from ŠGOP (2006) of 301 g C m^{-2} year^{-1} is much lower than the estimate from eddy covariance (922 g C m^{-2} year^{-1}, on average; Table 7.2). Interestingly, estimate of average NPP with dendrometer bands without leaves (608 g C m^{-2} year^{-1}; Table 7.2) is almost at the middle of the previous two.

There are several reasons for this difference between NPP estimates from National forest management plan and eddy covariance, namely: data in ŠGOP (2006) are an average of all pedunculate oak forests (young and old), while data from Jastrebarsko are for a young stand (35–40 years) which is at the top of its productivity; ŠGOP (2006) does not include either increment of trees thinner that 10 cm in dbh or increment of thin (<7 cm) branches.

In calculation of the "average" sink from forest inventory data the soil is frequently not considered (i.e., it is considered as "carbon neutral"), which is probably not the case, and we can assume that soil probably also acts as carbon sink. Recent studies show that even old forests, which are usually considered carbon neural, still accumulate carbon (Luyssaert et al. 2008), indicating that soil likely acts as a sink.

In conclusion, pedunculate oak forests seem to be an important carbon sink at national level. Flux measurements at Jastrebarsko forest lead toward conclusion that true sink is probably even greater than the sink estimated from national forest management plan. But, climate change, i.e., alterations in temperature and precipitations pattern combined with local anthropogenic impacts are likely to cause negative effects on forests carbon balance enhancing interannual variability. Thus, long-term monitoring is needed to better understand the behavior of these forests and their response to different weather conditions. This would enable us to better assess the impact of climate change and take appropriate measures for preservation of those forests.

7.3 Carbon Stock and Organic Matter Production of Selected Hungarian Forest Ecosystems

7.3.1 Selection of the Measurement Sites

For the estimation of carbon stocks and dynamics of organic matter production, three experimental stands were selected with different climate characteristics, representative of beech, hornbeam-oak, and Turkey oak forestry climate classes. On the basis of temperature and precipitation data, each area can be distinctly characterized and distinguished. Other important criterion in selection is: growth of the experimental ecosystem should fall into the culmination section of the current increment, namely when the organic matter production is the most intensive and the utilization of site's potential is the highest. At that time, we can assume that the productivity of the habitat — considering both nutrients and water — is nearly completely utilized by the trees in the stand. Furthermore, in choosing a representative stand, the stand should not have limiting habitat factors at the site either from hydrological or from pedological aspects that could render the stand suboptimal for growth of a particular forest type.

In each of these ecosystems, 1,000 m^2 large sample plots were marked, where diameter at breast height (dhb) and height of all stems were measured, and then the average dbh, height, total basal area, and the aboveground standing volume were calculated (Table 7.3). After dendrometric survey, in 2003–2005, four trees per species with dendrometric parameters as close to the average as possible were selected and cut in each plot (see Chapter 5.4). Selected trees were used to determine the organic matter and carbon content of above and belowground biomass. During the calculation a dendrometric factor was used to take into account the difference between the calculated "average tree" and sampled trees.

The selected beech stand is situated near Bakonybél village, in the 13C forest subcompartment at Bakony Mountains (47.27°N, 17.17°E). The location is flat, plateau situated at the altitude of 460 m above sea level. Water-losing (i.e., free draining) hydrology and deep rootable depth on lessivated brown forest soil of clayey-loam soil-physical properties are present. The stand is of the first yield class, suitable for high-quality primary wood production. It is almost pure common beech (*Fagus sylvatica* L.) stand with 98% share of beech trees while remaining 2% are trees of wych elm (*Ulmus glabra* Huds.).

The hornbeam-pedunculate oak forest can be found near Bejcgyertyános village. The plot was established in the 5A forest subcompartment, of the Farkaserdő forest block (47.20°N, 17.00°E), in Vas-Zala county hill-ridge forest region. The location is flat with the altitude of 180 m above sea level. Water-losing hydrology, deep rootable depth on lessivated brown forest soil with clayey-loam soil physical properties are the main characteristics of the site. The stand is of the first yield class, suitable for high-quality primary wood production. It is a mixed forest of hornbeam (*Carpinus betulus* L.) and pedunculate oak (*Quercus robur* L.). Although hornbeam, with a share in number of trees per hectare of 62%, versus 35% of oak trees, is dominant in numbers, the volume of oak trees is almost four times greater than that

Table 7.3 Yield properties of experimental stands (Führer and Jagodics 2009)

	Age	H_g	D_g	Stems	G	V
	(years)	(m)	(cm)	(N ha^{-1})	(m^2 ha^{-1})	(m^3 ha^{-1})
Beech	70	31.1	28.9	660	43.3	740.2
Pedunculate oak	52	22.9	25.0	550	26.9	357.4
Hornbeam	52	17.2	11.6	940	9.9	92.0

H_g–average height; D_g–average diameter; G–total basal area; V–stand volume of hornbeam. Admixed to the forest are also trees of bird-cherry (*Prunus avium* L.) and common beech (*Fagus sylvatica* L.) with 2% and 1% share in number of trees, respectively.

The hornbeam-pedunculate oak forest type is most similar to the Croatian Jastrebarsko forest (compare Tables 7.1 and 7.3). The main difference is in the hydrological conditions. In a pedunculate oak dominated forest, abundance of hornbeam is indicative for sites that are not waterlogged (or at worst the waterlogging does not last long), while presence of black alder (*Alnus glutinosa* Gaernt.) and narrow-leafed ash (*Fraxinus angustiflolia* L.) indicates prolonged waterlogging conditions (as it is the case in Jastrebarsko forest). Oak trees, as they grow and mature, typically increase their water demand, while maintaining high demand for light. On wet sites, like it is the Jastrebarsko site, this usually means that further in the development stage, as the stand will probably become dryer, black alder and narrow-leafed will decline and will give way to shade tolerant, mesophyte hornbeam, particularly on microelevations. Eventually, the forest at Jastrebarsko site will probably become more and more similar to the Hungarian hornbeam-pedunculate oak forest type.

The Turkey oak forest ecosystem lies near village Veszprém-Szabadságpuszta. The plot was set in the 78C forest subcompartment, in the Veszprém plateau (47.06°N, 17.90°E). The location is flat with the altitude of about 240 m above sea level. Water-losing hydrology, deep rootable depth on brown forest soil with mainly loam soil physical properties describe the site characteristics. As in other two forests types, the chosen Turkey oak stand is also of the first yield class, suitable for high-quality primary wood production. Most abundant tree species is Turkey oak (*Quercus cerris* L.) with 93% share in number of trees per hectare, while the remaining 7% are trees in the understory, namely wild service tree (*Sorbus torminalis* L.) and field maple (*Acer campestre* L.).

7.3.2 *Organic Matter of the Aboveground Biomass*

In order to estimate the quantity of carbon in the aboveground biomass, representative trees of main tree species were selected and cut in each of the three different forest types (see Chapter 5.4). Aboveground part of the trees was partitioned into three biomass compartments: foliage, branches, and trunk. For each biomass compartment, dry mass per hectare, carbon content, and carbon stock per hectare was estimated (Table 7.4).

7.3.2.1 Foliage

Both stand density and tree genetics determine the shape of the tree crown, appearance of epicormic shoots, and consequently position of leaves along the tree. In the beech stand, 96% of the foliage was found in the crown and 4% in the stem space.

Table 7.4 Dry mass and carbon content of the average trees (Führer and Jagodics 2007; 2009)

Calculated mass	Species	Foliage	Branch	Trunk	Total
Above ground dry mass ($t\ ha^{-1}$)	Beech	3.17	107	371	481
	Pedunculate oak	3.18	57.9	189	250
	Hornbeam	1.54	16.9	59.7	78.1
	Turkey oak*	3.89	42.3	220	266
Above ground carbon content (%)	Beech	52.1	50.9	50.4	
	Pedunculate oak	51.9	51.0	50.7	
	Hornbeam	53.8	51.3	50.1	
	Turkey oak	53.6	53.0	52.3	
Above ground carbon stock ($tC\ ha^{-1}$)	Beech	1.65	54.3	187	243
	Pedunculate oak	1.65	29.6	96.1	127
	Hornbeam	0.83	8.65	29.9	39.4
	Turkey oak[a]	2.08	22.3	116	140

[a] Including admixed tree species and shrubs

In the hornbeam-oak stand, where the stems of the oaks are shaded by the hornbeams, 99% of the foliage was in the crown space. On the other hand, in the Turkey oak stand, 19% of the foliage was on the epicormic branches of the stem. Epicormic shoots occur in some tree species when higher parts of the plant are damaged, or light level on the tree stem is increased. Heavy damage by gypsy moth (*Lymantria dispar* L.) in the year 2004 might have significantly reduced the foliage mass of the Turkey oak and could be responsible, among others, for increased epicormic growth.

The average absolute dry mass of leafage (including the crop, i.e., seeds and cupules) per tree for the analyzed trees was 4.81 kg for beech, 6.24 kg for pedunculate oak, 2.33 kg for hornbeam, and 4.50 kg for Turkey oak.

7.3.2.2 Branch Structure

The whole branch absolute dry mass of the average tree of the beech plot amounts to 162 kg, of which 12% is in the upper, 30% in the middle, and 50% in the lower third of the crown, while 8% can be found in the stem space. The average absolute dry mass of the branch structure in the pedunculate oak was 111 kg, while for hornbeam was 24 kg. The total branch absolute dry mass of the average tree in the Turkey oak stand was 50 kg, of which 96% was in the crown, the rest in the stem space. The branch mass of the average tree of other tree species was 30 kg. The absolute dry weight of the branch structure, carbon content and stock is given in Table 7.4. In addition, in Turkey oak forest, admixed species contribute to the branch structure with 1.67 $t\ ha^{-1}$ (0.87 $t\ C\ ha^{-1}$), while additional 2.11 $t\ ha^{-1}$ of dry weight biomass (1.10 $t\ C\ ha^{-1}$) is in the shrub-layer.

7.3.2.3 Trunk

The most of the biomass is allocated in trunks of the trees, as could be expected for trees in high forests (Table 7.4). The share of aboveground biomass in trunks is in the range of 75.6% (pedunculate oak) to 82.7% (Turkey oak) of the total aboveground biomass. In harvesting, the stems of the trees are extracted, but significant share of biomass (branches, stumps, etc.) usually remain in the forest. The knowledge on the amount of biomass that is moving from one carbon pool (i.e., aboveground live biomass) into another (coarse/fine woody debris) after forest thinning or harvesting is gaining importance. This information is important for modeling forest carbon cycling, or for estimating the potential for additional quantities of biomass that could be extracted with harvesting.

7.3.3 Organic Matter of Belowground Biomass

7.3.3.1 Stump and Root Swelling (Root-Wood)

The absolute dry mass of the stumps and root swellings in beach forest is 40.0 t ha^{-1}, and in the hornbeam-oak forest it is 50.2 t ha^{-1}, of which 39.5 t ha^{-1} comes from pedunculate oak and 10.7 t ha^{-1} from hornbeam. On the other hand, in the Turkey oak stand, despite a higher stem-number per hectare of Turkey oaks than the corresponding pedunculate oaks in hornbeam-oak forest, the total mass of stump and root swelling coming from Turkey oak is only 27.1 t ha^{-1}, with addition of 0.65 t ha^{-1}coming from other tree species. Carbon stock shares of stump and root swelling are given in Table 7.5.

7.3.3.2 Root System

The absolute dry mass of the root system in the beech stand is 72.8 t ha^{-1}, in the hornbeam-oak stand it is 64.6 t ha^{-1}, while in the Turkey oak stand it is 92.5 t ha^{-1}. Here it is interesting to note that Turkey oak forest has significantly more biomass in the root system compared to hornbeam-pedunculate oak. Furthermore, in all of the three

Table 7.5 Carbon mass stored in the underground dendromass (tC ha^{-1}) (Führer and Jagodics 2007, 2009)

		Beech f.	Hornbeam-oak forest			Turkey oak forest		
		Beech	Ped. oak	Hornbeam	Total	Turkey oak	Other	Total
Stump and root swelling		20.1	19.7	5.31	25.0	14.2	0.34	14.5
Root system	0–2 mm	6.96			4.27			5.67
	2–5 mm	2.73			2.65			2.55
	5–20 mm	7.58			7.56			8.62
	20–50 mm	7.52			8.30			6.36
	>50 mm	4.28			2.75			12.9
	Subtotal roots	29.1			25.5			36.1
Total		**49.2**			**50.6**			**50.6**

ecosystems, a lower carbon concentration was measured in the belowground dendromass than in the aboveground dendromass. In the beech stand, carbon concentration of belowground dry weight dendromass was 40.0%, in the hornbeam-oak stand 39.5%, and in the Turkey oak stand, it was 39.3% on average.

The amount of carbon stored in the underground dendromass, i.e., the stumps and root systems of the beech, hornbeam-oak, and Turkey oak stands is shown in Table 7.5.

7.3.4 Organic Matter of the Total Dendromass

The comparison of the total carbon content of three forest types, beech, hornbeam-oak, and Turkey oak forest, shows some significant differences (Fig. 7.4). Beech stand has accumulated more carbon per hectare than the other two forest stands, while hornbeam-oak stand has accumulated more than Turkey oak stand. Although it is difficult to compare beech stand with that of hornbeam-oak or Turkey oak because beech stand is almost 20 years older, it is interesting to note the difference in carbon allocation in above and belowground dendromass.

Total of 292 tC ha^{-1} of carbon is fixed in the dendromass in beech stand, of which the foliage is 1%, the branch system is 18%, the stem is 64%, the stump and root swelling is 7% and the root system is 10%. In the hornbeam-pedunculate oak

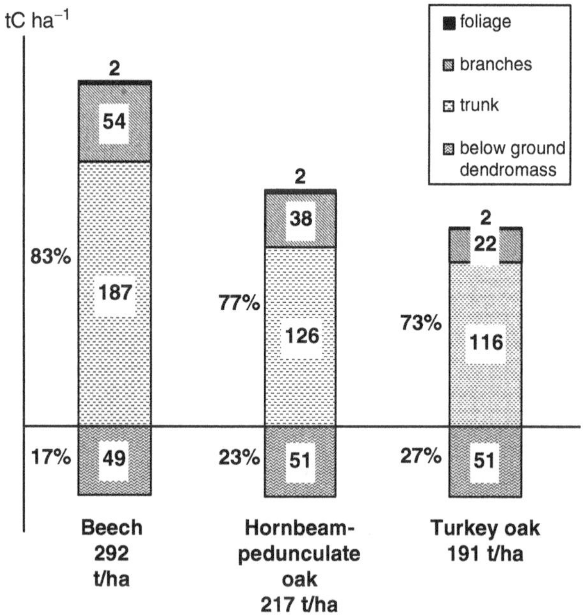

Fig. 7.4 Carbon stock of dendromass (in tC ha^{-1}) in the investigated ecosystems (Führer and Jagodics 2007, 2009)

stand 217 tC ha^{-1} carbon, of which is 1% foliage, 18% branch system, 58% stem, 11% stump and root swelling, and 12% root system. In the dendromass of the Turkey oak stand 191 tC ha^{-1}, carbon has been accumulated on one hectare, of which the share of the foliage, the branch structure, the stem, the stump, and root swelling and the root system are 1%, 11%, 61%, 8% and 19%, respectively.

Beech stand has the least amount of carbon allocated in the belowground part (stumps and root system) in both absolute and relative terms. On the other hand, Turkey oak stand has the greatest share of carbon allocation into belowground dendromass. Greater allocation of carbon in a tree root system usually occurs as a response to certain limiting factors like poor nutrient and/or water availability (Pallardy and Kozlowski 2007). Although comparison among different forest ecosystems cannot be so straightforward, observed difference between these three forest ecosystems can be an indicator of the potential for carbon storage in different forest types.

7.4 Carbon Fixing Capacity of the Forests in Hungary

7.4.1 Carbon Stock of the Forests in Hungary

Based on the data of forest management plan (ÁESZ 2002) and on the results of several decennial surveys (Führer et al. 1991; Führer 1995, 2004; Führer and Molnár 2003), approximately 377 Mt of carbon is stored in the forests of Hungary (Table 7.6). The average carbon storage per unit area is 206 tC ha^{-1}, from which 74 tC ha^{-1} can be found in the dendromass while 132 tC ha^{-1} is stored in the litter and in the soil (top 40 cm). The sandy skeletal soils have the lowest carbon content, about 60 tC ha^{-1}, while the brown (dark colored) forest soils can store more than 350 tC ha^{-1}.

Considering carbon stocks per tree species, the native hard broadleaved tree species (oak, beech, and Turkey oak) have a larger carbon stock than their corresponding share in the area, the hornbeam and other broadleaved tree species have a

Table 7.6 Carbon stock stored in the forests of Hungary in 1991–2000 (Führer 2004)

Carbon stock	MtC
Dendromass	
Foliage	3
Living stock above cutting surface	87
Stump, root system	46
Subtotal	*136 (36%)*
Soil	
Litter layer	5
Humus soil to 40 cm	236
Subtotal	*241 (64%)*
Total	**377 (100%)**

nearly same share, while a carbon stock in the locust, poplar, and coniferous stands is smaller than their corresponding share in the area (Führer and Molnár 2003).

7.4.2 *Annual Change of the Carbon Fixed in the Living Stock of Forest*

The quantity of carbon fixed in or released from the forest ecosystem equals difference between carbon gains (e.g., biomass increment) and carbon losses (e.g., harvesting, forest fire, heterotrophic respiration) (Führer 2004). The annual organic material forming, i.e., the increment of the tree volume (above cutting surface) in Hungary is 12 million m^3, which correspond to 3.20 MtC (Table 7.7). With the practice of sustainable forest management, only 75% (2.40 MtC) of that increment is harvested, while 25% (0.80 MtC) remains and accumulates every year in forests.

Furthermore, not everything that is harvested is extracted from the forest. Approximately 20% of the biomass of felled trees is the so-called harvesting loss (twig, branch, bark, sawdust, etc.), which remains in the forest after harvesting. The carbon content of the harvesting loss, entering the annual carbon cycle, begins to decompose and gradually returns back to the atmosphere. From the 1.92 Mt carbon stored in extracted round wood nearly 50% is in wood used in energy supply (mainly fuel wood), which is released into the atmosphere by burning. The remaining 50% of extracted round wood is used in wood processing industry, mainly for production of durable products. During production approx. 5% of wood

Table 7.7 Annual carbon balance of the growth stock of forests in Hungary (Führer 2004, Führer and Mátyás 2005)

Annual carbon balance	MtC year^{-1}	
(1) Annual carbon gain		
Foliage	2.40	
Increment of aboveground biomass	3.20	
Increment of stump and roots	1.30	
Subtotal	**6.90**	
(2a) Annual carbon loss		
Foliage		2.40
Harvesting losses (20% of the felling)		0.48
Stump and root from felling		0.97
Fuel wood (40% of felling)		0.96
Waste arising from processing of wood products (app. 5% of processed wood)		0.05
Subtotal		**4.86**
(2b) Stored carbon		
Products of wood industry (37.9% of felling)		0.91
Increment of aboveground biomass remaining in forest		0.80
Increment of stump and roots remaining in forest		0.33
Subtotal		**2.04**
Total	**6.90**	**6.90**

is woodworking waste which, coming to the carbon cycle, must be accounted as CO_2 source. The carbon fixed by photosynthesis from the atmosphere and stored in wood products does not return immediately to the annual cycle, or, more precisely, it is not accounted for because of the lasting character of the wood products. However, since all wood products have finite lifetime in total carbon accounting the average lifetime of products should be estimated, and appropriate values should be added to the annual carbon cycle.

Annual carbon uptake used in formation of the foliage amounts to approx. 2.40 MtC. It is assumed that an equal amount of carbon returns to the atmosphere every year through processes of decomposition.

It is difficult to determine the stock and annual change of carbon stored in the stump and root system. It is estimated that stump and root increment amounts approx. 40% (1.3 MtC) of the increment of the aboveground part, but this estimate comes with large uncertainty.

In summary, it can be stated that about 6.9 Mt carbon are stored into the living tree stock through the annual growth of forests in Hungary. From this, 70% goes to the annual carbon cycle and returns to the atmosphere, while the remaining 30% are partly stored in the trees in the forest, and partly stored in durable wood products.

7.4.3 Possibilities for Forestry and Wood Industry to Increase the Carbon Sequestration

As a consequence of the high carbon sequestration potential of forests, it could be possible, through measures of forest management, to influence the regional carbon cycle and help in reducing effects of anthropogenic GHG emissions in the following ways:

(a) *Increasing the carbon sequestration in wood products by improving timber quality*. The quality of the forest crop can be improved by a sustainable and professional forest management. As a result, the share wood, suitable for manufacturing durable wood products would increase. To achieve this, it is important that trees are harvested at the appropriate age before the quality of the wood is degraded from the point of view of wood processing. Postponing the felling, e.g., for reasons of nature conservation, while important for the maintaining biodiversity, usually has unfavorable influence on wood quality, and subsequently on carbon sequestration through wood products. An unmanaged forest during its development gradually reaches the limit of its carbon fixing capacity. In other words, most of the carbon dioxide uptake during the assimilation in such a forest will be eventually released by the respiration.

(b) *Increasing the productivity of the existing forests through species conversion and/or improving stand structure*. Today in Hungary in nearly 30% of the forest land area, there is a possibility to improve forest stands structure, which may result in higher yields and better wood quality.

(c) *Afforestations*. Afforestation of abandoned or low-productivity agricultural land would increase removal of atmospheric carbon. Removed carbon would be stored not only in dendromass, but also in the soil, which is usually much poorer in organic material than forest soil.
(d) *Increasing the share of energy wood in the energy mix*. Although carbon dioxide emissions from burning of fossil fuel or wood for energy are similar for equal amount of heat, by burning wood (unlike burning of fossil fuel) we do not add new carbon into the global carbon cycle. Therefore, increased use of firewood for heating, produced through sustainable forest management, would positively reflect on the GHG emission budget.

7.5 Soil Flux Measurements of Nitrous Oxide and Methane

Emission of non-CO_2 greenhouse gases could play significant role in global warming. Increase of temperature or change in precipitations will probably have an effect on the biological processes that regulate emissions of GHG, like nitrous oxide and methane, in natural ecosystems. The role of forests, under present or changed conditions, in net fluxes to non-CO_2 greenhouse gases, is still linked with great uncertainty. Some of the researches that have been undertaken in Hungary with the aim to quantify fluxes of non-CO_2 greenhouse gases are presented here.

7.5.1 *Sampling Sites*

Researches of fluxes of nitrous oxide were conducted in two regions of Hungary, Bodrogköz region and Mátra region. Researches of methane flux were carried out only in Bodrogköz region. Typical forest ecosystems in those regions are: lowland pedunculate oak-hornbeam forest, riverine oak-elm-ash woodland, and riverine poplar woodland in Bodrogköz region, while mountain sessile oak-hornbeam forest and Norway spruce forest plantation are typical for Mátra region.

7.5.1.1 Bodrogköz Region, Vajdácska (Lowland Pedunculate Oak - Hornbeam Forest)

Location: Long-forest (Long-erd , 48.34°N, 21.65°E, 97 m above sea level), near Vajdácska-Sárospatak village, Bodrogköz region, (Vajdácska, *Circaeo-Carpinetum* Borhidi 2003). The stands are situated mostly on the places of dried out oxbows at the inner parts of Bodrogköz. The canopy layer is dominated by pedunculate oak (*Quercus robur* L.), hornbeam (*Carpinus betulus* L.), and narrow-leaved ash (*Fraxinus angustifolia* L. ssp. pannonica), while in the shrub layer mostly field maple (*Acer campestre* L.) and narrow-leaved ash have the highest abundance. In the ground layer, *Circaea lutetiana*, *Convallaria majalis*, *Rubus caesius*, and *Viola sylvestris* are dominant or subdominant. On locations with cooler microclimate conditions the specialty of this stand is the well-developed common beech (*Fagus sylvatica* L.) which, although stands is located at only 90–95 m altitude above sea level, gives a certain mountain character to the place (Tuba 1995; Gál et al. 2006; Horváth et al. 2008; Czóbel et al. 2010).

7.5.1.2 Bodrogköz Region, Tiszacsermely (Riverine Oak-Elm-Ash Woodland)

Location: near Tiszacsermely (48.24°N, 21.85°E, 96 m above sea level) village at the flood-plain of Tisza River (Tiszacsermely, *Fraxino pannonicae-Ulmetum* Soó in Aszód 1935 corr. 1963). The stands lay mainly on the highest points of the flood-plain (Tuba 1995). The investigated stands are the consociation of narrow-leaved ash (*Fraxinus angustifolia* L. ssp. pannonica) and pedunculate oak (*Quercus robur* L.). The well-developed layer of shrubs is composed by *Sambucus nigra*, *Ulmus laevis*, *Fraxinus angustifolia* ssp. pannonica. Some invasive plants like *Amorpha fruticosa*, *Robinia pseudo-acacia* are present in the layer. In the ground layer, *Rubus caesius* and *Glechoma hederacea* are predominant, although *Aristolochia clematitis*, *Circaea lutetiana*, *Iris pseudacorus*, and *Sambucus nigra* can be abundant as well (Horváth et al. 2008; Czóbel et al. 2010).

7.5.1.3 Bodrogköz Region, Óbodrog (Riverine Poplar Woodland)

The studied riverside willow-poplar woodland site lays at the bank of Óbodrog oxbow (48.32°N, 21.57°E, 106 m above sea level), near the town Sárospatak. The scientific name of the association is *Senecioni sarracenci-Populetum albae* Kevey 1996. In the Bodrogköz section of Upper-Tisza, this community denotes the last point of the zone of flood-plain associations (Gál et al. 2006). In Bodrogköz region, the stands are mixed with gray poplar (*Populus* × *canescens*). In the canopy layer, the gray poplar is dominant. The shrub layer is rich is species, dominated by *Cornus sanguinea*, *Fraxinus pennsylvanica*, *Ulmus laevis*, and *Viburnum opulus*. In the ground layer, the *Rubus caesius*, *Cornus sanguinea*, and *Lythrum salicaria* are the most frequent. The protected *Maianthemum bifolium* can be found in the stand as well (Horváth et al. 2008; Czóbel et al. 2010).

The wetland poplar forests grown along the Óbodrog oxbow are common on the higher parts of the floodplain. In the upper canopy layer, only white poplar (*Populus alba* L.) and gray poplar (*Populus* × *canescens*) are dominant. The shrub layer is rich; *Cornus sanguinea*, *Vitis sylvestris*, *Rubus caesius* are dominant with some contribution of *Fraxinus pennsylvanica* and *Acer negundo*. In the ground layer, *Rubus caesius* is usually dominant and *Lysimachia vulgaris*, *Stachys palustris*, *Lycopus europaeus* are common. The soil is acidic Fluvisol, poor in ammonium but more abundant in carbon and nitrogen compared to the other wetland site.

7.5.1.4 Mátra Region, Tetves-Rét (Mountain Sessile Oak-Hornbeam Forest)

The studied sessile oak forest site lies in the forested area of Mátra Mountains, Northeast Hungary. The Ecology Research Station of Hungarian Forest Research Institute is also located in this forest. The surrounding low mountain range area is largely afforested mainly with broadleaf trees. Climate is temperate continental with mean annual air temperature of 5.7 C; mean annual precipitation is 780 mm, with a precipitation maximum in summer.

Geographical coordinates of the center of the 20 ha sessile oak stand (Tetves-rét) are 47.85°N and 19.97°E, with average altitude of 660 m above sea level. Exposure of the site is southeast. The stand is dominated by sessile oak (*Quercus petraea* L.) in community with hornbeam (*Carpinus betulus* L.) and some common beech (*Fagus sylvatica* L.) and Turkey oak (*Quercus cerris* L.) trees.

The scientific name of the association is *Carici pilosae-Carpinetum* Neuhäusl & Neuhäuslová-Novotná 1964, Borhidi 1996. In parts, herbaceous ground vegetation can be found, comprising, e.g., *Dentaria bulbifera, Galium odoratum,* and *Lathyrus vernus*.

The age of trees in the stand is 75–80 years, and leaf area index was estimated at 4.0 in summer months. The soil is a mollic leptosol, with 30.4% of sand, 53.8% silt, and 15.8% of clay, in the uppermost 5 cm of the soil mineral layer (Horváth et al. 2006; Rosenkranz et al. 2006; Czóbel et al. 2010).

For the mull-type organic layer, a soil organic carbon (SOC) content of 39.3%, a C/N ratio of 22.3, and a pH value of 5.4 (in 0.01 M $CaCl_2$) were determined. The upper 5 cm of mineral soil had a SOC content of 5.5%, a C/N ratio of 12.3, and a pH value of 3.1 (in 0.01 M $CaCl_2$). In the years 2001–2003, the forest received 17 kg N ha^{-1} nitrogen input through wet and dry deposition (Horváth et al. 2006).

7.5.1.5 Mátra Region, Nyírjes (Norway Spruce Forest Plantation)

The studied Norway spruce (*Picea abies* (L.) H. Karst.) forest site also lays in the forested area of Mátra Mountains, Northeast Hungary. Geographical coordinates of the 8 ha Norway spruce stand (Nyírjes) are 47.89°N and 19.96°E, with average altitude of 560 m above sea level. The age of the stand is 42–44 years, and leaf area index measured in summer, 1993, was 3.3. Mean annual precipitation is 780 mm, while mean annual air temperature is 5.7°C (Horváth et al. 2006).

7.5.2 Results

7.5.2.1 Bodrogköz Region

The results of individual measurements of methane soil flux can be seen in Fig. 7.5 for 2 years. The year 2006 was wetter then 2007, according to the observed yearly precipitation (580 and 660 mm $year^{-1}$, respectively, measured at the nearby Sárospatak

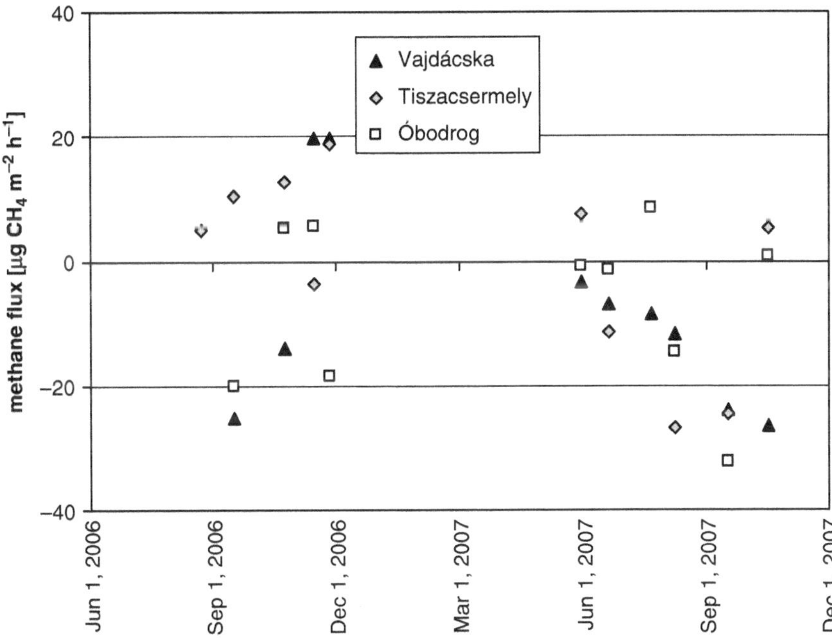

Fig. 7.5 Soil methane flux in forests of Bodrogköz region (Horváth et al. 2008)

meteorological station). Methane flux varies between –32 and 20 μg CH_4 m^{-2} h^{-1} during the 2 years. The magnitude and the direction (emission or uptake by soil) strongly depend on the soil water content (SWC) and on the temperature. Maximum uptake can be observed at high soil temperature and medium soil moisture. In extreme dry soil, methane uptake is inhibited by the water stress for methanotroph bacteria. As it was mentioned earlier (see Chapter 5.3), dry soils, especially well aerated forest soils, act as sinks for methane through the microbial oxidation by methanotroph bacteria (Prather et al. 1995; Steinkamp et al. 2001). On the other hand, at SWC higher than 50% the methane production dominates against the methane uptake (van den Pol-van Dasselaar et al. 1998). Parallel with the variation of SWC and temperature the direction of methane flux varies as well. As Table 7.8 demonstrates, methane fluxes in Bodrogköz region show high variability (high standard deviation) with low mean methane flux in the wetter year (2006). Depending on the forest type, the soil acted as net methane sink or net source. It is interesting to note the difference in methane flux dynamics during 2006 vegetation period in different forest types (Fig. 7.5). For example, oak–elm–ash woodland at Tiszacsemerly acted as a source during summer and became a sink at the beginning of autumn. On the other hand, hornbeam-oak forest at Vajdácska acted exactly opposite. Overall average of the measured methane fluxes during 2006 shows that methane emission and uptake are practically balanced. In the dryer year (2007), methane uptake of forest soils predominated in all three forest ecosystems. In spring, forests were close to being methane neutral, but during summer they all became strong methane sinks.

Table 7.8 Mean soil fluxes of nitrous oxide and methane in Bodrogköz region (Horváth et al. 2008)

	N_2O flux (μg N m^{-2} h^{-1}) Mean ± SD		CH_4 flux (μg CH_4 m^{-2} h^{-1}) Mean ± SD	
Location/Year	2006	2007	2006	2007
Vajdácska	1.2 ± 1.0	0.4 ± 2.4	0.0 ± 23.2	−13.5 ± 9.6
Tiszacsermely	0.1 ± 1.6	3.9 ± 7.1	8.6 ± 8.6	−10.0 ± 16.2
Óbodrog	0.8 ± 1.0	0.5 ± 4.2	−6.9 ± 14.3	−6.6 ± 14.6
Overall mean	0.7	1.6	0.6	−10.0
CO_2 equivalent[a] (t CO_2 ha^{-1} $year^{-1}$)	0.03	0.07	0.00	−0.02

[a] Calculated according to 100-year time horizon (Forster et al. 2007)

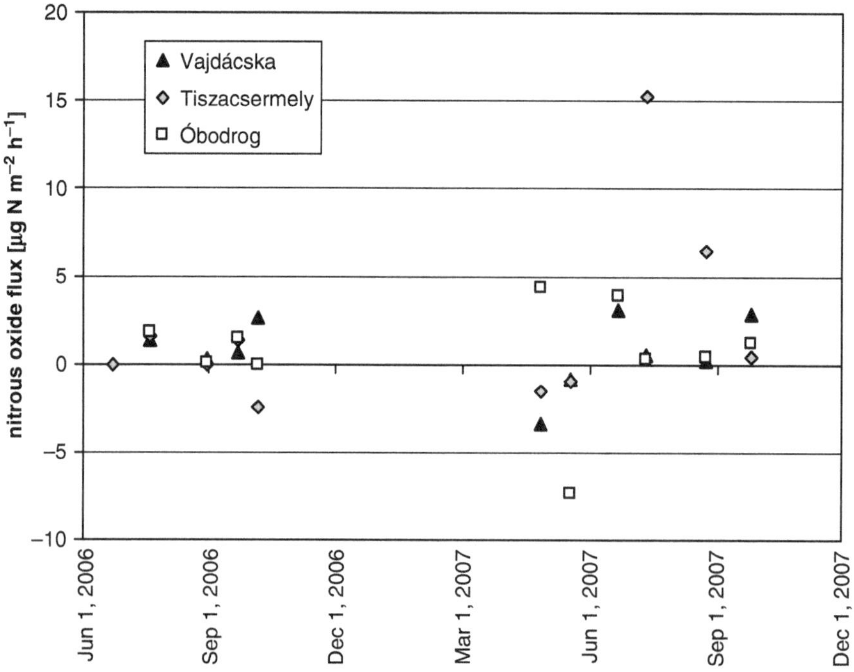

Fig. 7.6 Soil nitrous oxide flux in forests of Bodrogköz region (Horváth et al. 2008)

The results of individual measurements of nitrous oxide soil emission fluxes can be seen in Fig. 7.6. The nitrous oxide flux varied within a range of −7 and 15 μg N m^{-2} h^{-1} in 2006–2007. There is a difference in the measured mean flux between the wetter year (2006) and drier year (2007) (Table 7.8). In both cases, the N_2O flux is positive (emission). Measured mean fluxes for forested area (1.2 μg N m^{-2} h^{-1}) are generally lower by a factor of two if compared to the nitrous oxide emission measured above short vegetation in Bodrogköz region (2.4 μg N m^{-2} h^{-1}, Horváth et al. 2008).

Though measurements were conducted in summer half-year, if we suppose that most of nitrous oxide is emitted from the soil during warmer times (vegetation period), estimated annual emission rate for forests is 0.10 kg N ha^{-1} $year^{-1}$. Using global warming potential (GWP) for N_2O of 289 for a 100 years time horizon (Forster et al. 2007), emission of 1 μg N m^{-2} h^{-1} (in form of N_2O) is equivalent to 0.0410 t CO_2 ha^{-1} $year^{-1}$. Consequently, measured mean flux of N_2O is equivalent to the emission of 0.05 t CO_2 ha^{-1} $year^{-1}$. Similar calculation for CH_4 (GWP=25) gives that 1 μg CH_4 m^{-2} h^{-1} is equivalent to 0.0022 t CO_2 ha^{-1} $year^{-1}$. Hence, we obtain mean annual net flux of CH_4 of −0.01 t CO_2 ha^{-1} $year^{-1}$ (Table 7.8), which has practically negligible effect on (reducing) global warming.

Atmospheric deposition is generally the only N-source for nonfertilized ecosystems, and it is estimated to be around 15 kg N ha^{-1} $year^{-1}$ for Hungarian forests (Horváth et al. 2006; Kugler et al. 2008). Thus, only a negligible part (0.7%) of deposited nitrogen is emitted from the forest soil back into the atmosphere in the form of N_2O as an intermediate product of soil denitrification processes.

7.5.2.2 Mátra Region

Results of N_2O flux measurements in sessile oak-hornbeam forest and spruce plantation in Mátra region are described in Horváth et al. (2008) and summarized in Fig. 7.7 and Table 7.9. Average soil N_2O flux, measured between October 2002 and

Table 7.9 Soil flux of nitrous oxide at Mátra region (Horváth et al. 2006)

	Nyírjes (spruce)			Tetves-rét (oak)		
	N_2O flux (μgN m^{-2} h^{-1})	T_{soil} (°C)	w_{soil} [b] (V/V)	N_2O flux (μgN m^{-2} h^{-1})	T_{soil} (°C)	w_{soil} [c] (V/V)
October 2002	35.2	9.3	0.25	36.0	n.a.	n.a.
November 2002	18.1	6.4	0.27	4.7	7.3	0.27
December 2002	2.6	3.0	0.29	2.0	3.3	0.29
January 2003	3.7	0.9	0.30	1.1	1.1	0.31
February 2003	0[a]			0[b]		
March 2003	0[a]			30.9	4.3	0.31
April 2003	4.0	2.8	0.30	5.7	4.4	0.30
May 2003	11.3	8.1	0.27	16.8	11.0	0.28
June 2003	14.9	10.4	0.23	28.6	13.9	0.26
July 2003	22.5	11.6	0.22	35.8	14.8	0.23
August 2003	36.5	13.4	0.26	48.9	16.5	0.18
September 2003	28.2	11.1	0.15	32.8	12.8	0.15
Mean ± SE	14.8 ± 2.8	5.6	0.26	20.3 ± 2.9	8.2	0.26
CO_2 equivalent[d] (t CO_2 ha^{-1} yea^{-1})	0.61 ± 0.11			0.83 ± 0.12		

[a] Estimated (snow cover on soil)
[b] Saturation (i.e., water-filled pore space = 1.0) at w_{soil} = 0.32
[c] Saturation (i.e., water-filled pore space = 1.0) at w_{soil} = 0.34
[d] Calculated according to 100-year time horizon (Forster et al. 2007)

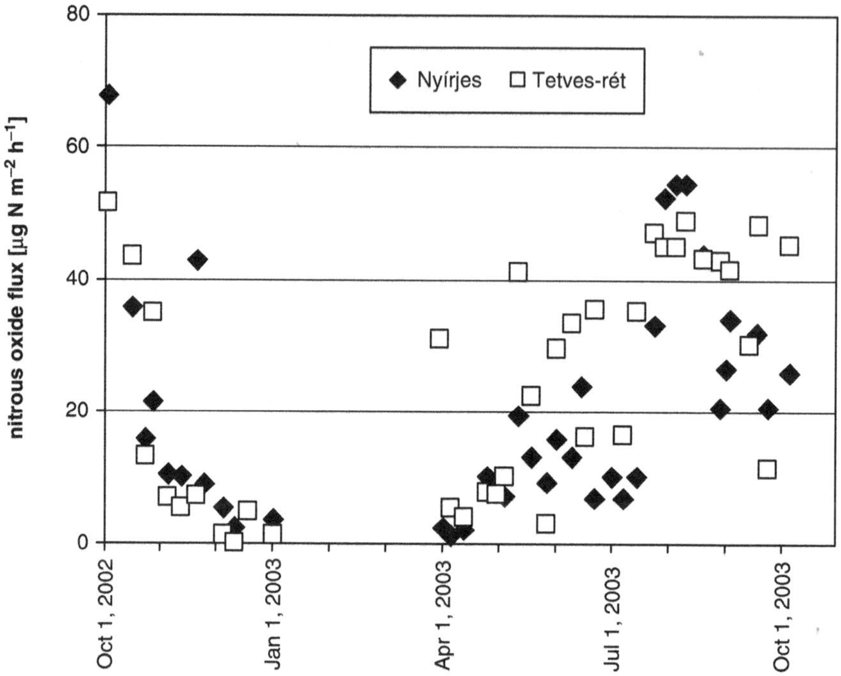

Fig. 7.7 Soil nitrous oxide flux in forests of Mátra region (Horváth et al. 2006) (spruce: Nyírjes station, oak: Tetves-rét station)

October 2003, were 15 and 20 µg N m^{-2} h^{-1} for spruce plantation and sessile oak-hornbeam stands, respectively. Fluxes were not measured from the beginning of January until the end of March 2003 because of snow cover. Yearly averages were calculated from the monthly means of 4–5 measurements per month, assuming zero flux for the winter period with snow cover.

Nitrous oxide is produced by nitrification and mostly by denitrification processes. Although many factors may control these processes and the subsequent N_2O production and exchange to the atmosphere, most of the research papers agree that there are three major factors that influence the emission of these gases, namely the soil moisture, the soil temperature, and the nitrogen availability (e.g., Meixner and Yang 2006).

Our results can be compared with those collected at 15 European forest sites surveyed by Pilegaard et al. (2006) as part of the European Union funded NOFRETETE project. The NOFRETETE locations represent different forest types (coniferous, deciduous) and span the typical nitrogen deposition rates found across Europe. In that survey, the N_2O emission rates were between 0.3 and 20.3 µg N m^{-2} h^{-1} with an average of 8.2 µg N m^{-2} h^{-1}, 2–3 times lower than our measurements.

The total atmospheric N-deposition to the canopy, calculated as the sum of wet + dry deposition, was estimated as 1.7 g N m^{-2} $year^{-1}$ (Horváth et al. 2006). Deposition to the soil was estimated from throughfall (TF) and stemflow (SF) measurements, and should equal nitrogen deposition to the canopy minus net canopy exchange (NCE). NCE was calculated to be 16% of total nitrogen deposition to the spruce forest canopy, and 6.5% for the oak forest. Though litter fall deposition is much higher than other deposition forms, it represents the internal circulation of nitrogen compounds inside the canopy. Emission of N_2O is around 10% of the atmospheric deposition to the soil (TF+SF). Thus, forest soils at this region play a role in the transformation of deposited nitrogen compounds from regional or continental pollution sources into a greenhouse gas that acts at on global scale.

7.6 Summary

Although there is a lack of continuous GHG flux measurements in Hungarian forests, we can draw conclusions from results of the existing observations and also by using data from similar forests in neighboring Croatia.

Hungarian forests store approximately 377 Mt carbon in dendromass and soil (Table 7.6). Their annual carbon uptake in dendromass (Table 7.7, converted to CO_2-eq) is equivalent to 16.5 Mt CO_2 year^{-1} (i.e., 25.3 Mt CO_2 year^{-1} if leaves included). Significant part of the carbon is released back to atmosphere, but less than the uptake. Depending on the method of calculation, Hungarian forests act as net sink for 4.1 Mt CO_2 year^{-1}, but if we would take into account that some of the carbon remains stored in durable wood products, the sink is 7.5 Mt CO_2 year^{-1}.

Since forest area in Hungary is 1.832 Mha, we obtain that the average uptake is 9.01 t CO_2 ha^{-1} year^{-1} (13.81 t CO_2 ha^{-1} year^{-1} including leaves), while net sink, after harvest removals, and without taking into account soil carbon changes, is 2.26 t CO_2 ha^{-1} year^{-1} (4.08 t CO_2 ha^{-1} year^{-1} including carbon in durable wood products).

Similar approach (Section 7.2.3, values are converted to CO_2-eq) for the pedunculate oak forests in Croatia, which are considered among the most productive, gives the average carbon uptake in trees of 11.04 t CO_2 ha^{-1} year^{-1} (not including leaves and thin branches). Net carbon sink, after harvest removals, and without taking into account soil carbon changes, in those forests is estimated to be at least 4.57 t CO_2 ha^{-1} year^{-1} (7.03 t CO_2 ha^{-1} year^{-1}, also considering durable wood products). Difference between the two countries with regards to carbon uptake by forests is within expectations (considering that oak forests are more productive than average). On the other hand, greater difference between carbon sink per unit area of forest is a result of different harvesting intensity in the two countries (75% of increment is usually harvested in Hungary, while only 50% is common in Croatia).

Results of comparison of carbon sink in forests in the two countries are consistent, but forest inventory method used did not take into account the true path of carbon in all pools. This method assumes that all carbon from biomass of the felled trees that remained in forest after harvesting is released within the same year.

Eddy covariance measurements at Jastrebarsko forest provide estimates of net ecosystem exchange of carbon dioxide, i.e. of the actual forest carbon sink which is 17.75 tCO_2 ha^{-1} year^{-1}, on average. But, this result is difficult to compare with the result obtained by forest inventory approach. Forest inventory approach considers only part of carbon uptake (e.g., carbon storage in leaves and thin branches is not taken into account) and only part of carbon release (e.g., release from heterotrophic respiration is not taken into account). On the other hand, calculations from forest inventories consider harvest as carbon release, while eddy covariance cannot measure carbon release from the harvested wood that was removed from forest. Therefore it is better to compare estimates of carbon uptake (NPP).

NPP estimate using data from ŠGOP (2006) of 11.04 tCO_2 ha^{-1} $year^{-1}$ is much lower than the estimate from eddy covariance (33.79 tCO_2 ha^{-1} $year^{-1}$, on average) while estimate of NPP from dendrometer bands (without leaves) is 22.28 tCO_2 ha^{-1} $year^{-1}$. This value is still much higher than the average for Croatian pedunculate oak forests (ŠGOP estimate) but we have to consider that ŠGOP estimate does not take into account carbon storage in thin branches and thin trees. Apart from that, Jastrebarsko forest is still quite young (35-40 years old) and it is gaining carbon at a higher rate than the average (rotation for pedunculate oak is 140 years).

In order to perform precise carbon accounting it is necessary to have the knowledge of size and changes of different carbon pools. Recently, progress was made in research of carbon share in each segment of tree (foliage, branches, stem, stump and root swelling, roots) in different forest types with the aim to better quantify the carbon stocks.

The contribution of the other greenhouse gases (CH_4 and N_2O) in the greenhouse balance between atmosphere and forest ecosystem (expressed in CO_2 equivalent) is negligible in the case of lowland forests of Bodrogköz region (average uptake of CH_4 is 0.01 t CO_2-eq ha^{-1} $year^{-1}$ and emission of N_2O is 0.05 t CO_2-eq ha^{-1} $year^{-1}$). However, for the forests in Mátra mountain region emissions of N_2O are not negligible with average emission from soil of 0.61 and 0.83 t CO_2-eq ha^{-1} $year^{-1}$ for spruce plantation and sessile oak-hornbeam forest, respectively.

References

ÁESZ (2002) Forest resources of Hungary, 2001. State Forest Service, Hungary, Budapest, in Hungarian with English summary

Aubinet M, Grelle G, Ibrom A, Rannik U, Moncrieff J, Foken T, Kowalski AS, Martin PH, Berbigier P, Bernhofer C, Clement R, Elbers J, Granier A, Grunwald T, Morgenstern K, Pilegaard K, Rebmann C, Snijders W, Valentini R, Vesala T (2000) Estimates of the annual net carbon and water exchange of European forests: the EUROFLUX methodology. Adv Ecol Res 30:113–175

Balboa-Murias MA, Rojo A, Álvarez JG, Merino A (2006) Carbon and nutrient stocks in mature *Quercus robur* L. stands in NW Spain. Ann For Sci 63:557–565

Bartholy J (2006) The possible climate consequences of the global climate change in Hungary. AGRO-21 Füzetek 48:12–18 (in Hungarian with English summary)

Bates BC, Kundzewicz ZW, Wu S, Palutikof JP (eds) (2008) Climate change and water. Technical paper VI of the intergovernmental panel on climate change. IPCC Secretariat, Geneva, 210 pp

Bornkamm R, Bennert HW (1989) Gehalt und Vorrat organischer Inhaltsstoffe in der Baumschicht von *Luzulo-Fagetum*-Beständen im Solling (BRD). Flora 183:133–148

Böswald K (1996) Zur Bedeutung des Waldes und der Forstwirtschaft in Kohlenstoffhaushalt, eine Analyse am Beispiel des Bundeslandes Bayern. Forstliche Forschungsberichte München, Bd 159, 147 pp

Brown S (2002) Measuring carbon in forests: current status and future challenges. Environ Pollut 116:363–372

Burschel P, Kürsten E, Larson BC (1993) Die Rolle von Wald und Forstwirtschaft im Kohlenstoffhaushalt – Eine Betrachtung für die Bundesrepublik Deutschland. Forstliche Forschungsberichte München, Bd 126, 135 pp

Cairns MA, Brown S, Helmer EH, Baumgardner GA (1997) Root biomass allocation in the world's upland forests. Oecologia 111:1–11

Cannel MGR, Milne R (1995) Carbon pools and sequestration in forest ecosystems in Britain. Forestry 68:361–378

Cava D, Contini D, Donateo A, Martano P (2008) Analysis of short-term closure of the surface energy balance above short vegetation. Agric For Meteorol 148:82–93

Cestar D, Kovaćić Đ (1982) Wood volume tables for Black Alder and Black Locust. Radovi Šumarski institut Jastrebarsko 49:1–149, in Croatian with English summary

Cestar D, Kovaćić Đ (1984) Wood volume tables for Narrow-leaved Ash (*Fraxinus parvifolia* Auct.). Radovi Šumarski institut Jastrebarsko 60:1–178, in Croatian with English summary

Ciais P, Schelhaas MJ, Zaehle S, Piao SL, Cescatti A, Liski J, Luyssaert S, Le-Maire G, Schulze ED, Bouriaud O, Freibauer A, Valentini R, Nabuurs GJ (2008) Carbon accumulation in European forests. Nat Geosci 1:425–429

Czóbel Sz, Horváth L, Szirmai O, Balogh J, Pintér K, Nemeth Z, Ürmös Zs, Grosz B, Tuba Z (2010) Comparison of N_2O and CH_4 fluxes from Pannonian natural ecosystems. European Journal of Soil Science , 61:671–682. doi: 10.1111/j.1365-2389.2010.01275.x.

Delle Vedove G, Alberti G, Peressotti A, Inglima I, Zuliani M, Zerbi G (2007) Automated monitoring of soil respiration: an improved automatic chamber system. Ital J Agron 4:377–382

Dewar RC (1991) Analytical model of carbon storage in the trees, soils, and wood products of managed forests. Tree Physiol 8:239–258

Dewar RC, Cannel MGR (1992) Carbon sequestration in the trees, products and soils of forest plantations: an analysis using UK examples. Tree Physiol 11:49–71

Foken T, Wichura B (1996) Tools for quality assessment of surface-based flux measurements. Agric For Meteorol 78:83–105

Forster P, Ramaswamy V, Artaxo P, Berntsen T, Betts R, Fahey DW, Haywood J, Lean J, Lowe DC, Myhre G, Nganga J, Prinn R, Raga G, Schulz M, Van Dorland R (2007) Changes in atmospheric constituents and in radiative forcing. In: Solomon S, Qin D, Manning M, Chen Z, Marquis M, Averyt KB, Tignor M, Miller HL (eds) Climate change 2007: The physical science basis. Contribution of Working Group I to the fourth assessment report of the Intergovernmental Panel on Climate Change. Cambridge University Press, Cambridge, United Kingdom and New York, NY, USA

Führer E (1995) The effect of weather on the productivity and health state of the forests. Erdészeti Lapok 130:176–178, in Hungarian

Führer E (2004) Carbon fixing capacity of the forests in Hungary. Hung Agric Res 13:4–7

Führer E, Jagodics A (2007) Ecological correlation between the climate and the organic material production of climate indicator tree species. In: Mátyás Cs, Vig P (eds) Erdő és klíma V. Conference proceeding, University of west Hungary, Sopron, pp. 269–280 (in Hungarian with English summary)

Führer E, Jagodics A (2009) Carbon stocks in the stands of climate indicator tree species. "KLÍMA-21" Füzetek 57:43–55, in Hungarian with English summary

Führer E, Járó Z (1992) Auswirkungen der Klimaänderung auf die Waldbestände Ungarns. Österreichische Forstzeitung 9:25–27

Führer E, Mátyás Cs (2005) Effect of climate change on carbon sequestration and stability of the Hungarian forest cover. Magy Tud 166(7):837–841, in Hungarian

Führer E, Molnár S (2003) The carbon quantity stored in the living stock of forests in Hungary. Faipar 51(2):16–19, in Hungarian with English summary

Führer E, Járó Z, Márkus L (1991) The carbon fixing capacity of forests in Hungary and climatic impacts on the growth of trees of long growing time. In: Faragó T, Iványi Zs, Szalai S (eds) The changeability and change of the weather II. Report to the Ministry of Environmental

Protection and Regional Development. Hungarian Meteorological Service, Budapest, pp 1.5./1–1.5./13 (in Hungarian with English summary)

Gál B, Szirmai O, Czóbel Sz, Cserhalmi D, Nagy J, Szerdahelyi T, Urmos Zs, Tuba Z (2006) Grassland and forest plant communities of the Hungarian Bodrogköz. Folia Historico Naturalia Musei Matraensis 30:43–62, in Hungarian

Goulden ML, Munger JW, Fan S-M, Daube BC, Wofsy SC (1996) Exchange of carbon dioxide by a deciduous forest: response to interannual climate variability. Science 271:1576–1578

Hanson PJ, Edwards NT, Garten CT, Andrews JA (2000) Separating root and soil microbial contributions to soil respiration: a review of methods and observations. Biogeochemistry 48: 115–146

Harrison AF, Howard PJA, Howard DM, Howard DC, Hornung M (1995) Carbon storage in forest soils. Forestry 68(4):335–348

Horváth B (2006) C-accumulation in the soil after afforestation: contribution to C-mitigation in Hungary? Forstarchiv 77:63–68

Horváth L, Führer E, Lajtha K (2006) Nitric oxide and nitrous oxide emission from Hungarian forest soils; linked with atmospheric N-deposition. Atmos Environ 40:7786–7795

Horváth L, Grosz B, Czóbel S, Nagy Z, Péli E, Szerdahelyi T, Szirmai O, Tuba Z (2008) Measurement of methane and nitrous oxide fluxes in Bodrogköz, Hungary; preliminary results. Acta Biologica Szegediensis 52(1):119–122

IPCC (2003) Good practice guidance for land use, land-use change and forestry. In: Penman J, Gytarsky M, Hiraishi T, Kruger D, Pipatti R, Buendia L, Miwa K, Ngara T, Tanabe K, Wagner F (eds). IPCC/IGES, Hayama, Japan

IPCC (2006) 2006 IPCC guidelines for national greenhouse gas inventories. In: Eggleston HS, Buendia L, Miwa K, Ngara T, Tanabe K (eds). IGES, Japan

Járó Z (1968) The fundamental principles of the site-typological systematisation. Kísérleti Közlemények 61/D(1–3):11–25, in Hungarian

Járó Z (1990) The circulation of organic matter and nutrient in beech stands. Erdészeti Kutatások 80–81:83–98, in Hungarian with English and Russian summary

Keeland BD, Young PJ (2006) Construction and installation of dendrometer bands for periodic tree-growth measurements. National Wetlands Research Center, Lafayette, LA. http://www.nwrc.usgs.gov/Dendrometer/. Accessed 29 Mar 2006

Kugler S, Horváth L, Machon A (2008) Estimation of nitrogen balance between the atmosphere and Lake Balaton and a semi natural grassland in Hungary. Environ Pollut 154:498–503

Livingston GP, Hutchinson GL (1995) Enclosure-based measurement of trace gas exchange: applications and sources of error. In: Matson PA, Harriss RC (eds) Biogenic trace gases: measuring emissions from soil and water. Blackwell Science, Cambridge, pp 14–50

Luyssaert S, Schulze E-D, Börner A, Knohl A, Hessenmöller D, Law BE, Ciais P, Grace J (2008) Old-growth forests as global carbon sinks. Nature 455:213–215

Luyssaert S, Ciais P, Piao SL, Schulze ED, Jung M, Zaehle S, Schelhaas MJ, Reichstein M, Churkina G, Papale D, Abril G, Beer C, Grace J, Loustau D, Matteucci G, Magnani F, Nabuurs GJ, Verbeeck H, Sulkava M, van der Werf GR, Janssens IA (2010) The European carbon balance. Part 3: Forests. Glob Change Biol 16:1429–1450

Magnani F, Mencuccini M, Borghetti M, Berbigier P, Berninger F, Delzon S, Grelle A, Hari P, Jarvis PG, Kolari P, Kowalski AS, Lankreijer H, Law BE, Lindroth A, Loustau D, Manca G, Moncrieff JB, Rayment M, Tedeschi V, Valentini R, Grace J (2007) The human footprint in the carbon cycle of established temperate and boreal forests. Nature 447:848–850

Marjanović H (2009) Modeling tree development and elements of stand structure in young stands of Pedunculate oak (Quercus robur L.). Dissertation, Faculty of forestry, University of Zagreb (in Croatian with English summary)

Marjanović H, Benndorf A, Indir K, Paladinić E, Peressotti A, Schweiger H, Vuletić D (2007) CO_2 and forests in Croatia. In: Peressoti A (ed) Local strategies for land use management according to Kyoto protocol. Forum Editrice Universitaria Udinese srl, Udine, Italy, pp 121–128

Matese A, Alberti G, Gioli B, Vaccari FP, Toscano P, Zaldei A (2008) Compact EDDY: a compact, low consumption eddy covariance logging system. Comput Electron Agric 64:343–346

Mátyás Cs (2005) Climate change, carbon sequestration and instability of forests. AGRO-21 Füzetek 43:80–86, in Hungarian with English summary

Mátyás Cs (2010) Forecasts needed for retreating forests. Nature 464(7293):1271

Mayer B (1996) Hydropedological relations in the region of lowland forests of Pokupsko basin. Radovi Šumarskog instituta Jastrebarsko 31:37–89, in Croatian with English summary

Meixner FX, Yang WX (2006) In: D'Odorico P, Porporato A (eds) Biogenic emissions of nitric oxide and nitrous oxide from arid and semi-arid land in: dryland ecohydrology. Springer, Dordrecht, Berlin, Heidelberg, New York, pp 233–255

Nabuurs GJ, Schelhaas MJ, Mohren GMJ, Field CB (2003) Temporal evolution of the European forest sector carbon sink from 1950 to 1999. Glob Change Biol 9:152–160

Ochsner TE, Sauer TJ, Horton R (2007) Soil heat storage measurements in energy balance studies. Agron J 99:311–319

Pallardy SG, Kozlowski TT (2007) Physiology of woody plants. Academic Press, San Diego, 454 p.

Pilegaard K, Skiba U, Ambus P, Beier C, Brüggemann N, Butterbach-Bahl K, Dick J, Dorsey J, Duyzer J, Gallagher M, Gasche R, Horvath L, Kitzler B, Leip A, Pihlatie MK, Rosenkranz P, Seufert G, Vesala T, Westrate H, Zechmeister-Boltenstern S (2006) Factors controlling regional differences in forest soil emission of nitrogen oxides (NO and N_2O). Biogeosciences 3:651–661

Prather M, Drewent D, Enhalt P, Fraser E, Sanhueza E, Zhou X (1995) Other trace gases and atmospheric chemistry. In: Houghton J, Meira-Fulho LG, Callender BA (eds) Climate Change 1994. Cambridge University Press, Cambridge, pp 77–126

Raich JW, Schlesinger WH (1992) The global carbon dioxide flux in soil respiration and its relationships to vegetation and climate. Tellus 44B:81–99

Rauš Đ (1996) Lowland forests of the Pokupsko basin. Radovi Šumarskog instituta Jastrebarsko 31:17–36, in Croatian with English summary

Reichstein M, Rey A, Freibauer A, Tenhunen JD, Valentini R, Banza J, Casals P, Cheng Y, Grünzweig JM, Irvine J, Joffre R, Law BE, Loustau D, Miglietta F, Oechel WC, Ourcival J-M, Pereira J, Peressotti A, Ponti F, Qi Y, Rambal S, Rayment MB, Romanya J, Rossi F, Tedeschi V, Tirone G, Xu M, Yakir D (2003) Modeling temporal and large-scale spatial variability of soil respiration from soil water availability, temperature and vegetation productivity indices. Global Biogeochem Cy 17(4):Art No. 1104

Reichstein M, Falge E, Baldocchi D, Papale D, Aubinet M, Berbigier P, Bernhofer C, Buchmann N, Gilmanov T, Granier A, Grünwald T, Havránková K, Ilvesniemi H, Janous D, Knohl A, Laurila T, Lohila A, Loustau D, Matteucci G, Meyers T, Miglietta F, Ourcival J-M, Pumpanen J, Rambal S, Rotenberg E, Sanz M, Tenhunen J, Seufert G, Vaccari F, Vesala T, Yakir D, Valentini R (2005) On the separation of net ecosystem exchange into assimilation and ecosystem respiration: review and improved algorithm. Glob Change Biol 11:1424–1439

Rosenkranz P, Brüggemann N, Papen H, Xu Z, Horváth L, Butterbach-Bahl K (2006) Soil N and C trace gas fluxes and microbial soil N turnover in a sessile oak (Quercus petraea (Matt.) Liebl.) forest in Hungary. Plant Soil 286:301–322

Schumacher FX, Hall FDS (1933) Logarithmic expression of timber-tree volume. J Agric Res 47:719–734

ŠGOP (2006) Šumskogospodarska osnova područja za razdoblje 2006–2015 – Knjiga I (Forest management area plan for the period 2006–2015 – Book I, in Croatian language). Hrvatske šume d.o.o., Zagreb, 566 p

Somogyi Z (2008) Recent trends of tree growth in relation to climate change in Hungary. Acta Silvatica et Lignaria Hungarica 4:17–27

Somogyi Z, Horváth B (2006) On carbon sequestration of forests planted since 1930. Erdészeti Lapok 151:257–259, in Hungarian

Špiranec M (1975) Wood volume tables. Radovi Šumarski institut Jastrebarsko 22:1–262, in Croatian with German summary

Steinkamp R, Butterbach-Bahl K, Papen H (2001) Methane oxidation by soils of an N limited and N fertilized spruce forest in the Black Forest Germany. Soil Biol Biochem 33:145–153

Tang JW, Baldocchi DD (2005) Spatial-temporal variation in soil respiration in an oak-grass savanna ecosystem in California and its partitioning into autotrophic and heterotrophic components. Biogeochemistry 73:183–207

Tuba Z (1995) Overview of the flora and vegetation of the Hungarian Bodrogköz. Tiscia 29:11–17

Twine TE, Kustas WP, Norman JM, Cook DR, Houser PR, Meyers TP, Prueger JH, Starks PJ, Wesely ML (2000) Correcting eddy-covariance flux underestimates over a grassland. Agric For Meteorol 103:279–300

Valentini R (2003) Fluxes of carbon, water and energy of European forests. Ecological Studies 163. Springer-Verlag, Berlin, Heidelberg, 270 p

Valentini R, Matteucci G, Dolman AJ, Schulze ED, Rebmann C, Moors EJ, Granier A, Gross P, Jensen NO, Pilegaard K, Lindroth A, Grelle A, Bernhofer C, Grünwald T, Aubinet M, Ceulemans R, Kowalski AS, Vesala T, Rannik Ü, Berbigier P, Loustau D, Gudmundsson J, Thorgeirsson H, Ibrom A, Morgenstern K, Clement R, Moncrieff1 J, Montagnani L, Minerbi S, Jarvis PG (2000) Respiration as the main determinant of carbon balance in European forests. Nature 404:861–865

van den Pol-van Dasselaar A, van Beusichem ML, Onema O (1998) Effects of soil moisture content and temperature on methane uptake by grasslands on sandy soils. Plant Soil 204:213–222

Vukelić J, Baričević D (2000) Development of vegetation in localites of penduculate oak dieback in Croatia. Glasnik za šumske pokuse 37:277–293

Webb EK, Pearman GI, Leuning R (1980) Corrections of flux measurements for density effects due to heat and water vapour transfer. Q J R Meteorol Soc 106:85–100

Weber M, Burschel P (1992) Wald und Holz als Kohlenstoffspeicher. Der Wald 42:148–151

Wilson K, Goldstein A, Falge E, Aubinet M, Baldocchi DD, Berbigier P, Bernhofer C, Ceulemans R, Dolman H, Field C, Grelle A, Ibrom A, Law BE, Kowalski A, Meyers T, Moncrieff J, Monson R, Oechel W, Tenhunen J, Valentini R, Verma S (2002) Energy balance closure at FLUXNET sites. Agric For Meteorol 113:223–243

Ziegler F (1991) Die Bedeutung des organischen Kohlenstoffes im Unterboden. Vorratsberechnungen an Waldböden. Zeitschrift für Umweltchemie und Ökotoxikologie 3(5):276–2771From the world map of eddy covariance sites that are members of the FLUXNET network it can be seen that the majority of sites are located in Western Europe and USA. (http://www.fluxnet.ornl.gov/fluxnet/graphics.cfm, retrieved on 5 Feb 2010).

Chapter 8
Arable Lands*

Eszter Tóth, Zoltán Barcza, Márta Birkás, Györgyi Gelybó, József Zsembeli, László Bottlik, Kenneth J. Davis, László Haszpra, Anikó Kern, Natascha Kljun, Sándor Koós, Györgyi Kovács, Attila Stingli, and Csilla Farkas

Abstract An overview of carbon dioxide (CO_2) exchange and soil respiration measurements is given for representative Hungarian agroecosystems. We present results of long-term atmosphere/biosphere CO_2 exchange measurements carried out at the Hungarian tall tower site (Hegyhátsál). Tall tower net ecosystem exchange (NEE) measurements provided consistent estimates of landscape-wide carbon dioxide dynamics. During 1997–2008, the region mostly behaved as a net CO_2 sink on annual scale. Year-round NEE was in the range of -352 ± 49 g C m^{-2} $year^{-1}$ and 43 ± 9 g C m^{-2} $year^{-1}$ (negative values indicate uptake). The measurements are

*Citation: Tóth, E., Barcza, Z., Birkás, M., Gelybó, Gy., Zsembeli, J., Bottlik, L., Davis, K.J., Haszpra, L., Kern, A., Kljun, N., Koós, S., Kovács, Gy., Stingli, A., Farkas, Cs., 2010: Measurements and estimations of biosphere-atmosphere exchange of greenhouse gases – Arable lands. In: Atmospheric Greenhouse Gases: The Hungarian Perspective (Ed.: Haszpra, L.), pp. 157–197.

C. Farkas (✉)
Bioforsk, Norwegian Institute for Agricultural and Environmental Research,
Frederik A. Dahlsvei 20, Ås 1432, Norway
e-mail: csilla.randria@gmail.com

E. Tóth and S. Koós
Research Institute for Soil Science and Agricultural Chemistry of the Hungarian Academy of Sciences, H-1022 Budapest, Herman O. út 15., Hungary

Z. Barcza and Gy. Gelybó
Department of Meteorology, Eötvös Loránd University, H-1117 Budapest, Pázmány P. s. 1/A, Hungary

M. Birkás, L. Bottlik and A. Stingli
Institute of Crop Production, Szent István University, H-2103 Gödöllő, Páter K. u. 1., Hungary

J. Zsembeli and G. Kovács
Karcag Research Institute, Research Institutes and Study Farm, Centre for Agricultural and Applied Economic Sciences, University of Debrecen, H-5300 Karcag, Kisújszállási út 166., Hungary

K.J. Davis
Department of Meteorology, The Pennsylvania State University, University Park, PA 16802, USA

L. Haszpra
Hungarian Meteorological Service, H-1675 Budapest, P.O. Box 39, Hungary

L. Haszpra (ed.), *Atmospheric Greenhouse Gases: The Hungarian Perspective*,
DOI 10.1007/978-90-481-9950-1_8,

A. Kern
Adaptation to Climate Change Research Group, Hungarian Academy of Sciences, H-1117 Budapest, Pázmány P. s. 1/A, Hungary
and
Department of Meteorology, Eötvös Loránd University, H-1117 Budapest, Pázmány P. s. 1/A, Hungary

N. Kljun
Department of Geography, Swansea University, Singleton Park, Swansea SA2 8PP, UK

representative to a mixture of arable lands with small contribution from other biome types. Effects of different soil tillage methods – applied in two long-term tillage experiments (Józsefmajor, Karcag) and in a peach orchard (Vác) – on soil CO_2 emission are also introduced. Soil respiration rate highly depended on the depth of soil disturbance and on the date of the measurements; CO_2 emissions measured immediately after tillage applications and during the vegetation period showed contradictory tendencies. Results obtained from a newly developed laboratory CO_2 emission measurement method indicated strong coherences between soil carbon dioxide fluxes and soil water potential values.

Keywords Eddy covariance method • NEE • Soil carbon dioxide efflux • Tillage • Undisturbed soil cores

8.1 Introduction

This chapter is dedicated to greenhouse gas measurements on Hungarian arable lands, which concern the assessment of carbon dioxide (CO_2) balance in agroecosystems and the measurements of soil CO_2 efflux in various soil management systems on different soil types.

We introduce the unique, tall-tower-based eddy covariance (EC) method that provides data for calculating the net ecosystem carbon dioxide exchange (NEE). These measurements are applied on a silt soil in a mixed agricultural landscape, which is typical for the Hungarian arable land use. We present results from an 11-year-long measurement period. The relationship between NEE and various meteorological parameters, as well as the spatial representativeness of the measured data were evaluated. Further data analysis concerned derivation of crop type maps and applying the footprint model to derive crop-specific NEE time series.

The EC measurements provide important information about the contribution of various agroecosystems to the carbon dioxide exchange at larger scales. However, for a better understanding of the processes standing behind the formation of the carbon balance in the soil-plant-atmosphere system, specific elements of this balance should be identified.

In Hungary, we make efforts on developing sustainable, soil structure and moisture conserving soil management systems that facilitate carbon sequestration in a given region, besides given production needs. From this aspect, soil carbon dioxide emission measurements on arable lands play an important role.

An essential feature of soils is that they are both a source and a sink of greenhouse gases (Ussiri and Lal 2009) strongly influenced by land use and soil management. In simple terms, land use means the entirety of technologies applied in using a piece of land and the crops for which it is used. Land use – whether conventional, extensive, intensive, or integrated – has different impacts on the soil carbon balance, and thereby on the quantity of CO_2 released into the atmosphere (Birkás et al. 2008). It is beyond doubt, however, that sustainable cropping is not possible without a positive carbon balance (Lal 2009). Accordingly, if carbon is lost from the system for any reason, at least as much of it has to be recycled (West and Post 2002).

Soil CO_2 flux is an important link in the carbon cycle between soil and atmosphere that also influences atmospheric CO_2 concentration. Deforestation or breaking up land for cropping are interventions well known for their contribution to carbon loss. According to Mann (1986) during the first 5 years after breaking up the soil of what once was a forest or a pasture, some 20% of the original organic carbon content (1,500 g m^{-2}) of the top 30 cm soil layer was lost through respiration. Tillage, irrigation, the application of manure or lime, as factors of land use, are organic material decomposing, i.e., carbon loss increasing practices (Read et al. 2001), though the intensity of such interventions can alter a lot. The factors affecting soil respiration have been identified and clarified by now. These include the level of soil disturbance, the quantity and quality of soil organic matter (SOM), field residue management, root respiration, aerobic microbial activity, soil moisture, looseness, porosity fractions, the soil carbon–nitrogen (C:N) ratio, the soil pH, the soil and the ambient air temperature, wind speed, and the CO_2 concentration gradient between the soil and the atmosphere (Al-Kaisi and Yin 2005; Reth et al. 2005; Reicosky et al. 2005, 2008; Chatskikh and Olesen 2007; Ussiri and Lal 2009).

The CO_2 flux data measured shortly or long time after tillage vary within a narrower or a wider range, as determined by the above factors. Since land use and soil management induced changes in soil respiration and carbon balance have a strong soil- and site-specific nature, measurement results of soil carbon balance elements cannot be supplemented by estimated values.

We present soil carbon dioxide emission studies carried out in Hungary through results of in situ chamber measurements and laboratory soil CO_2 efflux investigations. The in situ measurements were carried out in different regions and under various tillage systems, representing various levels of soil disturbance. For the laboratory analysis of soil CO_2 emission, a newly developed methodology was applied. The emission measurements were carried out on large, undisturbed soil columns in a climatic room at constant air temperature. The samples were collected from two soil management systems of a peach plantation. The water content in the soil columns was set up on the basis of different soil water potential values.

8.2 Carbon Dioxide Balance Measurement with EC Method on Mixed Arable Land

8.2.1 Climate and Instrumentation

The EC measurements have been performed on a tall tower, 82 m above the ground, at Hegyhátsál, West Hungary (46°57′N, 16°39′E, 248 m asl) since April, 1997 (Haszpra et al. 2001). The measurements were temporarily interrupted in 2000 for technical reasons. The climate of the Hegyhátsál region is temperate continental. Although certain meteorological parameters are measured at the station, comparable climate data are available only from the two closest meteorological observatories (Farkasfa, 46°55′N, 16°19′E, 312 m asl, and Rábagyarmat, 46°57′N, 16°25′E, 211 m asl; 26 and 17 km from Hegyhátsál, respectively). At Farkasfa the long-term (1961–1990) mean annual temperature is 8.9°C and the mean annual precipitation is 759 mm (at Rábagyarmat no such data are available). The recent years covered by the net ecosystem exchange (NEE) measurements were somewhat warmer than the multiannual mean, which might affect the carbon balance of the region. The climate data are summarized in Table 8.1.

Initially, the system consisted of an ultrasonic anemometer (Solent Windmaster, Gill Instruments Ltd., Lymington, United Kingdom), an aspirated fine wire thermocouple (Model ASPTC, Campbell Scientific Ltd., Loughborough, United Kingdom), and a fast response infrared CO_2/H_2O analyzer (LI-6262, Li-Cor Inc., Lincoln, Nebraska, USA). In 2002, the ultrasonic anemometer was replaced with a new model (Solent Research R3–50, Gill Instruments Ltd., Lymington, United Kingdom) also capable to measure the virtual temperature, thus the thermocouple was removed from the system. Measurements are made at 4 Hz. Air is sucked through the sampling tube and the analyzer at about 15 L min^{-1}, resulting in turbulent flow in the whole system. Pressure fluctuations generated by the pump are damped by means of a 6-L buffer volume (Haszpra et al. 2001). For details on the ancillary measurements and on the calibration procedure applied at Hegyhátsál, see Haszpra et al. (2001, 2005).

8.2.2 Footprint Climatology and Spatial Representativeness of the Tall Tower

Footprint analysis of EC measurements provides information about the source area of the measured vertical fluxes (Schmid 2002; Wang et al. 2006a; Vesala et al. 2008). To analyze the source of the turbulent flux signal and to attribute the measured signal to different surface elements we need to provide a so-called footprint climatology for the tower-based measurements. For this purpose, we used the LPDM-B model-based parameterized footprint model (Kljun et al. 2004) described in Section 5.2.

Table 8.1 Climate data for the Hegyhátsál region. Sums of photosynthetically active photon flux density (PPFD) and average daytime vapor pressure deficit (VPD) are measured locally at Hegyhátsál. 1997–2008 air temperature and precipitation amount data are from the nearby Rábagyarmat climate station, while the 1961–1990 means are from Farkasfa regional meteorological observatory. No net ecosystem exchange (NEE) measurements were performed in 2000 for technical reasons

Period		1961–1990	1997	1998	1999	2001	2002	2003	2004	2005	2006	2007	2008
Average temperature (°C)	Vegetation period (March–October)	13.2	13.6	14.7	15.2	15.4	15.1	15.3	13.9	14.1	14.8	15.3	15.1
	Whole year	8.9	9.7	10.5	10.5	10.7	11.3	10.6	9.8	9.6	10.3	11.3	11.4
Precipitation amount (mm)	Vegetation period (March–October)	584	523	693	530	466	469	381	564	595	604	628	543
	Whole year	759	692	837	735	550	598	487	705	801	738	786	661
Sum of PPFD (mol m^{-2})	Vegetation period (March–October)	n.a.	7,531	7,201	7,263	7,611	7,359	8,020	7,086	7,540	7,261	8,018	7,888
	Whole year	n.a.	8,507	8,262	8,240	8,530	8,120	9,047	7,931	8,494	8,258	8,935	8,870
Average daytime VPD (kPa)	Vegetation period (March–October)	n.a.	0.882	0.842	0.817	0.979	1.022	1.322	0.858	0.897	0.874	1.087	1.125
	Whole year	n.a.	0.684	0.651	0.615	0.735	0.773	0.954	0.639	0.668	0.649	0.820	0.854

The applicability of the LPDM-B based parameterization is restricted to high u_* classes. This means that during calm conditions the model is not applicable. This feature foreshadows that the footprint climatology will be somewhat biased. Calm winds frequently occur at the site during nighttime, when the situation is even more complicated since the EC level may lie outside the nighttime stable layer, and can fully decouple from the surface (Haszpra et al. 2005). The flux source area is generally located further away from the measurement point during stable conditions (Vesala et al. 2008), so our calculated region of representativeness will most probably be biased closer to the tower than in reality. This problem should be kept in mind in the interpretation of the results.

Figure 8.1 shows sample footprint climatology for the tall tower NEE measurement for the years 2004 and 2007. The plot was constructed using the maximum source weight distance, x_{max} calculated by means of Eq. 5.5 in Chapter 5. It can be seen that the estimated source region is well localized. There are two regions that contribute frequently to the measured signal: one is located at a distance of about 400–1,000 m to the north from the tower. The second area is located at the opposite side, to the south. Of course, this pattern is the consequence of the prevailing wind directions at the site (Haszpra et al. 2001). There are localized source regions at larger distances from the tower, but their number is low (cf. exclusion of stable nighttime cases, mentioned above). The footprint climatology for the other years reveals similar distributions. Note that there is a weird pattern in the maximum locations at 180° wind direction, which is the consequence of the tower's shadow effect and is an artifact of the wind direction correction method (Barcza et al. 2009).

The seasonal distribution of x_{max} shows that the footprint distribution is generally well localized except for spring (Barcza et al. 2009). This latter fact might be explained by the larger variability in wind direction and atmospheric stability.

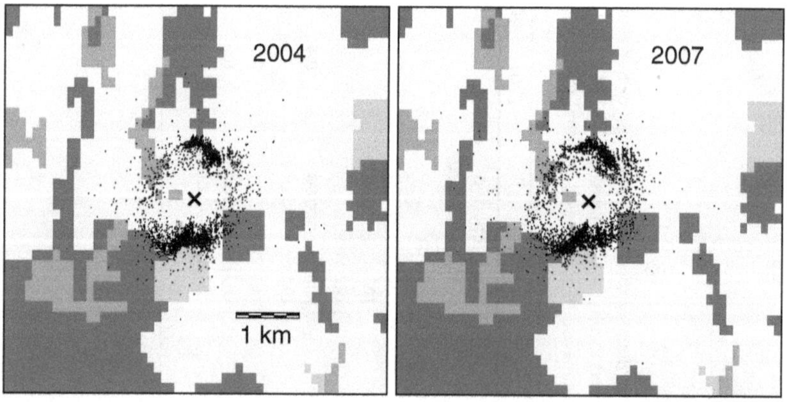

Fig. 8.1 Footprint climatology of the tall tower for 2004 and 2007, respectively. Maximum source strength locations (x_{max}) are plotted as dots. X symbol marks the location of the tower. The CORINE2000 land cover map is also shown (~100 m resolution). White: agricultural fields; light gray: pastures; dark gray: forests; mid-gray: other land use (settlements, transitional woodland–shrubland, marshes, etc.). Note the similarity between the plots

Summer x_{max} values are somewhat closer to the tower than winter ones because of the more effective thermal turbulence, but the differences are not remarkable.

The characterization of the footprint climatology can be used to attribute the NEE signal to different plant functional types (Barcza et al. 2009). To obtain the latter information, we used the CORINE Land Cover 2000 (CORINE2000) classification scheme (Büttner et al. 2002). The land cover and footprint information was discretized to a common grid to maintain consistency with remote-sensing-based studies discussed later in this chapter. The target grid is the ~250 m grid of remotely sensed vegetation index introduced later. For consistency, all the 250 m pixels were decomposed into CORINE2000 categories, taking into account the spatial extent of the pixels and the center points of the CORINE2000 grid. Using this approach, 1–3 CORINE2000 categories were assigned to each pixel. We distinguished between entirely agricultural pixels, mixed pixels with some agricultural presence, and other pixels without agricultural signature (for details see Barcza et al. 2009).

We estimated the relative importance of the different land cover types based on the CORINE2000 classification by assigning one grid cell to each hourly calculated maximum source weight location. Using this discretized footprint climatology (250 m grid) and the CORINE2000 categories that were assigned to the 250 m pixels, the relative importance of the different land cover categories was estimated for 6 years (2003–2008). The results showed that the majority (75–80%) of the measured signal originated from agricultural land (croplands). The contribution of pastures (grasslands) was also significant (15–20%). Pastures are located south of the tower and this 15–20% contribution could essentially be attributed to the same large pasture. Although there is a larger forested region southeast of the tower, its contribution was very small. The relative importance of urban areas was also small; it was generally below 2%. Transitional woodland–shrublands did not play significant role either, and the contribution of wetlands and water bodies was zero.

The relative importance of the different land cover types was different from the area-averaged land cover distribution within 10 km around the tower (Barcza et al. 2009). The contribution of agricultural land and grasslands to the EC measurements was considerably higher than their area-average fractional presence. Thus, simple upscaling cannot be used to estimate the carbon dioxide and water exchange of larger areas based on the tower measurements. A similar study was performed by Wang et al. (2006b) based on the WLEF tall tower site in the USA.

8.2.3 Net Ecosystem Exchange

As described in Section 5.2, NEE at 82 m height is calculated as the sum of the eddy flux plus the rate of change of storage (the mass flow term has been neglected). Storage plays an important role in the morning transition from stable to unstable conditions during the breakup of the nighttime stable stratification (inversion). Moreover, in many cases with low wind speed situation, the turbulent flux is close to zero during nighttime, and the storage term rules NEE.

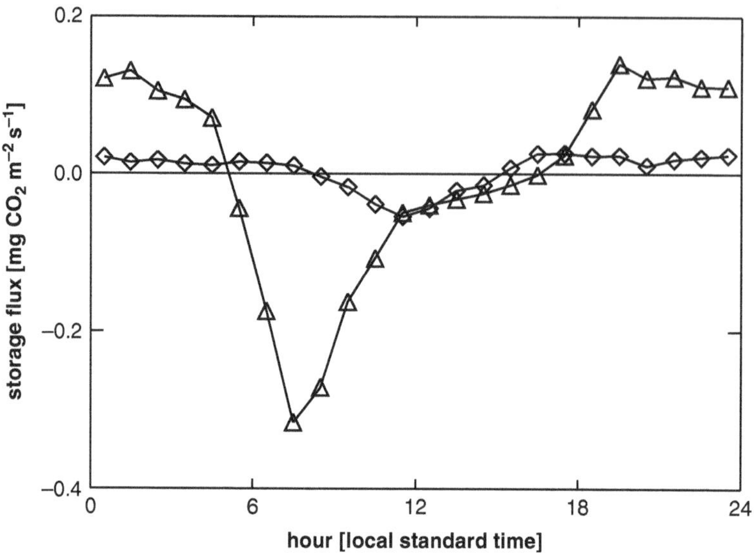

Fig. 8.2 Seven-year (1997–2004, without 2000) mean diurnal variation of the rate of change of CO_2 storage below the 82 m measurement level at Hegyhátsál. Triangles: plot compiled from all data measured during May–August. Diamonds: plot compiled from all data recorded during November–February

Figure 8.2 shows the 7-year (1997–2004, without 2000) mean daily course of the storage flux for Hegyhátsál. The nighttime contribution to NEE and the morning transition effect can clearly be detected in the plot. It is evident from the figure that storage plays an important role all day. The magnitude of the storage highlights that adequate estimation of the storage term is crucial in case of tall tower flux measurements in order to estimate NEE that is consistent with the environmental forces (radiation, temperature, etc.).

Daytime NEE reveals close coupling with photosynthetically active photon flux density (PPFD) during the active period of the vegetation. Maximum daytime CO_2 uptake reaches −1.4 to −1.5 mg CO_2 m^{-2} s^{-1} during the most active phase of the vegetation (1 mg CO_2 m^{-2} s^{-1} = 22.7 μmol CO_2 m^{-2} s^{-1}). Generally, daytime CO_2 uptake reaches its maximum value around noon, and decreases as PPFD declines. Mean nighttime respiration is around 0.1 to 0.3 mg CO_2 m^{-2} s^{-1} during the vegetation period.

Figure 8.3 shows the temporal variation of the daily net carbon dioxide exchange between the area surrounding the tower (soil + vegetation + anthropogenic activities) and the atmosphere for 1994–1999 and 2001–2008. The area is a net source of CO_2 during the first 3–4 months of the year. Typical daily net exchange is around 1 g C m^{-2} but NEE can reach 3–4 g C m^{-2} on some days. In spring, within a couple of weeks, the rapid growth of the vegetation changes the sign of NEE to negative (i.e., uptake). Typically, the vegetation takes up 2–4 g C m^{-2} day^{-1}. For short periods, the uptake rate can reach 5–6 g C m^{-2} day^{-1}, with an

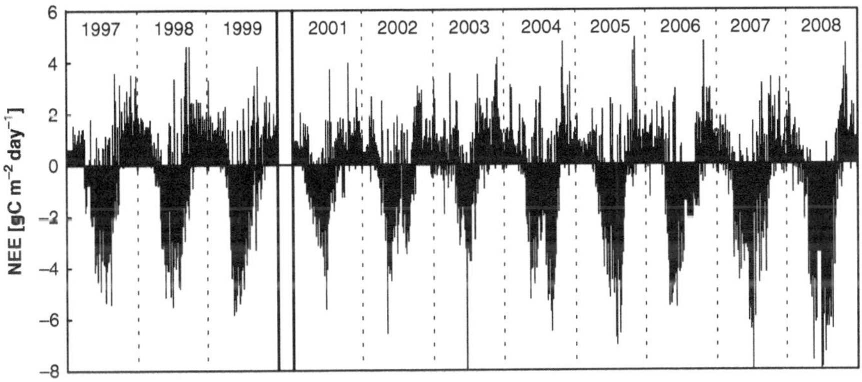

Fig. 8.3 Daily net ecosystem exchange (NEE) for the 11-year long measurement period (1997–1999, 2001–2008). Negative values indicate carbon dioxide uptake by the vegetation

absolute maximum of 6–8 g C m^{-2} day^{-1}. Around the beginning of October, the sign of NEE changes back to positive and the region remains a net source of carbon dioxide during the rest of the year. Carbon dioxide is released from the soil even during winter with snow cover, and fossil fuel emission released by domestic heating, traffic, etc. within the source area is expected to contribute to the measured upward CO_2 transport.

The length of the CO_2 uptake period determined from the sign change of the smoothed NEE time series is given in Table 8.2 for each year.

8.2.4 Gap Filling and Flux Partitioning of NEE

Gap-filled NEE time series as well as its components, gross primary production (GPP, which is assumed to be equal with gross ecosystem production) and total ecosystem respiration (R_{eco}; NEE = R_{eco} − GPP by definition) were calculated using all *three methods* described in Section 5.2 (referred to as STE, RH and NRH). Although it has been concluded in Stoy et al. (2006) that the NRH method gives the most realistic estimations, we did not set up any preference order in the lack of comparative analysis of the absolute performance of these methods for mixed arable lands. In addition, methods used in this study are not exactly the same as those presented in Stoy et al. (2006). We assumed those three methods to be equally efficient, and accepted the averages of NEE, GPP, and R_{eco} values obtained from the three different methods as the best estimates. Annual sums of NEE are presented in Table 8.2.

The three flux partitioning/gap filling methods have been compared and evaluated against independent data reported by Stoy et al. (2006). Their results for multiyear average of annual NEE sum suggested for an old field site (covered with herbaceous plants, therefore, it is the most similar to our site) that the nighttime data

Table 8.2 Annual net ecosystem exchange (NEE), gross primary production (GPP) and total ecosystem respiration (R_{eco}) at Hegyhátsál. The length of the carbon dioxide uptake period as well as the uncertainties of NEE, GPP and R_{eco} are also presented (see text for details)

Period	1997	1998	1999	2001	2002	2003	2004	2005	2006	2007	2008
Length of carbon uptake period (days)	183	179	153	138	174	106	168	154	169	202	184
NEE (g C m^{-2})	−167±40	−92±40	−68±9	−24±52	−27±65	43±9	−170±24	−188±12	−253±62	−193±11	−352±49
GPP (g C m^{-2})	1,007±56	998±112	1,092±31	823±45	891±76	848±82	1,056±109	1,075±50	1,320±82	1,137±72	1,476±104
R_{eco} (g C m^{-2})	840±70	906±95	1,025±37	799±97	864±37	892±81	886±85	887±50	1,067±43	945±74	1,125±71

based STE method (see Reichstein et al. 2005) provided less negative estimations of NEE than the daytime data based RH and NRH methods. Our calculations also supported this order as the 11 year (1997–2008, without 2000) average of NEE annual sums for thc STE method provided lower (less negative) estimate (−104 g C m^{-2} $year^{-1}$) than the RH (−142 g C m^{-2} $year^{-1}$) and the NRH (−160 g C m^{-2} $year^{-1}$) methods. According to the results, the RH method gives an intermediate estimation of annual sums of NEE at the arable land site that is in contrast with the findings of Stoy et al. (2006) where – for woodlands and grassland – the lowest estimation was provided by the RH method. Note that the methods used in Stoy et al. (2006) and in Reichstein et al. (2005) were not exactly the same as those used in this study, although the basic concepts were identical.

Results from STE method show that GPP is lower and R_{eco} is higher compared to the NRH method. The most important conclusion is that our study showed systematic differences between nighttime data based and light-response curve based estimations for arable land. Further analysis using independent measurements is required to evaluate the absolute performance of the methods.

According to the results, the Hegyhátsál region mostly acted as a net sink of CO_2 on annual scale. The year 2003 was an exception, when the region behaved as a carbon dioxide source. The calculations showed that during the 11 years period of the measurements (1997–2008, without 2000) the annual NEE ranged from −352 g C m^{-2} $year^{-1}$ to +43 g C m^{-2} $year^{-1}$, and each square meter of the region removed a total of 1,491 g carbon from the atmosphere (Table 8.2). Obviously, the removal of carbon is only meant from the atmospheric point of view: the carbon *was not* sequestered by the biosphere (presumably only a small part by the forests), but it was partly removed from the region by harvesting. Since the arable fields around the tower belong to individuals, it is hard to obtain yield information. Using ancillary information (county-level crop yield data), allocation estimations and residue management assumptions we will present the first results about the net biome production (NBP) of the Hegyhátsál region in Chapter 12.

NEE is the small residual of two large terms of GPP and R_{eco}. The effect of climate can be scrutinized via relating the monthly environmental factors to GPP and R_{eco}. The correlation between GPP and R_{eco} is rather high on the annual time scale (R^2=0.8), but this is the consequence of the calculation methodology of GPP and R_{eco} (GPP and R_{eco} are not independent parameters, and the annual NEE sums are relatively low; see Vickers et al. 2009).

GPP/R_{eco} varies between 0.95 and 1.31 during the 11-year interval. The 0.95 ratio occurred in 2003, which was an extremely warm and dry year, while the 1.31 GPP/R_{eco} value was reached in 2008, which was a warm and wet year, though the precipitation amount presented in Table 8.1 does not reflect it (see the discussion below).

The year-to-year variability of annual sums of NEE, GPP, and R_{eco} are important features of the parameters, which can partly be explained by climatic fluctuations. The results show that 2006, 2007 and 2008 (and to a smaller extent 1999) appears to be somewhat extreme compared to the other years: GPP and R_{eco} were larger than in the other years, while Table 8.1 does not show any obvious climate anomaly.

In the lack of precipitation measurements at Hegyhátsál the data were taken from a nearby (17 km) meteorological observatory and, taking into account the high spatial variability in precipitation amount, they may not always represent properly the local conditions at the measuring site. If we consider the mean VPD values presented in Table 8.1, it is clear that 1999 was not as dry as the other years. Thus, the large GPP and R_{eco} might be attributed to the larger local precipitation or to its different temporal distribution. Toward the end of the measurement period, 2006, 2007, and 2008 can also be considered as unusual regarding GPP and respiration values. It is obvious from Table 8.1 that in 2006 and 2007 the higher precipitation amount provided favorable conditions for vegetation growth. However, in 2008, the annual precipitation amount is lower than the average, for both vegetation period and the whole year. In this case, the higher activity of the ecosystems can be explained by the temporal distribution of precipitation events through the year (less than 100 mm precipitation was measured in the dormant season, against the 583 mm in vegetation period).

On the annual time scale, considering the environmental variables, the relationship between the annual precipitation and the ecosystem respiration ($R^2=0.07$), and between the precipitation amount and GPP ($R^2=0.19$) were examined (note that the relationship was somewhat stronger between the annual NEE and the precipitation amount: $R^2=0.26$). Although the results showed poor correlations with precipitation in case of both R_{eco} and GPP, previous results for a subset of measurement years (until 2004) showed stronger correlation (Haszpra et al. 2005). The reason why recent years made the correlation weaker could be found in the change of the temporal distribution of precipitation through the years. The correlation between the annual mean temperature and R_{eco}, as well as between the temperature and GPP were also low ($R^2=0.1$ and 0.2, respectively; the correlation between the annual temperature and NEE was zero). GPP is practically independent from the annual PPFD sum ($R^2=0.01$). These findings somewhat contradict the relations found on shorter time scales, which emphasize the complexity of the forcing factors on large-scale ecosystem processes.

8.2.5 Uncertainty Analysis

The uncertainties were evaluated based on the three estimations (using the STE, RH, and NRH gap filling/flux partitioning methods described in Section 5.2) of NEE, GPP, and R_{eco} on daily and annual time scales. We described the uncertainties as half of the difference between maximum and minimum estimations for each variable. Note that uncertainties arising from sampling errors and low turbulent conditions during nighttime are not addressed here (those issues are discussed in Lenschow and Stankov 1986; Hollinger and Richardson 2005; Richardson et al. 2006). Potential errors indicated by the lack of energy balance closure (Haszpra et al. 2005) are also ignored here. Typical magnitudes of sampling errors are described in Richardson et al. (2006).

Table 8.2 presents the annual sums of R_{eco} and GPP together with the respective uncertainty intervals. Note that since the NEE values also have uncertainty, and since GPP and R_{eco} calculations are based on NEE, the error intervals of GPP and R_{eco} are only meaningful for the presented, best NEE estimate, so the overall uncertainties of GPP and R_{eco} are higher.

A potential systematic uncertainty arises from anthropogenic CO_2 emission in the tower region. As it was pointed out by Soegaard et al. (2003), anthropogenic emission results in a shift in the calculated NEE toward more positive value (i.e., more carbon loss). As a rough estimation from the available 20×20 km^2 emission inventory (Inst. for Environmental Protection, Budapest) the annual anthropogenic contribution in the source area is expected to be less than 20 g C m^{-2}.

The uncertainty of the long-term, small-scale flux studies are generally believed to be about ±50 g C m^{-2} year^{-1} (Baldocchi et al. 2001). Taking into account all the above-mentioned factors and the results of the gap-filling related Monte-Carlo experiment presented in Haszpra et al. (2005), the overall error in the year-round estimate of NEE, GPP, and R_{eco} is around ±80 g C m^{-2} year^{-1}.

8.2.6 *Flux Separation into Crop-Specific Parts*

8.2.6.1 Crop Classification in the Region

Climate variables (especially amount and timing of precipitation) are important factors that control the phenology, productivity, and NEE of plants. Over heterogeneous landscape, where the measured flux is a mixture of fluxes from different ecosystems with different phenological cycles and human influence, it is difficult to evaluate the effect of fluctuations in environmental factors. As an example, a large precipitation event that occurs after harvest and plowing of winter crops will likely have larger effect on summer crop NEE due to increased photosynthesis. To overcome these difficulties in interpretation, it is necessary to get information specifically on different crops. The crop-specific estimates can also help us to adapt ecosystem models in order to simulate carbon balance in mixed agricultural regions where both C3 and C4 plants are present (see Chapter 12).

As it is difficult to get information about sown crops from local farmers, it is essential to determine the crop types planted in the region using alternative methods. Having the crop coverage information and footprint calculations, it is possible to separate fluxes specifically for the predefined crop types.

The phenological cycle of crops is not simple due to the human intervention like sowing, harvesting, plowing, application of fertilizers/weed killers, etc. None of the currently available land cover maps provide information on the crop types that are cultivated in the agricultural regions (e.g., Jung et al. 2006). This is due to the large temporal variability of crop types and the small spatial extent of the individual fields. Satellite-based vegetation indices may provide an alternative and cost-efficient method for crop type classification and characterization of their phenological cycle (Viña et al. 2004; Sakamoto et al. 2005).

We used the official MOD13 vegetation index product (Huete et al. 1999) of National Aeronautics and Space Administration (NASA, USA) that had been constructed from the multispectral information provided by the MODerate resolution Imaging Spectroradiometer (MODIS) sensor onboard satellites Terra and Aqua. The applied MOD13 product (calculated with the version 005 algorithm) provides information in 250 m spatial resolution in 16 days time intervals. The Normalized Difference Vegetation Index (NDVI) is calculated from the red and near infrared radiations taking into account the atmospheric attenuation.

In order to gain information about the agricultural crop types around the tower and the phenological cycle of crops, we analyzed the temporal course of NDVI individually in each pixel. Based on aerial photos taken in the vicinity of the tower and on local surveying, it can be assumed that mainly winter crops (typically winter wheat) and summer crops (mostly maize) are grown by the farmers in the region.

We thus chose a simple classification, distinguishing between two separate crop types: (1) crops with peak development in spring, with maximum greenness in early summer and harvest in summer (typical phenological cycle of winter wheat); (2) crops with peak development in summer, maximum greenness around August, and harvest in autumn (typical attributes for maize). This two-group classification might seem to be a rough oversimplification, but it is supported by the fact that winter wheat (a C3 plant) and maize (a C4 plant) comprises by far the largest portion of the arable land in the region, and also in Hungary (Hungarian Central Statistical Office 2010).

Wavelet decomposition, reconstruction, and derivative analysis of NDVI time series were applied to estimate NDVI signatures of summer and winter crops and to estimate summer and winter crop fractions using the 250 m resolution NDVI grid. The applied methodology is fully described in Barcza et al. (2009). Figure 8.4 shows the signatures of winter and summer crops for 2004 and 2007. For winter crops, there are two typical curves. One of them is characterized with a second greening after the harvest in summer. The magnitude of the second greening also exhibits interannual variability. These three typical signatures were used to estimate crop type distribution in the mixed MODIS pixels (Barcza et al. 2009).

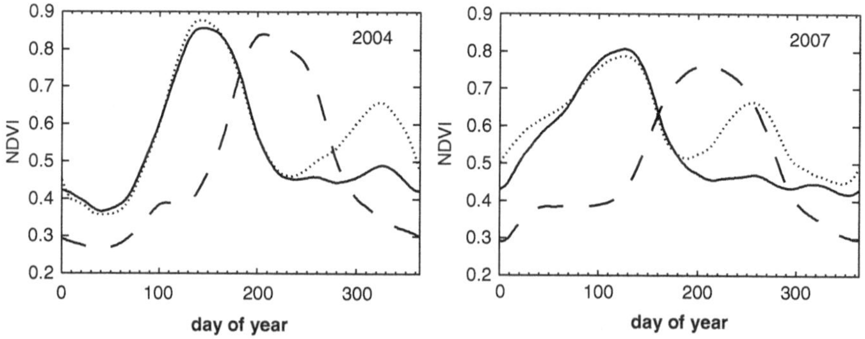

Fig. 8.4 Typical annual Normalized Difference Vegetation Index (NDVI) courses of winter crops (*solid line*), winter crops with a secondary greening (*dotted line*) and summer crops (*dashed lines*) in the investigated region for 2004 (*left*) and 2007 (*right*)

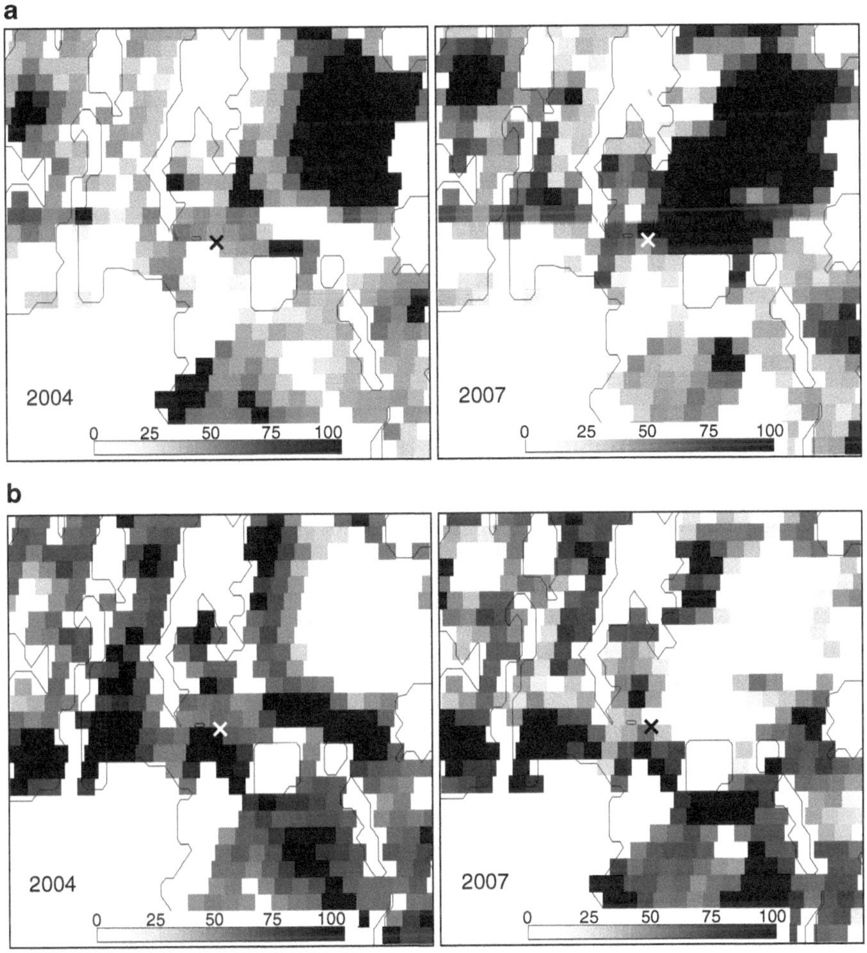

Fig. 8.5 Fraction of winter (*upper panels*) and summer (*lower panels*) crops (expressed in %) inside the individual MODIS pixels in the vicinity of the tower (X) in 2004 and 2007. The pixel size is ~250 m. The sum of summer and winter crop fractions is 100%

Figure 8.5 shows the distribution of winter and summer crops in the Hegyhátsál region in 2004 and 2007, respectively. The reconstruction was verified for a few large homogeneous plots using field evidence and aerial photos taken in the region. It was found that the method performed considerably well for clear signatures.

8.2.6.2 Contribution of the Different Crop Types to the Measured NEE

The crop-type distribution maps were used to estimate the contribution of the different crop cultivars to the measured NEE signal. Calculations also take into account the rasterized footprint for Hegyhátsál (Barcza et al. 2009).

Table 8.3 shows the results of the attribution of fluxes to different crop types. The percentage values presented describe the crop type contributions to the total

Table 8.3 Relative contribution of the main crop types to the measured NEE signal *inside* the agricultural pixels

Period	2003	2004	2005	2006	2007	2008
Winter crop	28.4%	43.7%	44.3%	70.2%	57.6%	44.9%
Summer crop	68.1%	51.7%	51.7%	26.1%	38.6%	51.8%
Other/unclassified crop	3.5%	4.6%	4%	3.7%	3.8%	3.3%

agricultural contribution. In order to estimate the total contribution of, for example, winter crops to the signal, the values presented should be multiplied with the fraction of agricultural land for the year considered.

8.2.7 *Crop-Specific NEE and Its Main Components*

8.2.7.1 Construction of Crop-Specific NEE Time Series

In order to gain information about specific crop types (e.g., to describe the dependence of summer crop NEE on the environmental factors, or to estimate dark respiration), the measured data have to be decomposed based on hourly footprint information. Since the relative importance (Table 8.3) as well as the spatial distribution of summer and winter crops are variable among the years, they may influence the measured signal as a result of the different timing of the strong carbon uptake period and harvest.

In order to separate the total NEE flux to winter and summer crop-specific NEE, selection was performed based on the fractional crop coverage map (Fig. 8.5) and the discretized peak footprint location (MODIS grid), for both winter and summer crops. The daytime hourly NEE was assigned to one of the two generic crop types if the contribution of the specific crop type was more than 80%. No other data selection was performed. The resulting crop-specific dataset was further analyzed.

The gap filling of crop-specific NEE data, and estimation of GPP and total ecosystem respiration are not trivial because of the increased number of data gaps. Crop-specific flux separation cannot be performed in calm atmospheric conditions during the night due to the characteristics of the footprint model (Section 5.2). Therefore, crop-specific nocturnal NEE data are not available, while occurrence of daytime data gaps is also enhanced by excluding data when the contribution of the specific crop type is less than 80% in the source area (represented by the ~250 m pixel). Gap-filling methods such as the mean diurnal variation method (Falge et al. 2001) and the STE flux partitioning method described in Section 5.2, used in case of the full (not crop-specific) dataset, require nighttime NEE data. Light-response curve methods (described also in Section 5.2) are also general enough to be applicable for gap filling and flux partitioning of the crop-specific dataset, and they have the advantage of not requiring nighttime data (Wang et al. 2006b).

Daily sums of crop-specific NEE and its main components were determined, and annual sums were calculated. Crop-specific annual sums and gap-filled time series are not available for every year (note that MODIS NDVI data are only available from 2003). Both data availability and data gaps vary significantly from year to year in case of crop-specific NEE according to the abundance of summer and winter crops, and they are also affected by footprint changes. Hence, there might be years when filling gaps in crop-specific data is not possible.

The number of data gaps together with the available data is influenced by the criteria we use to accept a data point as a signal of pure winter or summer crop. As we mentioned, a limit of 80% crop-specific contribution is set. Increasing or decreasing this limit can lead either to significant changes in number of data gaps, or changes in the contamination of the crop-specific dataset by signals from other crop types representing the leftover percentage. There is a trade-off between data availability and pureness of NEE time series. We decided to use the 80% limit, as it provided sufficient data availability with reliable pureness, and increasing this number only changed the result within its uncertainty.

8.2.7.2 Results of the Crop-Specific Analysis

The NEE gap filling, as well as GPP and R_{eco} calculations were performed by the same light-response curve methods on all available datasets (full dataset and crop-specific datasets). Having the estimations using the same approach ensures the comparability of the full and crop-specific time series. Two estimations for the crop-specific datasets based on light-response curve methods are available (RH and NRH methods in Section 5.2). Annual sums of crop-specific NEE, GPP, and R_{eco} are shown in Table 8.4 from 2003 to 2008 (for years when data coverage was sufficient). The determination of the uncertainty intervals associated is not addressed here, but the uncertainties are most likely exceeding those of the full dataset due to frequent data gaps.

The relative order of GPP and R_{eco} estimations in case of crops-specific data are similar to that of the full dataset (see above), but the difference between the two

Table 8.4 Crop-specific total fluxes for years when MODIS data were available and data coverage was sufficient to estimate the annual sums of net ecosystem exchange (NEE), gross primary production (GPP) and total ecosystem respiration (R_{eco})

Period	2003	2004	2006	2007	2008
Winter crop					
NEE (g C m^{-2})	–	−72	−247	−256	–
GPP (g C m^{-2})	–	1,190	1,583	1,221	–
R_{eco} (g C m^{-2})	–	1,118	1,336	965	–
Summer crop					
NEE (g C m^{-2})	8	−255	–	–	−643
GPP (g C m^{-2})	937	1,120	–	–	−1,275
R_{eco} (g C m^{-2})	945	865	–	–	632

methods are more significant due to the high number of data gaps that increases the uncertainty in the estimations.

The single positive NEE annual sum was estimated for summer crops in 2003 that was an extremely dry and hot year (Table 8.1). Although the crop-specific dataset for summer crop (maize) is the most complete in 2003, it does not reflect the typical characteristics of summer crops throughout the year due to the extreme weather. Taking into account that annual NEE sum from the full measured dataset was 43 ± 9 g C m^{-2} $year^{-1}$ and the fact that the CO_2 uptake of winter crops (typically winter wheat, see the crop classification earlier in this chapter) was usually shifted toward more positive (less negative) values compared to summer (typically maize) crops (Barcza et al. 2009), annual sum of NEE of nearly 8 g C m^{-2} $year^{-1}$ for maize was realistic. Summer and winter crop-specific data availability was nearly equal in 2004. Year 2004 is a valuable source of information being the only year when gap filled and partitioned data series for both summer and winter crops are available.

In order to check the consistency of our results with independent information sources, we used yield statistics data for maize (dominant summer crop around the tower and also in Hungary) and winter wheat (dominant winter crop). According to the data of the Hungarian Central Statistical Office, yield of maize (6,070 kg ha^{-1}) indeed exceeded that of winter wheat (4,390 kg ha^{-1}) for 2004 on the territory of Vas county, Hungary, where the tower is located. The relationship between annual sums of crop-specific NEE and yield of summer and winter crops is depicted in Fig. 8.6 for

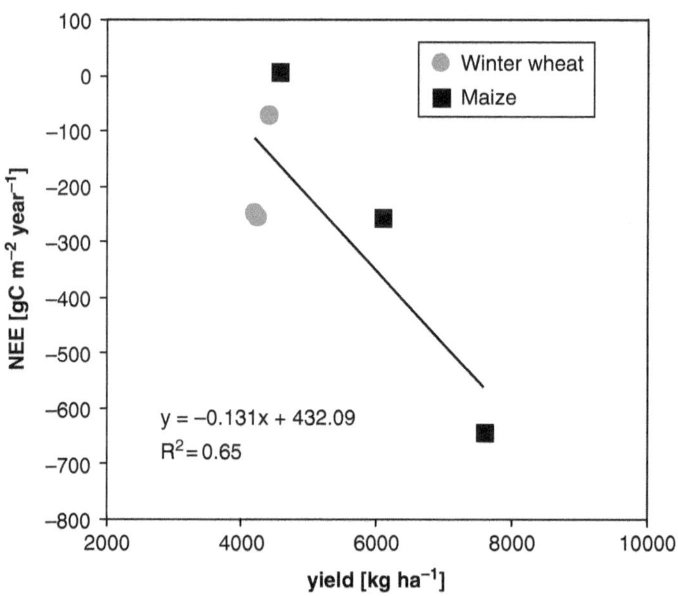

Fig. 8.6 Relationship between winter wheat and maize yields in Vas county (where the tower is located), and winter and summer crop-specific annual NEE sums. Yield data are provided by the Hungarian Central Statistical Office

all years when crop-specific annual sums are available due to sufficient data coverage and MODIS NDVI data availability. The overall square of correlation (combining all winter and summer crop-specific annual NEE sums and corresponding yield data) is 0.65 between NEE and yield data that suggests that the crop-specific results are realistic. Interestingly, summer crop NEE sums are in better correlation with maize yield data than winter crop sums, though the number of available data points is low. The contrasting correlation values can be caused by the replacement of winter wheat especially by rapeseed in recent years. Hence, the winter crop-specific NEE time series and winter wheat yield are not highly correlated, as the NEE dataset is not representative exclusively for winter wheat anymore. At present, we do not have spatially explicit crop map, which includes rapeseed to support this finding.

Note that the vegetation period of winter wheat is shorter than a complete year, which means that the winter wheat specific NEE data should be truncated in order to be comparable with the yield (and probably NEE from the previous year should be taken into account for a complete growing period). More research and longer data series are needed to provide more robust crop-specific results.

Differences between peaks of NEE in time between summer and winter crops can easily be detected in Fig. 8.7, where hourly gap filled crop-specific NEE is shown. NDVI time series for winter and summer crops show similar features regarding the time and number of peaks in NEE (Fig. 8.4). The crop-specific NDVI and NEE temporal courses are highly correlated in spite of the fact that in many cases carbon dioxide uptake is constrained by the availability of incoming solar radiation and/or soil water content that cannot be detected in the NDVI time series. It is clear from Fig. 8.7 that the NEE time series significantly differ for summer and winter crops. Maximum carbon dioxide uptake occurs in late spring for winter crops, which is followed by a decrease (most probably caused by harvest). Carbon dioxide is also taken up after harvest, which is supported by the second greening that is detected using the NDVI time series (Fig. 8.4). As environmental stress can inhibit photosynthesis the

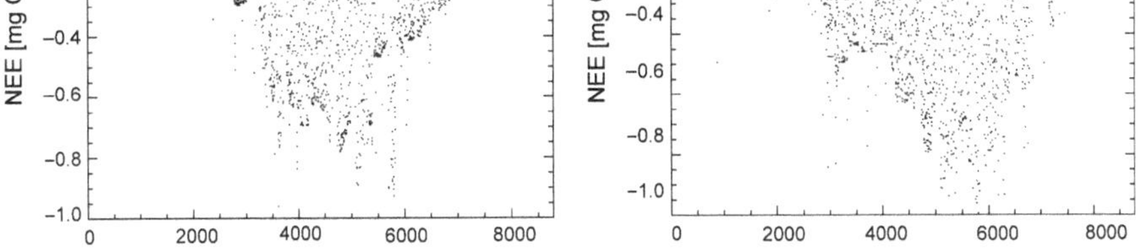

Fig. 8.7 Hourly gap filled winter and summer crop-specific NEE data for 2004. Smoothed curve is boxcar average using 200-h data window

time lag between the second greening and intensive carbon uptake is plausible. For summer crops, maximum CO_2 uptake occurs in summer or in autumn. The annual NEE course reflects the effect of late sowing and autumn harvest. Note, on average, carbon dioxide uptake is stronger for summer crops (maize) than for winter crops (winter wheat), which is in accordance with the higher yield of maize compared to that of winter wheat.

The crop-specific estimates will be further evaluated in the future as data from additional years will be available. The applied flux footprint decomposition offers an approach for relating data from complex, tall tower footprints to a larger regional context (Wang et al. 2006b; Barcza et al. 2009). The resulting dataset can be used to validate remotely sensed productivity estimates (see Section 12.3) and also to calibrate biogeochemical models in order to provide better estimates about the carbon dynamics of managed landscapes.

8.3 Soil Carbon Dioxide Emission Measurements

8.3.1 Measurements with Closed Chamber Method

In Hungary, the closed chamber dynamic method is the most common one for measuring soil carbon dioxide emission in agricultural ecosystems. The critical aspects of such measurements are (i) the choice of chamber height above the surface and the incubation time as well as their proportion (Livingston et al. 2006; Venterea and Baker 2008), and (ii) the effects of temperature changes during incubation.

This section is dedicated to soil CO_2 emission measurements on arable lands. The Józsefmajor (JM – 47°42′N, 19°38′E, 105 m asl) and the Karcag (K – 47°17′N, 20°53′E, 89 m asl) (see Fig. 1.1 in Chapter 1) experimental stations represent two different agroecological regions in Hungary. The long-term mean annual temperature is 9.7°C and 10.5°C at Józsefmajor and Karcag, respectively. The mean annual precipitation amount and potential evapotranspiration (PET) are 580 and 660 mm at Józsefmajor and 500 and 700 mm at Karcag, respectively. An overview of the two experimental sites, where long-term soil tillage experiments are being carried out, is given in Table 8.5.

At both experimental sites, the relationship between the level of soil disturbance and the soil carbon dioxide emission was studied at different grades of soil water status. Since soil water content at field conditions is uncontrolled and can vary a lot, the measurement results were evaluated for selected soil water content intervals. The soil was classified as very dry, dry, moist, and wet if its moisture content was below 0.10 g g^{-1}, from 0.10 to 0.17, from 0.17 to 0.23, and above 0.23 g g^{-1}, respectively.

For the Józsefmajor experiment, the short-term (within the following hours and days after tillage operations) and the long-term (seasonal) effects of soil mechanical disturbance on CO_2 efflux were evaluated. The most detailed experiments concerned short-term effects, where numerous soil tillage systems were involved

Table 8.5 The soil and soil management characteristics at the Józsefmajor and Karcag experimental sites

	Location of the experimental sites					
	Józsefmajor (JM)			Karcag (K)		
	Soil conditions in the 0–10 cm layer					
Soil types	**Chernic Calcic Chernozem**			**Luvic Chernozem**		
Soil texture	**Sand** 23%	**Silt** 42%	**Clay** 35%	**Sand** 7%	**Silt** 51%	**Clay** 42%
Soil hydraulic properties	**Total porosity** 0.46 $m^3\ m^{-3}$	**Field capacity** 0.37 $m^3\ m^{-3}$	**Wilting point** 0.10 $m^3\ m^{-3}$	**Total porosity** 0.42 $m^3\ m^{-3}$	**Field capacity** 0.34 $m^3\ m^{-3}$	**Wilting point** 0.16 $m^3\ m^{-3}$
Topsoil properties	**Bulk density** 1.28 g cm^{-3}	**Humus content** 3.17%	**$CaCO_3$** 0.0%	**Bulk density** 1.53 g cm^{-3}	**Humus content** 3.09%	**$CaCO_3$** 0.0%
	Soil tillage systems in focus					
Name	Direct drilling	Loosening + disking	Plowing (autumn)	Direct drilling	Reduced tillage	Conventional Plowing
Code	**DD**	**LD**	**P**	**DD**	**RT**	**P**
Depth of tillage	2–3 cm	35–40 cm + 16–20 cm	28–32 cm	2–3 cm	15–20 cm	25–30 cm

(see description in Section 8.3.1.2). These measurements were coordinated by the Institute of Crop Production of the Szent István University, Gödöllő, between 2002 and 2009. Studies concerning seasonal effects of soil disturbance on soil carbon dioxide efflux were coordinated by the Research Institute of Soil Science and Agricultural Chemistry of the Hungarian Academy of Sciences, Budapest. These measurements were carried out in 2005 during the vegetation period for three selected soil tillage systems, representing conventional (plow-based – P), minimum (direct drilling – DD), and soil root zone improving (deep loosening [DL] combined with disking – LD) tillage systems.

Depending on the data availability, the results were compared with those obtained at the Karcag experimental site, where the effects of three different tillage systems on soil respiration were examined between 2003 and 2007. These measurements were coordinated by the Karcag Research Institute in plots representing conventional (plow-based – P), reduced (RT), and minimum (direct drilling – DD) tillage systems.

8.3.1.1 Site Description and Soil Management Systems in the Karcag Long-Term Tillage Experiments

The main goal of the soil cultivation experiment introduced in the Karcag Research Institute in 1997 was to elaborate and refine new soil tillage techniques in order to prevent or mitigate soil degradation or even to improve soil properties and sustain the improved soil state in an economically beneficial and environment (soil-) protective way (Forgács et al. 2005). The supposed solution was the application of minimum tillage adapted to the given ecological conditions that were anything but favorable:

- Shortage of water (annual precipitation is 500 mm, while PET is 700 mm)
- Unfavorable distribution of precipitation in time
- Heavy textured, compacted soils usually endangered by secondary salinization or
- Originally salt affected soils

The experimental site is divided into 14 plots; seven of them (0.5 ha each) with conventional (plow-based – P) tillage and the other seven plots (1.5 ha each) with minimum tillage (DD). Besides conventional and minimum tillage practices, a reduced tillage (RT) system is being introduced and tested for its effects on soil water, heat, and carbon regimes. The RT consists of applying mulch tiller and finisher after leaving plant residues on the surface, and disturbing the soil the smallest level possible. Crop rotation characteristic for the region is applied (wheat or barley, maize, sunflower, peas or chickling vetch, millet or canary grass).

The soil type – Luvic Chernozem (WRB 2006) – of the experimental site is characteristic for the Trans-Tisza Region of the Great Hungarian Plain.

CO_2 emission measurements were carried out between 2003 and 2007 in the three different soil tillage systems (DD, P, and RT) of the Karcag experimental area, using an ANAGAS CD 98 infrared gas analyzer.

8.3.1.2 Site Description and the Soil Management Systems Applied in the Józsefmajor Long-Term Tillage Experiment

The long-term tillage experiment at Józsefmajor, nearby the city of Hatvan, was established in 2001 on a Chernic Calcic Chernozem (WRB 2006) developed on loamy clay. This soil type is common in this part of Hungary and gives sensitive response to tillage interventions. The clay and organic matter contents of the upper 40 cm layer vary from 46% to 49% and from 3.0% to 3.3%, respectively.

Considering, that the long-term goal was to reduce carbon wasting tillage practices, we also involved some so-called problematic tillage situations frequently encountered in the region. Soil left undisturbed for a long time was used as control.

Shortly after establishing the experimental plots in six replicates, tillage-induced soil respiration was studied in situ, at different soil moisture status. Soil carbon dioxide emission measurements were carried out between 2002 and 2009 shortly after summer (in July) and autumn (in September or October) tillage under a variety of weather conditions resulting in a wide range of soil water status at the time of measurements. In the top 40 cm soil layer, the soil moisture content varied between 0.08 and 0.32 g g^{-1}. The lowest and highest air temperatures at 16 cm above the soil surface were 8–36°C, respectively, but most of the readings fell in the 22–28°C range.

This was the first study of this sort, applying such wide range of parameters, in the Pannonian region. We examined the following factors in relation to the soil respiration: tillage depth and tillage systems most frequently applied in the region (deep and medium deep plowing – P, direct drilling – DD, loosening – L, disking – D, as well as tillage with cultivators – C in the way of primary tillage), along with and without surface forming. The 8-year-long measurement period between 2002 and 2009 and the timing of tillage operations (July and September) enabled gathering knowledge on short-term soil responses to a wide range of disturbances under varying weather conditions.

Tillage operations took place in July (so-called summer tillage) and September (so-called autumn tillage). The most frequent depths of autumn plowing were between 28 and 32 cm, while the typical depths of summer plowing were between 22 and 25 cm. The loosening (L) depths suitable for breaking up the plow pan in the soil varied from 35 to 40 and from 40 to 45 cm. From the aspect of tillage soil loosening is a technique aimed to break up tillage pan blocking normal water transport and crops rooting, and to deepen the root zone. From a microbiological aspect, it is an intervention eliminating toxic anaerobic conditions, helping the soil to rid itself of toxic substances (Szabó 2008). The so-called medium-DL (35–40 cm or 40–45 cm), widely applied in the Pannonian region, is deeper than the "regular" plowing, leaving a large surface, like plowing, which may be reduced by secondary tillage. Supplementary disking entails simultaneous crumb forming, flattening and mixing field residues in the soil.

The depth of disking (D) as primary tillage was between 16 and 20 cm. For stubble treatment, it was 6–10 cm in case of flat plate disks, and 10–14 cm if conventional disks were applied.

The cultivators were loosening the soil to a depth of 10–12 cm in the case of stubble treatment and up to 22–25 cm or 30–32 cm in the case of primary tillage. Concerning no-till (DD), the depth of soil disturbance varied from 2 to 3 cm.

In order to evaluate short-term responses of soil respiration on soil physical disturbance, carbon dioxide emission measurements were carried out in situ, using the closed chamber dynamic method. We measured the air concentration in the headspace of the 16 cm high chambers after 0.5, 0.75, 1.0, 1.5, 4, 24, and 48 h of the tillage operation. The method is described in Section 5.4.2.

Seasonal effects of soil tillage operations on soil carbon dioxide efflux were studied monthly during the vegetation period of 2005 in the plowing (P), no-till (NT), and disking with DL treatments. Soil CO_2 fluxes were measured using 20 cm tall chambers according to the measurement method described in Section 5.4.1. The incubation intervals were 0.5 and 1 h.

Soil respiration investigations carried out shortly after tillage interventions and during the vegetation period were complemented with soil temperature and soil water content measurements.

8.3.1.3 Evaluation of the Measured Data

A common database was constructed and analyzed, using the Statistica software package (StatSoft 1996) for evaluating the following relations:

- The short-term effects of two tillage systems (P and L) on soil CO_2 emission at two different soil temperature and water content levels, using the results of frequent measurements carried out shortly after the tillage operations
- The seasonal effects of four different soil tillage systems (P, LD, DD, and RT) on soil CO_2 emission
- The influence of soil disturbance level on soil CO_2 efflux by coupling data from the overall database

8.3.1.4 Results and Discussion

Short-Term Effects of Soil Management Systems on Soil CO_2 Emission

Figures 8.8 and 8.9 demonstrate soil CO_2 flux in the next 4 and 48 h after tillage operations in two different soil tillage systems of the Józsefmajor experiment (plowing: JM_P and loosening: JM_L). In Fig. 8.9, fluxes measured 1 day after tillage operations at Karcag site are also given. Air temperature measured 16 cm above the soil surface and soil water content status are indicated next to the legends. If more than three measurements were available, a second-order polynomial was fitted to the data. For such cases, determination coefficients can be found nearby the legends.

The effect of tillage systems was the most remarkable in the first few hours after tillage operations, when the measured fluxes increased around three times in the case of L treatment (from 2.0 to 6.6 g CO_2 m^{-2} day^{-1} and from 2.1 to 6.5 g CO_2 m^{-2} day^{-1}

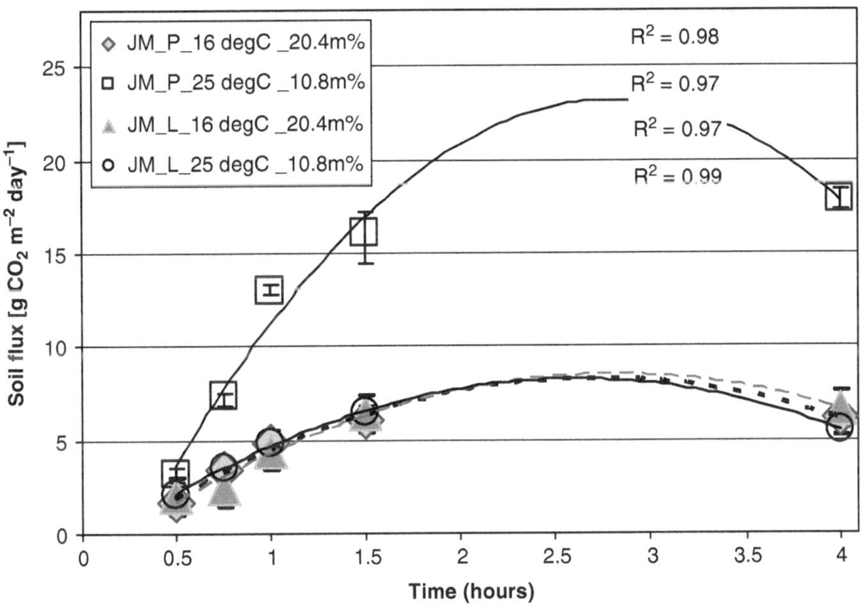

Fig. 8.8 Soil carbon dioxide emission as a function of time shortly after tillage operations in two different soil tillage systems (P – plowing, L – loosening) in the Józsefmajor experiment at various soil temperature (in degC) and water content (in m%) values. Symbols and error bars indicate the average of three replicates and standard deviations, respectively

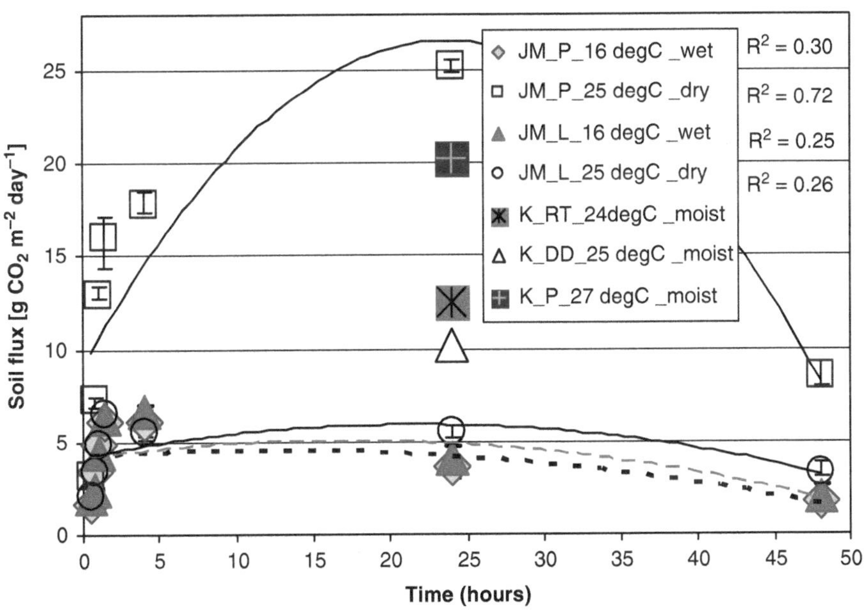

Fig. 8.9 Soil carbon dioxide emission as a function of time up to 2 days after tillage operations in different soil management systems in the Józsefmajor and Karcag experiments at various soil temperature (in degC) and water content (in m%) values. Symbols indicate average of three replicates. P – plowing, L – loosening, RT – reduced tillage, DD – direct drilling

within 4 and 1.5 h in case of wet and dry soils, respectively), almost four times (from 1.6 to 6.2 g CO_2 m^{-2} day^{-1} within 4 h) in the P treatment under wet and cold conditions, and almost eight times (from 3.2 to 25.2 g CO_2 m^{-2} day^{-1} within 24 h) in the warm and dry plowed soil. Plowing treatment on dry soil at air temperature of 25°C had the strongest and longest effect on soil CO_2 emission.

For all the examined tillage systems, increase in soil respiration until the end of the 4-h measurement fitted to a second-order polynomial (Fig. 8.8). According to its shape, the initial rapid increase in carbon dioxide fluxes stopped around 1.5–2 h after the mechanical disturbance. The increased soil fluxes started falling after approx. 4 h, except for the JM_P experiment under dry and warm conditions (Fig. 8.9). In this treatment, soil respiration showed increasing tendency until around 24 h after tillage operations.

In all the cases, the emission rates decreased to the starting level 48 h after tillage operations. Many authors reported about similar decline directly after tillage operations. For example, Omonde et al. (2007) observed the greatest CO_2 emission immediately after tillage and a sharp decline within hours after tillage operations. Al Kaisi and Yin (2005) registered the maximum CO_2 emission from all the examined treatments (deep rip, chisel plow, and moldboard plow) immediately after tillage operations, and CO_2 emission from these tilled treatments decreased sharply by 52–68% within the first 2 h.

Soil respiration rates from wet and cold soils of the P plots were similar to those observed in the loosening treatments at any of the examined soil conditions (Figs. 8.8 and 8.9). We found significantly higher CO_2 fluxes in the plowing treatment when the soil was dry and 25°C warm. The intensive respiration levels (>25 g CO_2 m^{-2} day^{-1}) observed in freshly plowed dry and warm soil could be observed in each of the past 6 years.

Results obtained in the Karcag experiment 24 h after tillage applications showed the highest fluxes in the conventional P treatment, where the detected values were around 20.3 g CO_2 m^{-2} day^{-1}. The CO_2 emission rates in the minimum tillage (DD) plots (10.2 g CO_2 m^{-2} day^{-1}) were two times lower than in the plowed plots, while somewhat higher rates – up to 12.5 g CO_2 m^{-2} day^{-1} – were measured in the RT plots (Fig. 8.9). Hence, flux measurements in the Karcag experiment showed similar short-time tendencies as in the Józsefmajor experiment: significantly higher soil respiration rates were found in the P plots than in the other treatments.

Seasonal Response of Soil Respiration on Soil Management as Compared to Short-Term Effects

Figure 8.10 shows the soil CO_2 emission values at different soil water status. Tillage-induced seasonal changes in soil respiration rates, averaged over the first 48 h of measurements showed opposite tendencies compared to short-term changes.

Hence, in the short-term studies, higher level of soil disturbance caused higher soil carbon dioxide efflux at all the soil water content ranges at both locations: the highest and the lowest respiration rates were measured in the plowing and DD

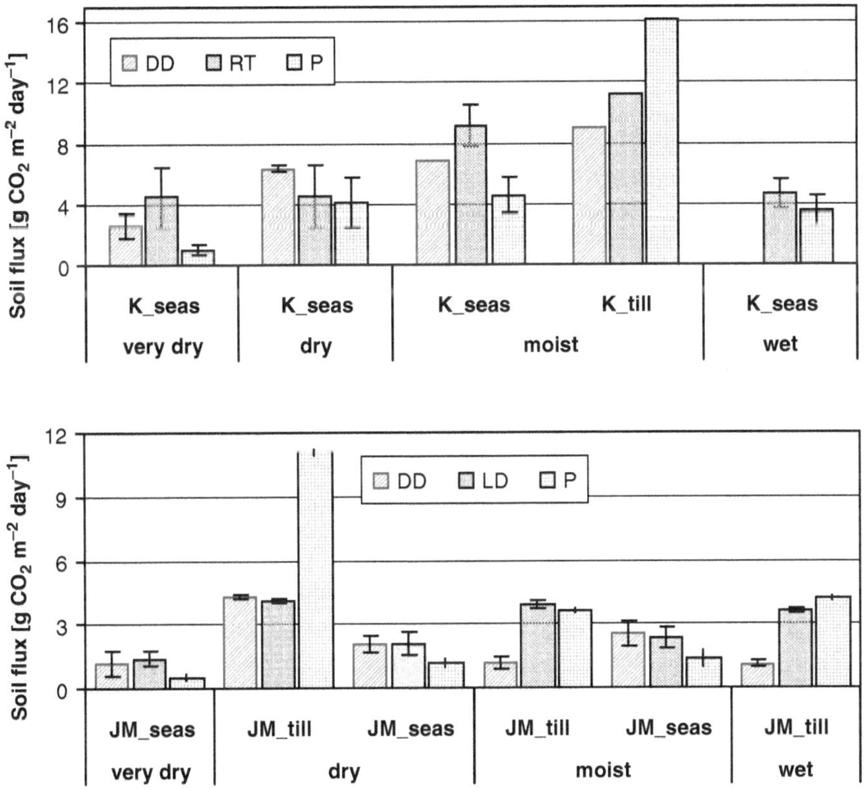

Fig. 8.10 Short-term and seasonal changes in soil respiration rates measured in the Józsefmajor (JM) and Karcag (K) experiments shortly after the tillage operations (_till) and during the vegetation period (_seas) at various soil water content ranges. *Columns* represent the average of the measurements. *Error bars* indicate the standard deviations

treatments, respectively. The emission rates were 1.5–3 times higher in the P treatments than in the DD ones.

Contrarily, seasonal fluxes were the lowest ones in the P treatment (Fig. 8.10) in both experiments, varying from 0.5 to 1.4 g CO_2 m^{-2} day^{-1} and from 1.0 to 4.2 g CO_2 m^{-2} day^{-1} at Józsefmajor and Karcag, respectively. In dry and very dry soils, the emission rates from the DD plots were at about 1.5–2.5 times higher than in the plowing treatment at both locations. Similar tendencies could be observed in moist soils: the average CO_2 fluxes were 1.4 (P) and 2.6 g CO_2 m^{-2} day^{-1} (DD) in the Józsefmajor, and 4.6 (P) and 6.8 g CO_2 m^{-2} day^{-1} (DD) in the Karcag experiments. La Scala et al. (2006) and Quincke et al. (2007) also observed that the tendencies of seasonal emission rates among tillage systems diversed from the fluxes measured immediately after tillage operation. Quincke et al. (2007) found that a flush of CO_2 loss occurred immediately after tillage, but the cumulative loss of carbon 30 day after tillage was similar for plowing and no tillage systems. Accordingly, they recorded significantly higher and lower soil respiration rates in the plowing and directly drilled plots within a few hours and 30 days after tillage operations, respectively. These results are in accordance with our measurements at the Karcag and Józsefmajor experimental sites.

Many studies, however, reported about reduced soil CO_2 emission in minimum or RT systems compared to traditional plowing (Aslam et al. 2000; Ussiri and Lal 2009). These results contradict to our observations on seasonal changes in soil respiration. There might be several reasons for that.

Firstly, the plots under minimum and reduced soil tillage involved in this study are relatively young and most probably reflect influences of a transition period on soil carbon cycle. It has been recognized that changing from traditional to conservation tillage practices can increase total soil organic carbon storage and can reduce overall CO_2 emission from a long-term (>10 year) perspective only (Havlin et al. 1990; Franzluebbers et al. 1995; Halvorson et al. 2002), while short-term (≤10 year) tillage effects on soil carbon dynamics are complex and often variable (Conservation Technology Information Center 2003).

Secondly, higher soil respiration rates in minimum tillage could occur due to higher soil water content in the DD plots compared to P plots and strong positive correlation between the soil water content and CO_2 fluxes. In our case, higher – by 2–7 g g^{-1} – soil water contents were recorded in the DD treatment than in the P treatment at all the measurement dates, even when the soil water status of both treatments belonged to the same soil water content range.

Besides the above-mentioned factors, more intensive soil biological activity detected in directly drilled soils compared to plowed ones (Tóth et al. 2009) could also lead to significant increase in soil respiration. In a 3-year study with 257 measurement days, Vinten et al. (2002) found higher CO_2 emission in minimum tillage system compared to plowing. Chavez et al. (2009) had similar investigations during a 30-day-long period on Rhodic Hapludox soil. Similarly, Lee et al. (2009) concluded that the relatively short-term use of minimum tillage did not appear to decrease soil respiration rates across the field. According to their measurements, seasonal CO_2 fluxes ranged from 16.8 to 171.8 kg CO_2 ha^{-1} day^{-1} for conventional tillage and from 17.6 to 192.0 kg CO_2 ha^{-1} day^{-1} for minimum tillage over a 3-year period from 2004 to 2006.

Soil respiration rates in the LD soil management system at Józsefmajor site did not differ significantly from those, measured in the DD treatment. On the other hand, RT plots at Karcag, in general, emitted more CO_2 during the vegetation period than DD plots (Fig. 8.10).

Comparing the identical soil tillage systems at the two sites directly after tillage or during the vegetation period, we found that the Luvic Chernozem soil (Karcag) emitted more CO_2 than the Calcic Chernozem soil (Józsefmajor) at all soil water status in the DD plots. Soil respiration in the plowed plots was also three or four times higher.

Soil respiration rates during the vegetation period showed increasing tendency with increasing soil water content at both experimental sites. These results are in accordance with those reported by Davidson et al. (2000) and Jabro et al. (2008).

The only exception was observed at Karcag when the emission from dry soil of the DD plot was higher than in the moist one.

Directly after tillage operations a contradictory tendency could be observed: the emission was higher on drier soil. The average emission rates, measured in the Józsefmajor experiment were 4.3 (dry), 1.1 (moist), and 1.2 g CO_2 m^{-2} day^{-1} (wet) in the DD, 13.8 (dry), 3.6 (moist) and 4.3 g CO_2 m^{-2} day^{-1} (wet) in the P and 6.8 (dry), 3.9 (moist) and 2.1 g CO_2 m^{-2} day^{-1} (wet) in the LD treatments. This indicates that if the surface layer is in furrows and contains holes, respiration – and consequently the decomposition of humus – is increased as a consequence of the microbial activity (Szabó 2008). Such circumstances can be avoided by surface forming simultaneously with plowing. A dry soil will continue to contain holes despite surface forming, causing intensive respiration in the first days.

When the water content of the top soil layer was higher (wet) than the range suitable for tillage, soil respiration remained lower. In dry soil after plowing, we found plenty of gaps helping ventilation among larger clods pressed into the soil. Deep plowing in the dry summer conditions leaving the surface in furrows proved to be extremely unfavorable from the aspect of CO_2 emission. The literature quoted shows that plowing is a tillage intervention type boosting CO_2 flux levels. Different types of plowing, however, should be distinguished from one another, since plowing in the suitable moisture range, followed by immediate surface forming, cannot be regarded as a tillage method leading to high carbon losses (Birkás et al. 2008).

Soil CO_2 Emission as Influenced by Depth of Soil Tillage

Figures 8.11 and 8.12 show the relationships between tillage depth and CO_2 fluxes observed in the Józsefmajor and Karcag experiments, respectively, at various soil moisture levels.

Strong coherences between soil CO_2 emission rates and average tillage depths were found at both experimental sites. These coherences were more expressed directly after tillage operations than during the vegetation period. Tillage systems, causing deeper soil disturbance (e.g., deep soil loosening – L) or higher level of turnover (e.g., plowing – P) resulted higher emissions either because oxygen reached deeper soil layers or because formerly covered layers appeared in the surface. In many cases, short-term effects (Fig. 8.11 – JM_till and Fig. 8.12 – K_till) of soil tillage on soil respiration showed contradictory tendencies compared to seasonal effects (Fig. 8.11 – JM_veg and Fig. 8.12 – K_veg). Thus, during tillage operations and immediately after them the highest and lowest emission rates were measured in dry and wet soils, respectively (Fig. 8.10 – JM_till). Nevertheless, these rates were usually higher in wet soils than in the dry ones during other times of the vegetation period.

Due to different circumstances, tillage may be carried out at soil moisture conditions far from the optimal. Evaluating the short-term soil respiration rates corresponding to different soil water content status, the soil moisture content ranges, most suitable for the various tillage techniques, can be derived. The optimum moisture

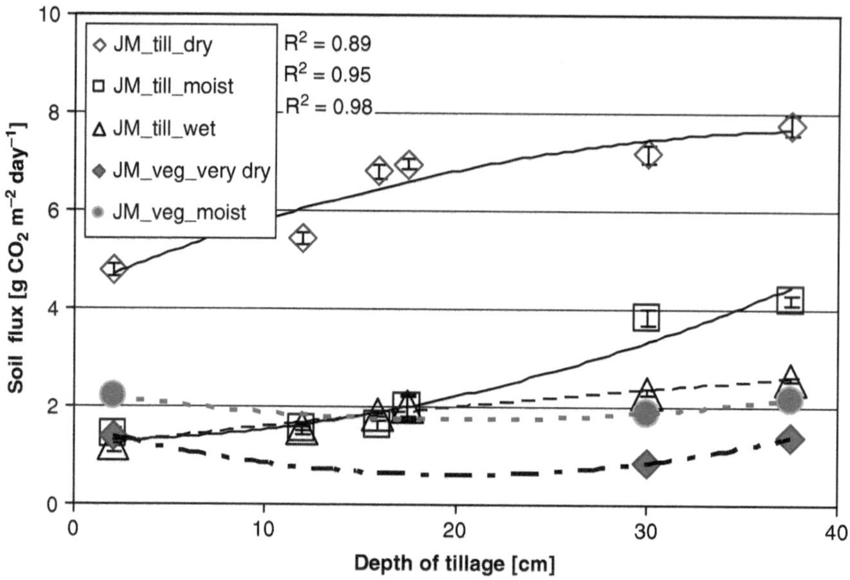

Fig. 8.11 Relationship between the average depth of tillage treatments and soil CO_2 emission at different soil moisture content ranges at the Józsefmajor experimental site. The empty symbols refer to average of fluxes measured during or a few days after tillage operations (_till), while the filled symbols stand for the average of fluxes measured any other times of the year (_veg) within the vegetation period. Note, that each symbol represents an average of at least 12–20 measurements

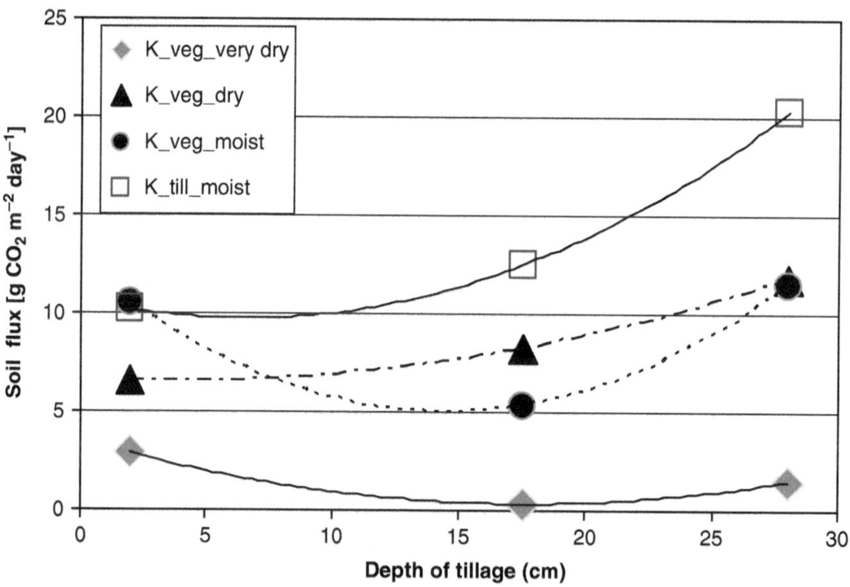

Fig. 8.12 Relationship between the average depth of tillage treatments and soil CO_2 emission at different soil moisture content ranges at the Karcag experimental site. The empty symbols refer to fluxes measured during or a few days after tillage operations (_till), while the filled symbols represent fluxes measured any other times of the year (_veg) within the vegetation period

range for tillage in the given Chernic Calcic Chernozem was found to be 0.08–0.16 g g^{-1} in the top 15 cm soil layer, and it is 0.16–0.26 g g^{-1} in the layer underneath.

8.3.2 Incubation in Climatic Room

8.3.2.1 Site Description and Soil Management Systems in a Peach Orchard

The experimental site was set up in a peach orchard (47°46′N, 19°09′E, 280 m asl), nearby the city of Vác, located 30 km north from Budapest (see Fig. 1.1). The yearly average temperature and annual precipitation sum in the region are 10.8°C and 570 mm, respectively.

The soil type of the area is Mollic-Cambisol (WRB 2006) developed on sandy loam. The sand, loam, and clay contents of the upper 20 cm layer varied from 58% to 59%, 21% to 23% and from 19.1% to 19.5%, respectively.

The orchard was planted in 1991, since then two different soil management systems have been tested at the experimental site. Every second row of the peach plantation is covered with undisturbed grass, while the other rows are disked once per 2 or 3 weeks from April to September, depending on weather conditions. The maximum depth of the disking treatment is 12–15 cm. The row spacing and the plant-to-plant distance are 6 and 4.5 m, respectively.

In 2009, 23 large undisturbed soil columns were taken from each treatment. The samples were transported into a laboratory. Soil CO_2 emission measurements were carried out 12 times at different soil water status on day 1, 16, 21, 28, 38, 46, 56, 63, 85, 105, 112, and 119 after the start of the experiment. A detailed description of soil sampling and laboratory incubation measurements is given in Section 5.4.4.

With soil samples originated from such an experiment, the effect of different soil management systems on carbon dioxide emission and soil properties could be examined. Soil carbon dioxide fluxes were calculated from detected changes in carbon dioxide concentration during the incubation time using equation reported by Tóth et al. (2005). Differences in means of soil CO_2 fluxes and bulk density values attributed to various treatments were analyzed by ANOVA using the STATISTICA software (StatSoft 2001).

8.3.2.2 Results and Discussion

Effect of Treatments on Soil Properties

We found considerable differences between almost all the examined soil hydraulic, soil chemical and soil biological properties measured in the disked (D) and grass-covered (G) rows (Table 8.6). Direct indicators of soil microbiological activity, like

Table 8.6 Soil chemical, biological, and hydraulic properties measured in different treatments

Soil properties	Treatment			
	Grass-covered row		Disked row	
Soil layer (cm)	0–5	5–10	0–5	5–10
pH (KCl)	7.76	7.96	7.30	7.29
pH (H_2O)	7.12	7.21	8.15	8.13
Total N (mg kg^{-1})	1.80		1.30	
K_2O (mg kg^{-1})	387		244	
P_2O_5 (mg kg^{-1})	382		337	
WEOC (μg C [g soil]$^{-1}$)	138.10	93.90	41.56	41.11
WEON (μg C [g soil]$^{-1}$)	10.58	7.01	1.48	3.16
Microbial biomass C	234.5	87.0	52.0	32.9
Microbial biomass N	50.0	17.0	9.1	8.8
Organic carbon (%)	1.32		0.98	
Humus (%)	2.28		1.69	
Bulk density (g cm^{-3})	1.18	1.43	1.35	1.47
Water content at saturation (v%)	57.3	48.1	51.3	47.6
Field capacity (v%)	38.1	33.2	31.2	32.5
Wilting point (v%)	10.3	10.4	9.7	10.6
Plant-available water content (v%)	27.8	22.8	21.5	21.9

water extractable organic carbon (WEOC) and nitrogen (WEON) contents, as well as microbial biomass carbon and nitrogen contents indicated higher microbiological activity in the nondisturbed grass-covered rows compared to the disked ones. Thus, WEOC and WEON values in the upper 5 cm layer of grass-covered treatment were more than three and seven times higher than in the same layer of the regularly disked treatment, respectively. Despite the fact that WEOC and WEON represent a minor part of SOM, they appear to be involved in many soil processes. Thus, it has been recognized that these molecules do influence soil biological activity (Chantigny 2003).

Soil chemical properties showed that the undisturbed, grass-covered rows were richer in nutrients. This also indicates that soil life, especially microbiological soil life, was more intensive under grass.

Soil bulk density and hydraulic properties, measured in the grass-covered raw, reflected the indirect influence of higher organic matter content and loosening effect of more dense root system on soil structural status (Table 8.6). Thus, bulk density values measured from the small 100 cm^3 undisturbed soil cores in the 0–5 cm soil layer were 1.18 and 1.35 g cm^{-3} in grass-covered and disked rows, respectively. In the 5–10 cm layer, this difference was much smaller: 1.43 and 1.47 g cm^{-3} in the grass-covered and disked treatments, respectively. Bulk density values, determined for the upper 10 cm soil layer from all the 20 large undisturbed soil columns differed significantly (at 5% significance level) between the two treatments (1.19 ± 0.07 g cm^{-3} – disking and 1.07 ± 0.06 g cm^{-3} – grass) and also reflected looser structure in the grass-covered rows.

In accordance with the measured bulk density data, soil hydraulic properties of the two treatments differed mostly in the upper 5 cm layer. Still, water content at saturation was high (above 50 v%) even in the disked rows. The field capacity and the plant-available soil water content values were above 30 and 20 v%, correspondingly, reflecting the good soil water retention properties of the soil studied. Nevertheless, all the measured characteristic points of the soil water retention curves showed higher porosity and water retention capacity in the grass-covered treatment compared to disking. We concluded that in the grass-covered nondisturbed rows more stable soil structure was formed, which supported better soil water regime in this treatment.

Effect of Treatments on Soil Carbon Dioxide Emission

One-way ANOVA results for comparing the means of soil CO_2 fluxes measured from the 20 large undisturbed soil columns are given in Fig. 8.13.

The mean emission values calculated from the overall dataset were significantly higher in the undisturbed soil as compared to the disked soil (Fig. 8.13). Significantly higher CO_2 emission values were measured in the grass rows compared to the disking treatment in 9 out of 12 measurement times (Fig. 8.13) and at

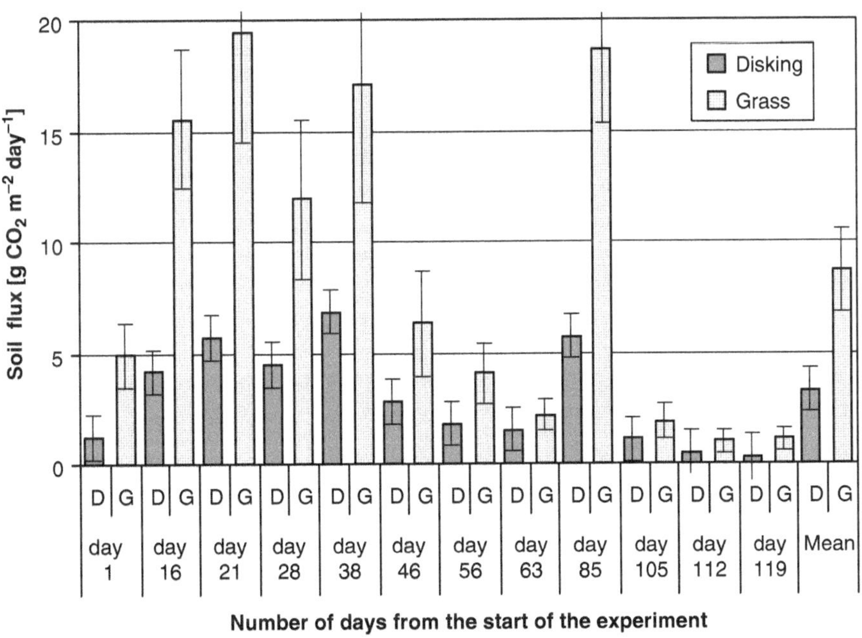

Fig. 8.13 Mean values of soil CO_2 fluxes – measured in selected times – in the disked (D) and grass covered (G) rows

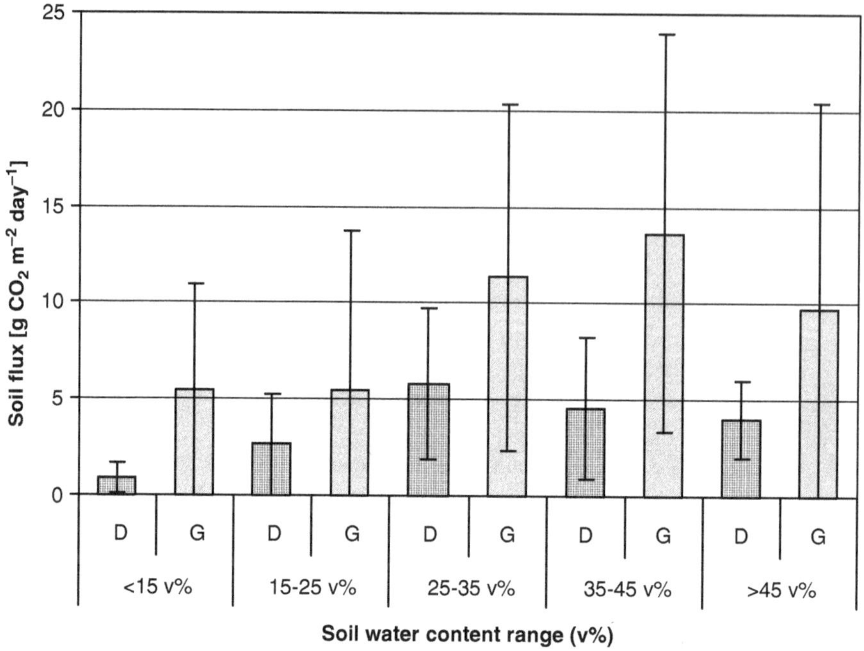

Fig. 8.14 CO_2 efflux measured in grass (G) and disking (D) treatments under various moisture conditions

various soil water contents (Fig. 8.14). Still, higher emission rates were measured in the grass-covered rows even for days when differences in emission values were not significant between the two treatments (data for 63rd, 105th, and 112th days). Our emission measurement results indicate that although tillage results in oxygen-rich state in the topsoil, regular soil disturbance destroys the soil microbiological life and reduces soil microbial activity. These conclusions are in good agreement with those derived from soil biological properties (Table 8.6).

Effect of Soil Water Content on CO_2 Emission

In the end of the experiment, volumetric soil water contents were calculated posteriorly for every incubation time and for all the 20 soil samples per treatment. Figure 8.14 illustrates the mean values and standard deviations of the measured CO_2 emissions for five different soil water content ranges. A one-way ANOVA was performed for comparing the means of the measured fluxes between the grass-covered (G) and disking (D) treatments. Statistically significant differences were found at 5% significance level for all water content ranges, except for the highest one (>45 v%). However, since only a few samples fall in this water content range, no sound conclusions can be drawn.

Depending on the soil water content, two to four times higher carbon dioxide fluxes were measured from samples, taken in the grass-covered rows compared to those originating from the disking rows (Fig. 8.14). In general, we found increase in soil respiration with increasing soil water content until the field capacity value was reached. Thus, the average field capacity values of the 0 to 10 cm topsoil layer (Table 8.6) were 32 and 37 v% in the disking and grass-covered treatments, respectively. Accordingly, in the D and G treatments, the highest CO_2 fluxes were measured in the 25–35 v% and 35–45 v% soil water content ranges. Furthermore, a slight decrease could be observed as soil water status approached toward saturation.

We believe that higher soil respiration rates reflected higher soil microbiological activity in the less disturbed grass-covered treatment compared to the disking treatment at any soil water content status. Our results indicate that in the given soil type water content near field capacity creates the most favorable conditions for soil microbial life, providing optimum proportion of water and air for microbes in the soil.

Coherences Between pF Values and CO_2 Emission

Soil properties, such as soil temperature and soil microbial life, affecting soil CO_2 emission depend on the state of water in the soil, which is not explicitly reflected by the soil water content itself. As it has been shown and discussed previously, in different soils, the same soil water content does not necessarily correspond to the same soil water potential. This might be true even for soils having similar mechanical composition (or texture) but differing in structure (Table 8.6). This means that similar soil water contents may represent very different water–air ratio in soil pores and, consequently, can provide rather altering conditions for soil microbial life. Moreover, for soils with marked textural differences, the same soil water content can correspond to totally different energetic status of water within soil pores. Thus, the exact way of characterizing the soil water status is using soil water potential values.

Our aim was to develop a measuring technique allowing to analyze the relationship between soil carbon dioxide emission and soil water potential. On the basis of preliminary estimations, we tried to set up soil samples to four different water potential values – pF2.0, pF2.3, pF3.2, and pF3.4 – by watering the samples. After the experiment ended and the bulk density of the undisturbed soil cores was determined, the real water potential value of each sample could be calculated precisely for each measurement day. The spectrum of water potential values, varying from pF1.0 to pF4.6, was much wider than it had been previously assumed.

At the first five measurement events (days 1, 16, 21, 28, and 38), coherences between soil fluxes and water potential values were not as strong as later during the experiment (data not shown). We suppose that soil sampling, transportation, and taking the samples into a completely new environment in the climatic room disrupted the microbiological balance state in the soil. Relationship between

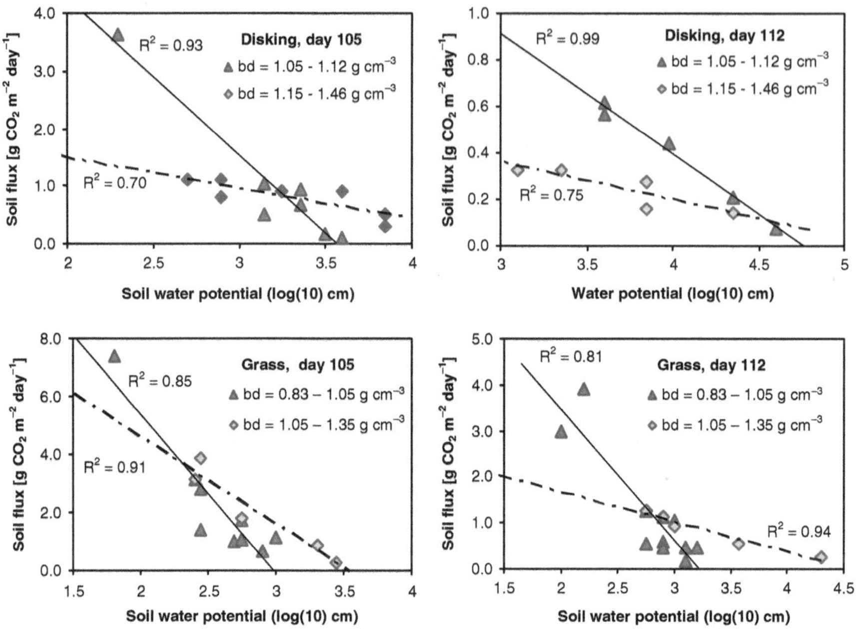

Fig. 8.15 Soil CO_2 efflux measured in grass covered (G) and disking (D) treatments as a function of soil water potential at various soil bulk density ranges

fluxes and water potential values can be observed after the fifth measurement procedure (from day 46); the coherence between these values became more distinct.

Figure 8.15 shows soil respiration rates measured on day 105 and 112 as a function of soil water potential in the two different treatments on different measurement days. Data are divided in two separate groups depending on bulk density of the soil samples. In the disking treatment, determination coefficients (R^2) varied from 0.70 to 0.75 and from 0.93 to 0.99 for higher and lower bulk density values, respectively. We achieved stronger relationship between the soil CO_2 flux and soil water potential for low bulk density (bd) values most probably because the range of this bd group was smaller. Without separation, the R^2 values were 0.60 and 0.24 on day 105 and 112, respectively.

Concerning the grass covered rows determination coefficients were 0.63 and 0.55 for the 105th and 112th measurement days, correspondingly, if no separation of data was performed according to bulk density values. In case of separation, R^2 values for samples having higher and lower bulk density reached 0.91 and 0.85 on the 105th, and 0.94 and 0.81 on the 112th days in the grass-covered row, respectively.

Our results clearly demonstrate the strong connection between soil carbon dioxide fluxes and soil water potential, and the structure-dependency of soil respiration rates.

8.4 Conclusion

On the base of our EC measurement agricultural lands in the study region take up CO_2 from the atmosphere on annual scale. The carbon content of crop yield (wheat, maize, etc.) is transported to villages and cities from arable lands, which causes a spatial separation between the uptake and release of atmospheric CO_2. Because of the horizontal transport, assuming equilibrium soil carbon stocks, there has to be a local sink of CO_2 in arable lands and a local source in cities. De facto, according to the tall-tower-based measurements performed over arable lands, we see a net CO_2 sink from the point of view of the atmosphere.

Nevertheless, EC method measures only atmospheric fluxes, so this method does not allow estimating changes in soil carbon stocks. Because the main aim is to enhance soil carbon content and mitigate soil CO_2 emission, soil tillage experiments have an increasing role to determine the CO_2 emission mitigating factors. Soil tillage systems significantly influence the main factors determining soil CO_2 efflux, and – depending on their effects on SOM content – they may, eventually, be carbon preserving, balance keeping but they may also contribute to increasing carbon losses from soil.

Due to the complexity of the system in focus and lack of knowledge about processes involved in soil carbon balance formation, the effect of different soil tillage systems on soil respiration and carbon sequestration cannot be unequivocally predicted. Therefore, knowledge-based solutions for sustainable, soil structure and moisture conserving tillage systems that facilitate carbon sequestration should be based on site-specific data measured over longer periods.

Investigating the influence of some conventional and newly implemented soil management systems on soil respiration in Hungarian agroecosystems, we found that shortly after tillage operations carbon dioxide flux levels are determined by the tillage depth, the resulting soil surface, and by soil water and temperature status. Seasonal changes in soil carbon dioxide efflux showed contradictory tendencies: tillage systems causing the smallest soil disturbance resulted in the highest emission rates due to increase in soil biological activity and higher soil water contents.

These results were further compared to those obtained from an undisturbed grass and disk treatments of a peach plantation. Significantly higher CO_2 fluxes were measured from the grass covered soil in a wide range of soil water contents at almost all the measurement times. Our results indicate that besides the favorable effect of soil tillage on soil aeration, regular soil disturbance reduces soil microbial activity.

Analyzing the relationship between the measured soil respiration rates and soil water potential, we confirmed that soil carbon dioxide emission is a structure dependent soil process. Therefore, flux measurements under laboratory conditions should preferably be carried out on large undisturbed soil cores, since using disturbed samples information can be lost and the results are more difficult to interpret.

We concluded that soil respiration could be regarded as a soil-state indicator and could be used for selecting the least damaging tillage mode under extreme weather conditions. Our results indicated that tillage systems causing high level of soil disturbance and leaving the soil surface open contributed to increased level of soil CO_2 emission and to soil carbon losses. Nevertheless, the circumstances of applying a certain tillage method determined soil respiration to a high extent. Thus, different types of plowing should be distinguished from each other, since plowing in the suitable moisture range, followed by immediate surface forming, could not be regarded as a tillage method leading to high carbon losses.

Acknowledgments Use of the TV-tower for EC measurements at Hegyhátsál is kindly provided by Antenna Hungária Corporation. The project was supported by the U.S.-Hungarian Scientific and Technological Joint Fund (J. F. no. 504), by the Hungarian National Research Fund (OTKA T23811, OTKA F026642, OTKA T32440, OTKA T42941, OTKA F047242, OTKA-NKTH CK77550), by the Hungarian Higher Education Support Program (FKFP-0168/1997), by the European Commission's Fifth R&D Framework Programme (AEROCARB and CHIOTTO, contract no. EVK2-CT-1999-00013 and EVK2-CT-2002-00163), by the European Commission's Sixth R&D Framework Programme (CARBOEUROPE-IP, contract no. GOCE-CT-2003-505572), and by the Hungarian State Eötvös Fellowship. Studies on soil respiration were supported by the Hungarian National Scientific Research Found (OTKA T62436; K6-0314; T048302), by the Hungarian Ministry of Education and Culture (OM-00381/2008) and by MEH-MTA No. VI.11 "Climate" project.

References

Al-Kaisi MM, Yin X (2005) Tillage and crop residue effects on soil carbon and carbon dioxide emission in corn-soybean rotations. J Environ Qual 34:437–445

Aslam T, Choudhary MA, Saggar S (2000) Influence of land-use management on CO_2 emissions from a silt loam soil in New Zealand. Agric Ecosyst Environ 77:257–262

Baldocchi DD, Falge E, Gu L, Olson R, Hollinger D, Running S, Anthoni P, Bernhofer Ch, Davis K, Fuentes J, Goldstein A, Katul G, Law B, Lee X, Malhi Y, Meyers T, Munger JW, Oechel W, Pilegaard K, Schmid HP, Valentini R, Verma S, Vesala T, Wilson K, Wofsy S (2001) FLUXNET: a new tool to study the temporal and spatial variability of ecosystem-scale carbon dioxide, water vapor and energy flux densities. Bull Am Meteorol Soc 82:2415–2435

Barcza Z, Kern A, Haszpra L, Kljun N (2009) Spatial representativeness of tall tower eddy covariance measurements using remote sensing and footprint analysis. Agric Forest Meteorol 149:795–807

Birkás M, Antos G, Neményi M, Szemők A (2008) Environmentally-sound adaptable tillage. Akadémiai Kiadó, Budapest

Büttner G, Feranec J, Jaffrain G (2002) Corine land cover update 2000. Technical Report 89 (European Environment Agency, 2002). http://reports.eea.eu.int/technical_report_2002_89/en

Chantigny MH (2003) Dissolved and water-extractable organic matter in soils: a review on the influence of land use and management practices. Geoderma 113:357–380

Chatskikh D, Olesen JE (2007) Soil tillage enhanced CO_2 and N_2O emissions from loamy sand soil under spring barley. Soil Tillage Res 97:5–18

Chavez LF, Amado TJC, Bayer C, La Scala NJ, Escobar LF, Fiorin JE, Costa de Campos B-H (2009) Carbon dioxide efflux in a Rhodic Hapludoc as affected by tillage systems in Southern Brazil. R Bras Ci Solo 33:325–334

Conservation Technology Information Center (2003). 2002 National crop residue management survey [Online]. http://www.ctic.purdue.edu/CTIC/CRM.html (accessed 28 May 2004; verified 9 Nov. 2004). Conserv. Technol. Info. Center, West Lafayette verified 9 Nov. 2004). Conserv. Technol. Info. Center, West Lafayette, IN

Davidson EA, Verchot LV, Cattânio JH, Ackerman IL, Carvalho JEM (2000) Effects of soil water content on soil respiration in forests and cattle pastures of eastern Amazonia. Biogeochemistry 48:53–69

Falge E, Baldocchi D, Olson R, Anthoni P, Aubinet M, Bernhofer C, Burba G, Ceulemans R, Clement R, Dolman H, Granier A, Gross P, Grünwald T, Hollinger D, Jensen N-O, Katul G, Keronen P, Kowalski A, Lai CT, Law BE, Meyers T, Moncrieff J, Moors E, Munger JW, Pilegaard K, Rannik Ü, Rebmann C, Suyker A, Tenhunen J, Tu K, Verma S, Vesala T, Wilson K, Wofsy S (2001) Gap filling strategies for defensible annual sums of net ecosystem exchange. Agric Forest Meteorol 107:43–69

Forgács L, Zsembeli J, Tuba G (2005) Examination of a soil protective cultivation method in the Research Institute of Karcag. In: Realizáciou poznatkov vedy a vyskumu k trvalo udrzatel'nému pol'nohospodárstvu. Michalovce, Slovakia, pp 64–68

Franzluebbers AJ, Hons FM, Zuberer DA (1995) Tillage and crop effects on seasonal soil carbon and nitrogen dynamics. Soil Sci Soc Am J 59:1618–1624

Halvorson AD, Wienhold BJ, Black AL (2002) Tillage, nitrogen, and cropping system effects on soil carbon sequestration. Soil Sci Soc Am J 66:906–912

Haszpra L, Barcza Z, Bakwin PS, Berger BW, Davis KJ, Weidinger T (2001) Measuring system for the long-term monitoring of biosphere/atmosphere exchange of carbon dioxide. J Geophys Res 106:3057–3070

Haszpra L, Barcza Z, Davis KJ, Tarczay K (2005) Long-term tall tower carbon dioxide flux monitoring over an area of mixed vegetation. Agric Forest Meteorol 132:58–77

Havlin JL, Kissel DE, Maddux LD, Claasen MM, Long JH (1990) Crop rotation and tillage effects on soil organic carbon and nitrogen. Soil Sci Soc Am J 54:448–452

Hollinger DY, Richardson AD (2005) Uncertainty in eddy covariance measurements and its application to physiological models. Tree Physiol 25:873–885

Huete A, Justice C, Leeuwen WV (1999) MODIS vegetation index – MOD13. Algorithm Theoretical Basis Document. http://modis.gsfc.nasa.gov/data/atbd/atbd_mod13.pdf

Hungarian Central Statistical Office (2010) http://statinfo.ksh.hu/Statinfo/themeSelector.jsp?&lang=en. Accessed February 2010

Jabro JD, Sainju U, Stevens WB, Evans RG (2008) Carbon dioxide flux as affected by tillage and irrigation in soil converted from perennial forages to annual crops. J Environ Manag 88: 1478–1484

Jung M, Henkel K, Herold M, Churkina G (2006) Exploiting synergies of global land cover products for carbon cycle modelling. Remote Sens Environ 101:534–553

Kljun N, Calanca P, Rotach MW, Schmid HP (2004) A simple parameterisation for flux footprint predictions. Bound-Lay Meteorol 112:503–523

La Scala N, Bolonhezi D, Pereira GT (2006) Short-term soil CO_2 emission after conventional and reduced tillage of a no-till sugar cane area in Southern Brazil. Soil Tillage Res 91:244–248

Lal R (2009) Soils and world food security (Editorial). Soil Tillage Res 102:1–4

Lee J, Hopmans JW, van Kessel C, King AP, Evatt KJ, Louie D, Rolston DE, Six J (2009) Tillage and seasonal emissions of CO_2, N_2O and NO across a seed bed and at the field scale in a Mediterranean climate. Agric Ecosyst Environ 129:378–390

Lenschow DH, Stankov BB (1986) Length scales in the convective boundary layer. J Atmos Sci 43:1198–1209

Livingston GP, Hutchinson GL, Spartalian K (2006) Living trace gas emission in chambers: a non-steady-state diffusion model. Soil Sci Soc Am J 70:1459–1469

Mann LK (1986) Changes in soil carbon storage after cultivation. Soil Sci 142:279–288

Omonde RA, Vyn TJ, Smith DR, Hegymegi P, Gál A (2007) Soil carbon dioxide and methane fluxes from long-term tillage systems in continuous corn and corn-soybean rotations. Soil Tillage Res 95:182–195

Quincke JA, Wortmann CS, Mamo M, Franti T, Drijber RA (2007) Occasional tillage of no-till systems: carbon dioxide flux and changes in total and labile soil organic carbon. Agron J 99:1158–1168

Read D, Beerling D, Cannell M, Cox P, Curran P, Grace J, Ineson P, Jarvis P, Malhi Y, Powlson D, Shepherd J, Woodward I (2001) The role of land carbon sinks in mitigating global climate change. Royal Society, London, UK, p 27

Reichstein M, Falge E, Baldocchi D, Papale D, Aubinet M, Berbigier P, Bernhofer C, Buchmann N, Gilmanov T, Granier A, Grunwald T, Havrankova K, Ilvesniemi H, Janous D, Knohl A, Laurila T, Lohila A, Loustau D, Matteucci G, Meyers T, Miglietta F, Ourcival J, Pumpan M (2005) On the separation of net ecosystem exchange into assimilation and ecosystem respiration: review and improved algorithm. Global Change Biol 11:1424–1439

Reicosky DC, Lindstrom MJ, Schumacher TE, Lobb DE, Malo DD (2005) Tillage-induced CO_2 loss across an eroded landscape. Soil Tillage Res 81:183–194

Reicosky DC, Gesch RW, Wagner SW, Gilbert RA, Wente CD, Morris DR (2008) Tillage and wind effects on soil CO_2 concentrations in muck soils. Soil Tillage Res 99:221–231

Reth S, Göckede M, Falge E (2005) CO_2 efflux from agricultural soils in Eastern Germany – comparison of a closed chamber system with eddy covariance measurements. Theor Appl Climatol 80:105–120

Richardson AD, Hollinger DY, Burba GG, Davis KJ, Flanagan LB, Katul GG, Munger JW, Ricciuto DM, Stoy PC, Suyker AE, Verma SB, Wofsy SC (2006) A multi-site analysis of random error in tower-based measurements of carbon and energy fluxes. Agric Forest Meteorol 136:1–18

Sakamoto T, Yokozawa M, Toritani H, Shibayama M, Ishitsuka N, Ohno H (2005) A crop phenology detection method using time-series MODIS data. Remote Sens Environ 96:366–374

Schmid HP (2002) Footprint modeling for vegetation atmosphere exchange studies: a review and perspective. Agric Forest Meteorol 113:159–184

Soegaard H, Jensen NO, Boegh E, Hasager CB, Schelde K, Thomsen A (2003) Carbon dioxide exchange over agricultural landscape using eddy correlation and footprint modeling. Agric Forest Meteorol 114:153–173

StatSoft (1996) Statistica 5.1 software. Statsoft, Tucksa

Stoy PC, Katul GG, Siqueira MBS, Juang J-Y, Novick KA, Uebelherr JM, Oren R (2006) An evaluation of models for partitioning eddy covariance-measured net ecosystem exchange into photosynthesis and respiration. Agric Forest Meteorol 141:2–18

Szabó IM (ed) (2008) A talajművelés mikribiológiája (Microbiology of tillage). In: Az általános talajtan mikrobiológiai alapjai, 3rd edn (Microbiological fundamentals of general soil science). Mundus, Budapest, pp 349–355

Tóth E, Koós S, Farkas Cs (2009) Soil carbon dioxide efflux determined from large undisturbed soil cores collected in different soil management systems. Biologia 64:643–647

Tóth T, Fórizs I, Kuti L, Wardell JL (2005) Data on the elements of carbon cycle in a Solonetz and Solonchak soil. Cereal Res Commun 33:133–136

Ussiri DAN, Lal R (2009) Long-term tillage effects on soil carbon storage and carbon dioxide emissions in continuous corn cropping system from an alfisol in Ohio. Soil Tillage Res 104:39–47

Venterea RT, Baker JM (2008) Effects of soil physical nonuniformity on chamber-based gas flux estimates. Soil Sci Soc Am J 72:1410–1417

Vesala T, Kljun N, Rannik Ü, Rinne J, Sogachev A, Markkanen T, Sabelfeld K, Foken T, Leclerc MY (2008) Flux and concentration footprint modelling: state of the art. Environ Pollut 152:653–667

Vickers D, Thomas CK, Martin JG, Law B (2009) Self-correlation between assimilation and respiration resulting from flux partitioning of eddy-covariance CO_2 fluxes. Agric Forest Meteorol 149:1552–1555

Viña A, Gitelson AA, Rundquist DC, Keydan G, Leavitt B, Schepers J (2004) Monitoring maize (Zea mays L.) phenology with remote sensing. Agron J 96:1139–1147

Vinten AJA, Ball BC, O'Sullivan MF, Henshall JK (2002) The effects of cultivation method, fertilizer input and previous sward type on organic C and N storage and gaseous losses under spring and winter barley following long-term leys. J Agric Sci 139:231–243

Wang W, Davis KJ, Ricciuto DM, Butler MP (2006a) An approximate footprint model for flux measurements in the convective boundary layer. J Atmos Oceanic Technol 23:1384–1394

Wang W, Davis KJ, Cook BD, Butler MP, Ricciuto DM (2006b) Decomposing CO_2 fluxes measured over a mixed ecosystem at a tall tower and extending to a region: a case study. J Geophys Res Biogeosci 111G:G02005. doi:10.1029/2005JG000093

West TO, Post WM (2002) Soil organic carbon sequestration rates by tillage and crop rotation: a global data analysis. Soil Sci Soc Am J 66:1930–1946

WRB IUSS Working Group (2006) World reference base for soil resources, 2nd edn. World Soil Resources Report No. 103. FAO, Rome

Part III
Modeling of Biosphere–Atmosphere Exchange of Greenhouse Gases

Chapter 9
Models and Their Adaptation*

Zoltán Somogyi, Dóra Hidy, Györgyi Gelybó, Zoltán Barcza, Galina Churkina, László Haszpra, László Horváth, Attila Machon, and Balázs Grosz

Abstract This chapter describes computer simulation models developed for estimating greenhouse gas fluxes of different ecosystems. In general, each model is used for the simulation of specific parts of the complex systems; therefore, merging the results of various kind of models can give a better insight. In this chapter, we describe four models used for estimating biospheric fluxes of greenhouse gases in Hungary. The Biome-BGC and the DNDC models are process-based ecological models. The adapted version of Biome-BGC is capable of describing the carbon,

*Citation: Somogyi, Z., Hidy, D., Gelybó, Gy., Barcza, Z., Churkina, G., Haszpra, L., Horváth, L., Machon, A., Grosz, B., 2010: Modeling of biosphere–atmosphere exchange of greenhouse gases – Models and their adaptation. In: Atmospheric Greenhouse Gases: The Hungarian Perspective (Ed.: Haszpra, L.), pp. 201–228.

B. Grosz (✉)
Department of General and Inorganic Chemistry, Eötvös Loránd University, H-1117 Budapest, Pázmány P. s. 1/A, Hungary
e-mail: grosz@mail.datanet.hu

Z. Somogyi
Hungarian Forest Research Institute, H-1142 Budapest, Stefánia út. 14., Hungary

D. Hidy, L. Haszpra, L. Horváth and A. Machon
Hungarian Meteorological Service, H-1675 Budapest, P.O. Box 39, Hungary

Gy. Gelybó and Z. Barcza
Department of Meteorology, Eötvös Loránd University, H-1117 Budapest, Pázmány P. s. 1/A, Hungary

G. Churkina
Leibniz-Centre for Agricultural Landscape Research, D-15374 Müncheberg, Eberswalder Strasse 84, Germany

A. Machon
Center for Environmental Science, Eötvös Loránd University, H-1117 Budapest, Pázmány P. s. 1/A, Hungary
and
Department of Botany and Plant Physiology, Szent István University, H-2103, Gödöllő, Páter K. u. 1., Hungary

L. Haszpra (ed.), *Atmospheric Greenhouse Gases: The Hungarian Perspective*,
DOI 10.1007/978-90-481-9950-1_9, © Springer Science+Business Media B.V. 2011

nitrogen, and water fluxes of the Hungarian arable lands and grasslands. This model is used to estimate the net primary production (NPP) and the net biome production. DNDC is used to predict the soil fluxes of nitrous oxide and methane, and simulates the biogeochemical cycles of carbon and nitrogen occurring in agricultural soil. MOD17 is a simple lightweight model that is based on remote sensing and ancillary meteorological information, and provides gross primary production (GPP) and NPP data. The CASMOFOR model was developed to estimate how much carbon can be accumulated in afforestation projects.

Keywords Biome-BGC • MOD17 • DNDC • CASMOFOR

9.1 Introduction

Greenhouse gases (GHGs) play a determining role in the formation of Earth's climate. Any change in their atmospheric concentration may result in climate change with potentially serious consequences on the biosphere and humankind. Atmospheric measurements have reported significant changes in the atmospheric GHG concentrations. Therefore, understanding and monitoring of processes influencing the atmospheric greenhouse gas concentrations are essential. Direct measurements of greenhouse gas concentrations and fluxes are important to obtain evidence, but they are expensive, and do not necessarily give information representative for large areas. Mathematical models synthesizing the available theoretical knowledge and conclusions of experimental programs may help us to estimate the processes where no measurements are available or to predict the behavior of these processes under changing environmental conditions. Numerical models have been developed for estimating the greenhouse gas exchange between the atmosphere and the vegetation–soil system. While measurement programs presented in Part II of this book, and in the scientific literature in general, can quantify the actual fluxes of GHGs on local scale, computer simulations can extend this ecological information in both time and space. In this chapter, we introduce four mathematical models used in Hungary for the estimation of biosphere–atmosphere fluxes.

Biome-BGC is a widely used biogeochemical model that can be used to simulate carbon, nitrogen, and water fluxes of different terrestrial ecosystems, such as deciduous and evergreen forests, grasslands, and shrubs. Biome-BGC is one of the few models that handle nitrogen cycle; thus the effect of varying nitrogen deposition and mineral nitrogen input can be scrutinized. The model can be used to simulate different carbon flux components between the atmosphere and the biosphere in daily time steps. We introduce the basic structure of the model, the procedure applied to calibrate it for the Hungarian conditions, and the developments made in the publicly available original source code in order to describe grassland management and to provide realistic estimates for herbaceous ecosystems.

The DNDC (denitrification–decomposition) model estimates the trace gas fluxes of soil. The fluxes of N_2O, NO, NH_3, and CH_4 are not constant either temporally or spatially: the continuous variation of these fluxes is due to the changes in the bio- and geochemical reactions, soil and meteorological conditions, and the complex transformations.

Satellite-based remote sensing techniques have opened new horizons in landscape-scale ecosystem modeling efforts in recent years. Data measured by instruments on polar orbiting satellites have made it possible to gain information on vegetation characteristics even where no in situ measurements are available. MOD17 is a simple light use efficiency model that is based on remote sensing data and ancillary meteorological information. MOD17 uses the measurements of the MODIS sensor onboard satellites Terra and Aqua. The MOD17 model can provide gross primary production (GPP) and net primary production (NPP) data with 8-day and annual temporal resolution, respectively, at 1-km spatial resolution. We present the basic logic of the model and the meteorological data used for the simulations.

The chapter also introduces a carbon accounting model for forests that is an original Hungarian development. This model, called CASMOFOR, has been developed to estimate how much carbon is sequestered in afforestation projects. The model is an MS Excel based framework of algorithms, and it can be applied in any country where the required parameters are available. For Hungary, the model includes all required data that are based on the results of the forestry research of recent decades. The model also includes relevant economic data to provide estimates for costs and revenues of the traditional forestry system; however, these data can also be used to estimate the costs and revenues of carbon sequestration by means of forestry. Thus, the model can also be considered as an ideal tool for decision makers.

9.2 The Biome-BGC Process-Oriented Ecosystem Model

Dóra Hidy, Zoltán Barcza, Galina Churkina

9.2.1 Model Structure

Biome-BGC v4.1.1 is a mechanistic biogeochemical model simulating the storage and fluxes of water, carbon, and nitrogen between the ecosystem and the atmosphere, and within the components of the terrestrial ecosystems. Biome-BGC was developed from the Forest-BGC family of models (Running and Coughlan 1988) and is an extension and generalization of the Forest-BGC model for the description of different vegetation types. Biome-BGC v4.1.1 uses daily time step, and is driven by daily values of maximum and minimum temperatures, precipitation amount, solar radiation, and vapor pressure deficit (VPD). Biome-BGC uses meteorological data, site-specific data (soil texture, effective soil depth, annual mean nitrogen deposition and biological nitrogen fixation, site elevation and latitude), and ecophysiological data (e.g. maximum stomatal conductance, canopy specific leaf area (SLA)) to simulate the biogeochemical processes of the given biome. The main processes assessed are photosynthesis, allocation, litterfall, carbon (C), nitrogen (N), and water dynamics in the litter and soil (Thornton et al. 2000). In the present work, Biome-BGC v4.1.1 is used with modifications described in Trusilova et al. (2009) to simulate the carbon and water fluxes of Hungarian grasslands and arable lands.

The most important blocks of the model are the carbon flux block, the phenological block, and the soil flux block. In the carbon flux block, GPP of the biome is calculated using Farquhar's photosynthesis routine (Farquhar et al. 1980). Autotrophic respiration is separated into maintenance and growth respirations. Maintenance respiration is the function of the nitrogen content of living material, while growth respiration is calculated proportionally to the carbon allocated to the different plant compartments. The phenological block calculates the foliage development; therefore, it affects the accumulation of carbon and nitrogen in leaf, stem, root, and litter. The soil block describes the decomposition of dead plant material (litter) (Running and Gower 1991).

In Biome-BGC, the main parts of the ecosystem are defined as plant, soil, and litter. Since Biome-BGC simulates water, carbon, and nitrogen cycles of C3 and C4 plants, the following main pools are defined: leaf (carbon, nitrogen, and water), fine root (carbon, nitrogen), soil (carbon, nitrogen, and water), and litter (carbon, nitrogen). Carbon and nitrogen pools have subpools, i.e. actual pools, storage pools, and transfer pools. Actual subpools contain the amount of carbon or nitrogen available on the actual simulation day. Storage subpools store the amount that will appear next year (like a core or bud), while the transfer subpools store the whole content of the storage pool after the end of the actual transfer period until the next one, and what will be transferred gradually into the leaf carbon pool (like a germ) in the next transfer period.

The model simulation has two phases. The first is the spin-up simulation, which starts with very low initial level of soil carbon and nitrogen and runs until a steady state is reached with the climate in order to estimate the initial values of the state variables (Thornton et al. 2000). For the spin-up phase, the 1901–2000 period was used for which the basic meteorological data were available from the CRU TS 1.2 database (Climatic Research Unit, University of East Anglia; New et al. 2002). This database contains monthly average temperature and diurnal temperature range, as well as monthly precipitation amount data. For the preparation of the daily meteorological data (daily maximum and minimum temperature, daytime average temperature, daily precipitation amount, daytime VPD, daytime average global radiation (GR), daylength) required by the model, a statistical weather generator (C2W) was used (Bürger 1997). The second phase, the normal simulation, uses the results of the spin-up simulation as initial values for the carbon and nitrogen pools. This simulation is performed for a given, predetermined time period. In this phase, mostly in situ measurements are used (VPD, maximum and minimum temperature, precipitation, GR). The missing daily meteorological data, not available from in situ measurements, are estimated by the MTCLIM model (Thornton et al. 2000).

9.2.2 Model Calibration

In order to set up model simulations, direct measurement of meteorological and soil-specific data are needed. The input parameters of the process-based models determine the model output. There are several input parameters for which values are difficult to obtain directly from experiments. The exact values of these parameters are not known; therefore, we refer to them as *unknown model parameters* (Kennedy and O'Hagan 2001; Van Oijen et al. 2005).

Owing to the uncertainty associated with the *unknown model parameters*, we can experience significant bias in the model results. In order to improve model performance, the *unknown model parameters* have to be estimated, or at least their uncertainty has to be constrained. The *unknown model parameters* can be estimated using inverse techniques based on measurement data (e.g. Wang et al. 2009). This process is called model calibration.

In Biome-BGC, the total number of ecophysiological parameters is 25 for grasses and arable lands. However, annual leaf and fine root turnover is set to 1 (because they are deciduous biomes, which means that the entire leaf and carbon pools are turned over every year), all-sided to projected leaf area index (LAI) ratio is set to 2 (because leaves are flat in these ecosystems), and shaded to sunlit SLA ratio are set to 2 (because in case of grasses and arable lands, it can be assumed that SLA in low irradiance is approximately twice the SLA in high irradiance, while Rubisco content is constant – see White et al. 2000). As burning of cropland residues or grasslands is forbidden within the European Union, we also set fire mortality to 0. Consequently, we still have 21 ecophysiological parameters that should be adjusted to the local conditions. This means that the model has a high degree of freedom, which is a disadvantage from the point of view of model calibration. Additionally, only few data are available for two site-specific parameters (effective soil depth and nitrogen fixation) from national and international literature (White et al. 2000; Horváth et al. 2005). Therefore, we also consider them as *unknown model parameters*. As a result, we have 23 *unknown model parameters* regarding the Hungarian sites where two of them are site-specific parameters, while the others are ecophysiological parameters. These 23 parameters (or a part of them, depending on the implementation) are estimated through model-data fusion. The *unknown model parameters* are listed in Table 9.1 together with their abbreviations, definitions, and units.

A state-of-the-art biogeochemical model calibration (model-data fusion) is frequently performed in Monte Carlo (MC) framework (Mosegaard and Tarantola 1995; Hollinger and Richardson 2005; Wang et al. 2009). In the MC method, the *unknown model parameters* are varied using pseudorandom numbers, and the model

Table 9.1 Abbreviation, definition, and unit of the unknown model parameters

Abbrev.	Model parameter	Unit
YSG	Yearday of start new growth	Day of year
YEL	Yearday of end litterfall period	Day of year
GP	Transfer growth period as fraction of grow. season	dim.less
LP	Litterfall as fraction of grow. season	dim.less
WPM	Annual whole-plant mortality fraction	dim.less.
FRC:LC	New fine root C: new leaf C	dim.less
CGP	Current growth proportion	dim.less
C:N(lv)	C:N of leaves	$kgC\ kgN^{-1}$
C:N(lit)	N of leaf litter, after retranslocation	$kgC\ kgN^{-1}$
C:N(r)	C:N of fine roots	$kgC\ kgN^{-1}$
CWIC	Canopy water interception coefficient	$LAI^{-1}\ day^{-1}$
CLEC	Canopy light extinction coefficient	dim.less
SLA	Canopy average specific leaf area (SLA)	$m^2\ kgC^{-1}$
PLNR	Fraction of leaf N in Rubisco	dim.less
MSC	Maximum stomatal conductance	$m\ s^{-1}$
CC	Cuticular conductance	$m\ s^{-1}$
BLC	Boundary layer conductance	$m\ s^{-1}$
WCRS	Soil water content (SWC) ratio: start of conduct. reduction	dim.less
WCSC	SWC ratio: complete conduct. reduction	dim.less
VDPS	Vapor pressure deficit (VPD): start of conduct. reduction	Pa
VDPE	VPD: complete conduct. reduction	Pa
SD	Soil depth	m
Nfix	Nitrogen fixation	$kgN\ m^2\ year^{-1}$

error is evaluated based on the measurement (reference) data using some kind of metrics. However, due to the high degree of freedom of the model, a huge number of model runs are needed to sample the parameter space accurately. For example, assuming that each (in our case 23) parameter is sampled ten times a total of 10^{23} model runs are necessary for MC analysis (global search), which is not realistic. Experience shows that the number of *unknown model parameters* that can be constrained during the inversion is generally low, which cause problems for models with large degree of freedom (Wu et al. 2009).

Sensitivity analysis is necessary to decrease the number of model parameters to be constrained (i.e. to decrease the degree of freedom) or to determine the *unknown model parameters*, which have the strongest effect on the output of the model. A possible solution for a sensitivity analysis is the use of the least-square linearization (LSL, Verbeeck et al. 2006), which divides output uncertainty into its sources using the results of the MC analysis. A multiple linear regression analysis is performed, which attempts to model the relationship between the variability of the model output and the value of the parameters. Every value of the independent variable (parameters values) is associated with a value of the dependent variable (likelihood value calculated from output data). As the result of the LSL method, the total variance of the model output and the *sensitivity coefficient* of each parameter can be determined. *Sensitivity coefficient* shows the percent of the total variance for which the given parameter is responsible. It can be approximated using regression coefficients and the variations of the parameter uncertainties (Verbeeck et al. 2006).

Once the sensitivity analysis has been performed, the model calibration can take place. Calibration can be accomplished by means of a simple Maximum Likelihood (ML) method or a more sophisticated Bayesian procedure. ML method simply seeks the best parameter set, which is associated with the best fit to the measurement results (Hollinger and Richardson 2005).

Bayesian-approach is another global optimization method to estimate the optimal values of the *unknown model parameters*. Bayesian procedure begins with quantifying the variability of the *unknown model parameters* in the form of a priori probability distribution. We assume that the *unknown model parameters* have uniform distribution between their possible minimum and maximum values (Mo and Beven 2004) that are generally estimated from the literature (White et al. 2000). Then we compare the measured data with the simulated ones calculating a likelihood function. Likelihood is the degree of goodness-of-fit between simulated and measured data (Mosegaard and Tarantola 1995; Van Oijen et al. 2005; Verbeeck et al. 2006).

The next step is the construction of a random walk in the parameter space based on the Metropolis-algorithm (Metropolis et al. 1953). The optimization process is a quasirandom walk through the parameter space, which attempts to find a parameter set that minimizes the model-data error defined in terms of likelihood function. Each *unknown model parameter* is varied *randomly* within its possible minimum and maximum value, and the model is run several times using different values for the *model parameters*.

Finally, we update the a priori distribution with model information (distribution of the likelihood function), which means the convolution of the a priori distribution with the likelihood function provided by the large number of model simulations. If the calibration procedure is successful, the uncertainties of *unknown model parameters* (confidence intervals) are decreased and the correlation between the model output data and the measured data is increased (see e.g. Trusilova et al. 2009).

Note that for arable lands and grasslands, the execution of the calibration procedure and the likelihood metrics are different, because in case of the arable lands only a simpler version of the Bayesian method is used, i.e. only ML estimation is performed. It means that we do not pay attention to the distribution of the calibrated parameters.

For grasslands, a complete Bayesian model calibration is performed. We implemented a novel, n × m-step, multi-objective calibration method[1] (n is the number of *unknown model parameters*, which determines the number of iteration sections; and m is number of iteration steps in an iteration section) with a special likelihood function. See Chapters 10 and 12 for details.

---

[1]Multi-objective optimization is the process of simultaneously optimizing two or more conflicting objectives, which means that more reference data are considered during the calibration procedure.

9.2.3 Model Developments

9.2.3.1 Management Modules

Originally, Biome-BGC was developed to simulate the carbon, water, and nitrogen cycles of unmanaged ecosystems; however, the agricultural fields and most of the forests and grasslands are managed in Central Europe. To simulate the effect of grassland management with Biome-BGC, we integrated a few new modules into the source code. We implemented the simulations of mowing, grazing, harvest, plowing, sowing, and fertilizing. To simulate the effect of management activities on the carbon, nitrogen, and water pools, we defined new fluxes between the pools and between the pools and the environment. From the management modules listed above, we describe mowing and grazing in details because the effect of these managements can be validated with local measurement data in Hungary.

The main effect of grazing and mowing is the removal of leaves. In both cases, the content of the actual carbon and nitrogen pools decrease (Herrmann et al. 2005). The rate of the plant material reduction is calculated from the reduction of LAI. LAI is the product of leaf carbon content (kg C m^{-2}) and SLA. SLA is one of the model's ecophysiological constants. It is a measure of the leaf thickness, the dry weight of the leaf of unit area. The decrease in the leaf carbon content can be calculated from the decrease in LAI. The decreases in the other pools (leaf nitrogen, canopy water) are proportional to the decrease of leaf carbon pool.

The rate of the plant material reduction depends on the management type: in case of mowing, calculations are based on the decrease of LAI. After mowing, the cut fraction of the aboveground biomass can be taken away from the field or can be left on site. In the first case, the removed plant material is excluded from the further calculations (the cut fraction is net loss to the system); in the second case, it goes partly to the litter pool (the ratio can be set in the initialization file of the model).

In case of grazing, a given amount of leaf material is consumed by animals every day when grazing occurs. An important parameter used to simulate the effect of grazing is the livestock unit (LSU). LSU is a unit used to compare or aggregate different animal species and 1 LSU is equivalent to 500 kg live weight (1 adult cattle = 1 LSU). In case of grazing, the following parameters determine the decrease of the plant material: the animal stocking rate regarding a unit area (LSUh; LSU ha^{-1}), the daily ingested dry matter regarding a unit LSU (IDM_{daily}; kg dry matter LSU^{-1}), the dry matter content of grass (DMCG; %), and the carbon content of dry matter (CCG; %).

Besides the defoliation effect of grazing (i.e. intake by animals), it is also important that a fixed proportion of the aboveground biomass flows to the litter compartment, as a result of trampling and excretal returns, increasing the content of the dead plant material's pool (litterC and litterN). The carbon and nitrogen contents of the returning plant material (CCM and NCM; %) from excretal to the litter pool are the function of IDM_{daily} and the proportion of the biomass returned to the litter pool for a unit LSU (PBRL; %):

$$\text{litterC+} = ((\text{IDMdaily} \cdot \text{LSUh}) \cdot \text{PBRL}) \cdot \text{CCM} \tag{9.1}$$

$$\text{litterN+} = ((\text{IDMdaily} \cdot \text{LSUh}) \cdot \text{PBRL}) \cdot \text{NCM} \tag{9.2}$$

The parameters of the management modules can be set by the user.

9.2.3.2 Improvement of Stomatal Conductance Calculation

Stomatal conductance (g_s) determines the rate of passage of either water vapor or carbon dioxide through the stomata. Stomatal conductance strongly depends on soil water status, which can be characterized by plant available water[2] (Tardieu and Simonneau 1998). As a general rule, plant available water is considered to be 50% of the water-holding capacity of the soil, which refers to the amount of water held between water field capacity[3] and wilting point[4] (Tardieu and Simonneau 1998). The water-holding capacity of the soil is the highest in silt soil and the lowest in heavy clay soil. The value of g_s has a strong effect on the main carbon and water fluxes: GPP, total ecosystem respiration (R_{eco}), and latent heat flux (LE). Biome-BGC calculates the stomatal conductance as the product of the maximum stomatal conductance (ecophysiological parameter; constant during the simulation) and limiting stress functions (based on minimum temperature, VPD, soil water potential).

Instead of the soil water potential (ψ), the volumetric soil water content (SWC) is widely used to calculate the limitation of stomatal conductance (Franks et al. 1999). A practical advantage of SWC as a limitation factor is that it is easy to measure in the field. The disadvantage is that SWC is not comparable among different soil types. A possible solution to calculate a limiting factor from SWC is to take the proportion of field capacity into account, because field capacity is depending on soil texture (Reichstein 2001):

$$\text{SWC ratio} = \frac{\text{SWC}_{\text{actual}}}{\text{SWC}_{\text{field capacity}}} \tag{9.3}$$

Using the original version of Biome-BGC, we experienced unrealistic peaks in stomatal conductance, and consequently, in the carbon and water fluxes. The reason for this phenomenon is that there are peaks in the limiting factors of stomatal conductance based on soil water potential. Soil water potential is the polynomial function of SWC: $\psi = f[(\text{SWC})^{-b}]$ (where b is the function of soil texture). The decrease of SWC causes small increase in soil water potential at high SWC. This means that

---

[2]Plant availabe water is the portion of the water field capacity that can be absorbed by plant.

[3]Field capacity is the soil water content after the soil has been saturated (all pores filled with water) and allowed to drain freely for about 24–28 h.

[4]Wilting point is the soil water content when plants have extracted all the water they can.

close to the saturation (high SWC values), soil water potential is not sensitive to the decrease in the SWC; therefore, stomatal conductance is not decreasing significantly despite the decrease in the SWC. At lower SWC values, a small decrease in SWC can cause huge increase in soil water potential. This can result in unrealistic peaks in the limitation factor based on soil water potential and – through these unrealistic peaks – in stomatal conductance. To avoid the occurrence of this phenomenon, we have modified the source code of the model to use SWC ratio as the limitation factor instead of soil water potential.

As a result of this modification, the list of ecophysiological parameters has changed: instead of the original model parameters called *leaf water potential: at the start of conductance reduction* (LWPS) and *leaf water potential: at the end of conductance reduction* (LWPE) we use *SWC ratio: at start of conductance reduction* (WCRS) and *SWC ratio: at end of the conductance reduction* (WCRE) parameters.

9.2.3.3 Improvement of Model Phenology

The phenological state of vegetation significantly affects the exchanges of carbon dioxide and latent heat between the ecosystem and the atmosphere. To determine the start of the growing season (SGS), the phenological state simulated by the model can be used. We experienced that the SGS calculated by the original model is unrealistically late in case of herbaceous ecosystems of Hungary; therefore, we have developed a new phenology module. We use the growing season index (GSI), which combines a set of variables into a variable for the estimation of the length of the growing season (Jolly et al. 2005). Minimum (suboptimal) temperature (TMIN), VPD, daylength (photoperiod; PHTP), and 10-day heatsum with 5°C basic temperature ($HTSM_{10}$) are combined. The first three are daily values, while the last one, heatsum, is calculated in 10-day long periods. For each variable, we set threshold limits (LIMIT1, LIMIT2), within which the relative phenological performance of the vegetation was assumed to vary from inactive (0) to unconstrained (1). The values of the limits regarding the different variables can be set by the user. We start to calculate the daily partial GSI index (GSI_i, i: TMIN, VPD, PHTP) on the 1st day of the year:

$$\mathrm{GSI_i = 0,\ if\ i \leq LIMIT1_i} \tag{9.4}$$

$$\mathrm{GSI_i = \frac{i - LIMIT1_i}{LIMIT2_i - LIMIT1_i},\quad if\ LIMIT1_i < i < LIMIT2_i} \tag{9.5}$$

$$\mathrm{GSI_i = 1,\ if\ i \geq LIMIT2_i} \tag{9.6}$$

On the 10th day of the year, we calculate the 10-day heat sum from the difference between the basic temperature and the maximum temperatures of the last 10 days.

$$HTSM_{10} = \sum_{day=1}^{10} \left(T_{max} - T_{basic}\right)_{10-day};\ \text{if}\ \left(T_{max} - T_{basic}\right)_{10-day} \geq 0 \qquad (9.7)$$

From the 10th day of the year, GSI_{TOTAL} is calculated from the 10-day average of the multiplication of daily partial GSI indexes and from the index of heat sum:

$$GSI_{TOTAL} = GSI_{HTSM_{10}} \cdot \left(\frac{1}{10} \sum_{day=1}^{10} (GSI_{TMIN} GSI_{VPD} GSI_{PHTP})_{10-day} \right) \qquad (9.8)$$

If on a given day, GSI_{TOTAL} is greater than a limit ($LIMIT_{SGS}$), we assume that the SGS is found. After finding the SGS, its last day is searched, and if on a given day (after SGS) GSI_{TOTAL} is less than a limit set by the user ($LIMIT_{EGS}$), we assume that the end of the growing season is reached.

9.3 The DNDC Process-Oriented Ecosystem Model

Balázs Grosz, Attila Machon, László Horváth

9.3.1 Description of the DNDC Model

The accurate knowledge of trace gas fluxes is important for both the ecosystem and the atmospheric environment. There are two ways to estimate these fluxes: direct measurement and modeling. Measurements are expensive, require sophisticated equipment, and there are limitations for extrapolating or applying these data for other regions. By means of mathematical modeling, we can also estimate the fluxes for larger regions; however, the models may need a huge amount of input data, which can only be obtained by measurements, and their calibration and/or validation also require direct measurements. In this section, we introduce the DNDC model that was applied for the estimation of soil fluxes of GHGs like CH_4 and N_2O. For this simulation, we employed the most recent model version: DNDC86K. There are two ways how this model can be applied. One can use it for modeling small areas in the so-called "spot mode" and one can simulate gas exchange of larger areas in "regional mode." The main difference between these two modes is that DNDC in regional mode requires fewer input parameters, since in "regional mode," certain parameters are estimated and derived using internal submodels requiring less direct measurements than the "spot mode." We used the model in regional mode except for the sensitivity analysis, where the model was operated in spot mode.

The model consists of two main parts. The first one includes the ecological drivers like crop type, anthropogenic activity, meteorological and soil conditions, and it estimates and/or quantifies the soil environmental variables like pH, redox potential, moisture, etc. The second model component contains the nitrification, denitrification, and fermentation algorithms, and computes certain soil properties and estimates the greenhouse gas fluxes. In addition to the common laws of physics, chemistry, and biology, empirical assumptions based on laboratory experiments are also applied. The model also offers the possibility for parameterization.

In the DNDC model, four major soil organic carbon pools are separated: the herbal residue, the microbial biomass, the humid (i.e. active humus), and the passive humus pool. All of these pools contain subpools with different decomposition rates. The daily decomposition rate is controlled by soil clay content, available nitrogen, soil temperature, and moisture. When one of the carbon subpools is composted, another part of the decomposed organic carbon is emitted in the form of CO_2, while the remainder is transferred into another carbon subpool. The dissolved organic carbon (DOC) is an intermediate, and the soil microbes consume it immediately.

During the decay of soil organic carbon, nitrogen compounds of the soil are also transferred into other repositories. Ammonium is produced in part by mineralization that takes place in the nitrification processes. The free ammonium keeps balance with the absorbed ammonium via clay and solved ammonia. The amount of evaporated ammonia depends on the ammonia content of soil water, soil temperature, and pH. During rainfall, nitrate ions are leached into deeper soil layers by soil water currents.

The model contains an "anaerobe balloon." This part of the model simulates the soil aeration status, and calculates the diffusion and reduction of oxygen in the soil composition. Based on the computed redox potential, the model divides the soil layers into aerobe and anaerobe sections, where nitrification and denitrification occur, respectively. The algorithms and equations are presented in Li (2000).

9.3.2 Sensitivity Analysis

In order to understand the sensitivity of the model to the meteorological conditions, several model runs were performed by varying the daily minimum and maximum temperatures and the amount of precipitation. The variations of these parameters yield minimal variations in the total soil nitrogen fluxes; however, they significantly affect the emission ratios of various nitrogen compounds (NH_3, NO, N_2O, and NO_2) (Table 9.2). In general, there is a linear relationship between these parameters and the meteorological conditions, but in some cases, we get exponential relationship what is expected because in chemical kinetics, reaction rates exponentially depend on ambient temperature. Overall, the model is sensitive to the changes in the meteorological factors. Next, we have also altered the soil properties. Parallel simulations have been performed by varying the initial parameters (e.g. pH, clay, organic

Table 9.2 Summary of DNDC model sensitivity analysis

	Flux (C or N g ha^{-1} year^{-1}), Bugac					
	N_2O	NO	N_2	NH_3	CO_2	CH_4
Original (0.75 cow/ha)	775	958	104	689	1,638	0
+1°C	440	1,048	50	736	1,772	0
−1°C	1,300	874	1,315	630	1,781	0
+2°C	80	1,130	13	791	1,924	0
−2°C	1,352	792	1,068	585	1,393	0
125% rainfall	780	1,096	102	730	1,914	0
75% rainfall	767	836	103	27	1,375	0
+10% clay	683	979	83	693	1,678	0
−10% clay	891	904	128	684	1,604	0
+ 0.5 pH	513	836	59	883	1,661	0
−0.5 pH	955	1,039	135	583	1,624	0
Microbiological index = 0.5	272	672	19	614	1,168	0
Non grazed	776	904	104	166	2,135	0
Grazed (0.5 cow/ha)	775	941	104	522	1,801	0

carbon, ammonia, or nitrate content). We concluded that the model is not sensitive to the amount of the initial biomass. The model does not give significantly different results for modifications of the initial, dry or wet deposited nitrate and ammonium, but it is sensitive to the microbial processes and variations in the surface carbon decomposition, as well as to clay content and pH (Table 9.2).

Finally, we examined the effects of the variations in the land management. The DNDC model is capable of handling various land use practices, e.g. cutting, irrigation, (organic or inorganic) fertilization, and the grazing by cattle, horse, sheep, etc.

For the sensitivity analysis, we used the measurements carried out at Bugac (46.69°N, 19.60°E, 111 m asl), which is one of the NitroEurope Supersites (for detailed site description, see Chapter 6). Land exploitation is grass pasture, grazed by Hungarian gray cattle. We set the respective input parameters of DNDC and altered the grazing time and the number of cattle per hectare. These variations yield significant influence on the gas emissions (Table 9.2). The ammonia emission depends on the number of cattle but N_2O or N_2 emissions are not affected. The model is also sensitive to the vegetation properties: for instance, grass classification determines the yearly amount of dry matter. Additionally, plants also affect other processes like decomposition of organic compounds, and affect the C/N ratio or the microbial activities. The sensitivity analysis has shown the most critical input parameters, which have to be measured accurately.

Table 9.2 shows that the model gives zero emission for methane in all cases. This is partly due to the internal rounding of the DNDC model: the actual meteorological and soil parameters at the experimental site yield small methane emission. The sensitivity analysis has shown that not even extreme modifications in the environmental parameters (±2°C, ±100 mm of rainfall) increase the methane emission significantly.

9.3.3 Input Dataset in Regional Mode

Like all ecological models, DNDC uses mathematical algorithms for the simulation of natural processes. If we simulate processes on larger areas, the results may be less precise. Uncertainties are derived from the limited spatial representativeness of the point measurements used, from the error of the spatial averaging of the measurements over the area considered, from the nonlinearity of certain processes, and from many other reasons influencing the input data of the model. The most affected properties are the land exploitation, the soil characteristics, and the meteorological data. The developers of the model are aware of these problems, thus much less data are needed for regional simulations than for the spot ones. Although data that are important for the spot mode are also important in regional mode, in the latter case, these are calculated by built-in algorithms from the limited dataset requested by the regional mode. Certainly, the reliability of the model results is much less in the case of regional simulations. The uncertainty of the input data and the ambiguity of the model parameters cause this unreliability. In order to minimize this type of errors in the simulations, we tried to choose a feasible grid size. A large grid size would increase the errors deriving from the spatial discretization, while a smaller one increases the data and computer resource demand.

The optimal grid size is found to be $1/6° \times 1/6°$, equivalent to an area of approximately 200 km^2. This grid size provides sufficient resolutions in practice but does not require an overwhelming computation time. All in all, 466 grid cells can cover the area of Hungary. The model contains the 50 most important plant types, defined by their physiological parameters, such as hydrological and nutrient uptake, C/N ratio, growing index, and so on. DNDC offers the reparameterization of plant information. The model also allows us to adjust the spring/autumn seed-time, the crop rotation, and the irrigation, grazing, or fertilization parameters. Meteorological data have been provided by the Hungarian Meteorological Service based on approximately 300 measuring sites. The required soil and land exploitation data have been provided by the Research Institute for Soil Science and Agricultural Chemistry of the Hungarian Academy of Sciences and the CLC-50 database (Büttner et al. 2000). Applying the regional simulations of DNDC, we could estimate the biospheric non-CO_2 greenhouse emissions (methane and nitrous oxide) over Hungary for the first time.

9.4 MOD17 Remote Sensing–Based Model

Györgyi Gelybó, Zoltán Barcza

9.4.1 Model Description

The MODerate resolution Imaging Spectroradiometer (MODIS) sensors onboard NASA EOS Aqua (launched in 2002, afternoon Equator crossing) and Terra (operating since 2000, morning Equator crossing) satellites perform measurements in 36 spectral bands. The sensors provide global coverage in every 1–2 days. There are several preprocessed products available for the scientific community based on measurements of the MODIS sensor. The MOD17 product provides estimations of GPP with 8-day, and NPP with annual temporal resolution. The spatial resolution of the model is 1 km for predefined geographical areas.

The GPP model (referred to as MOD17 model) of the MOD17 product is a light-use efficiency-based model (Monteith 1972, 1977). In this simple model, the theoretical maximum of light-use efficiency (ε_{max}) is decreased by abiotic stress factors according to environmental circumstances. GPP is determined by the actual value of this light-use efficiency and the amount of available photosynthetically active radiation (APAR):

$$GPP = \varepsilon * APAR, \tag{9.9}$$

where APAR is the fraction of incident photosynthetically active radiation (PAR) absorbed by the vegetation. APAR can be calculated as a product of incident PAR and FPAR, which shows the amount of radiation absorbed in percentages. Using the well-known assumption (e.g. Tsubo and Walker 2005) that PAR is a fixed portion of GR, we can estimate GPP with the following expression:

$$GPP = \varepsilon * GR * 0.45 * FPAR \tag{9.10}$$

The model can distinguish between 11 plant categories (Plant Functional Types, PFTs), i.e. evergreen needleleaf forest, evergreen broadleaf forest, deciduous needleleaf forest, deciduous broadleaf forest, mixed forest, closed shrubland, open shrubland, woody savanna, savanna, grassland, cropland. PFT specific model parameters, i.e. values of ε_{max} and parameters describing environmental (meteorological) stresses are stored in the biome properties look-up table (BPLUT) for each PFT. These parameters do not vary with geographical location, as they characterize biomes globally.

The model accounts for minimum temperature and moisture stresses; the latter is given by VPD. ε_{max} response to unfavorable circumstances is described by simple ramp functions, with values between 0 and 1. Temperature and VPD values when ε reaches its maximal value (ε_{max}), or decreases to zero are biome-specific parameters and stored in the BPLUT.

Net photosynthesis (GPP minus maintenance respiration of fine roots and leaves) calculations in the MOD17 model are based on exponential function of daily average temperature and the amount of biomass. The leaf biomass is represented by Leaf Area Index (LAI) and specific LAI, the latter is a BPLUT parameter. Further parts of autotrophic respiration that are necessary to determine Net Primary Production (NPP), such as growth respiration and annual maintenance respiration, cannot be calculated on daily basis with the algorithm logic used in the model (for details, see Running et al. 1999); therefore, NPP is given in annual data files. Annual maintenance respiration of live woody tissue, annual growth respiration of leaves, live wood, dead woody tissues, and fine roots are calculated using look-up table parameters and the annual maximum of leaf mass calculated previously on daily time step.

9.4.2 Input Data

The algorithm of the MOD17 model requires input databases from other MODIS products in case of FPAR and land cover information. Input data regarding spatial distribution of PFTs are provided by the MOD12 Land Cover product given for each year separately (Strahler et al. 1999). FPAR data for the model are available from the MOD15 LAI & FPAR product with 8-day temporal resolution.

Meteorological fields for the algorithm are provided by GMAO (NASA Goddard Global Modeling and Assimilation Office, former Data Assimilation Office (DAO)) as global reanalysis on 1° latitude × 1.25° longitude grid. Data requirements for GPP calculations include daily GR, VPD, and minimum temperature data. GPP calculations are performed on a daily basis, and then are composed to 8-day product.

LAI data for the NPP algorithm is also provided by the MOD15 product for calculations (Knyazikhin et al. 1999).

9.4.3 Adaptation to Hungary

The official MOD17 data product used for validation and country-scale calculations is the version 5.1 provided by Numerical Terradynamic Simulation Group (NTSG), University of Montana. Due to inconsistencies in the DAO/GMAO meteorological dataset, data are available only for the period from 2000 to 2006. Besides using the preprocessed GPP data, we also performed calculations using the algorithm described in Zhao et al. (2005).

Validation of the MOD17 product is carried out using tall tower eddy covariance measurements near Hegyhátsál (46°57′N, 16°39′E, 248 m asl – for location see Fig. 1.1). As the MOD17 model is sensitive to inaccuracies in meteorological fields (Zhao et al. 2006), we performed model calculations using both GMAO and local meteorological measurements carried out at the tower. The quality control and gap filling of FPAR product, as well as the interpolation of GMAO meteorology to the MODIS grid are based on Zhao et al. (2005).

9.5 CASMOFOR Model

Zoltán Somogyi

The methods and models for the description of the forest carbon cycle depend on the objective of the modeling. In Hungary, a specific model called CASMOFOR was developed to enable one to analyze appropriate afforestation scenarios to sequester carbon. The need to implement afforestation projects to sequester carbon has been growing in order that human-induced greenhouse gas emissions are offset to the extent possible. Afforestation projects may be of climate policy relevance, and when considering afforestation options, decision makers need scientifically sound information, as well as tools and capabilities to develop different options and comparing them quickly and in an easy manner.

In Hungary, afforestations and reforestations have 8 decades of history. Since 1930, about 800,000 ha have been afforested. Further large potentials for afforestations and reforestations were predicted for Hungary (Somogyi 2000). In order to estimate the amount of carbon that were sequestered in afforestations of the past decades, to estimate the amount of carbon that could be potentially sequestered in future afforestations, to help planning further afforestations to sequester carbon, and to better understand the carbon cycle of the Hungarian forests, a country-specific forest carbon cycle model was developed.

When designing this model, it was considered that several forest carbon models had been developed (e.g. CO2FIX: Nabuurs et al. 2001; GORCAM: Graz/Oak Ridge Carbon Accounting Model – http://www.joanneum.ac.at/GORCAM.htm) that could be used for similar purposes. However, in order that these models can be used in any specific situation, they must be parameterized for the local conditions, which may require a specific data management framework and/or specific data types. When local data is available, especially with respect to the ecology and the management of the local forests, country-specific models could sometimes be better used because such models could be designed to accommodate the specific framework and types of available data. In Hungary, it seemed paramount to develop such a country-specific model because of the rich country-specific information base available. It is expected that this specific model can better be used to arrive at accurate carbon sequestration estimates for the Hungarian conditions, and to assist decision makers than any other model currently available internationally.

This country-specific model is called CASMOFOR, which stands for CArbon Sequestration MOdel for FORestations. The model can be accessed at http://www.scientia.hu/casmofor.

As noted above, CASMOFOR is specific in the sense that it is fully parameterized for Hungary; however, it is also a flexible modeling framework that could be used in any country where appropriate data is available to model basic forest processes like tree growth and decomposition, as well as to model forest management. CASMOFOR was developed in an MS Excel based system of spreadsheets with MS Visual Basic macros, and has a user-friendly interface to facilitate its calibration to standard forestry situations. Thus, the model could easily be adapted to other countries' needs.

CASMOFOR models the whole carbon cycle of forests under management by considering all major processes that affect the size of the carbon pools (Fig. 9.1). CASMOFOR is an accounting type model. The accounting functions correspond to internationally agreed methods, i.e. those in the various IPCC (2003, 2006) guidelines, and other relevant developments in the methodology of biomass-related forest carbon estimation (Somogyi et al. 2007). Carbon stock changes of all pools are calculated in 1-year-steps considering all major processes that increase or reduce the carbon stocks of the pools. The processes are mainly nonlinear ones.

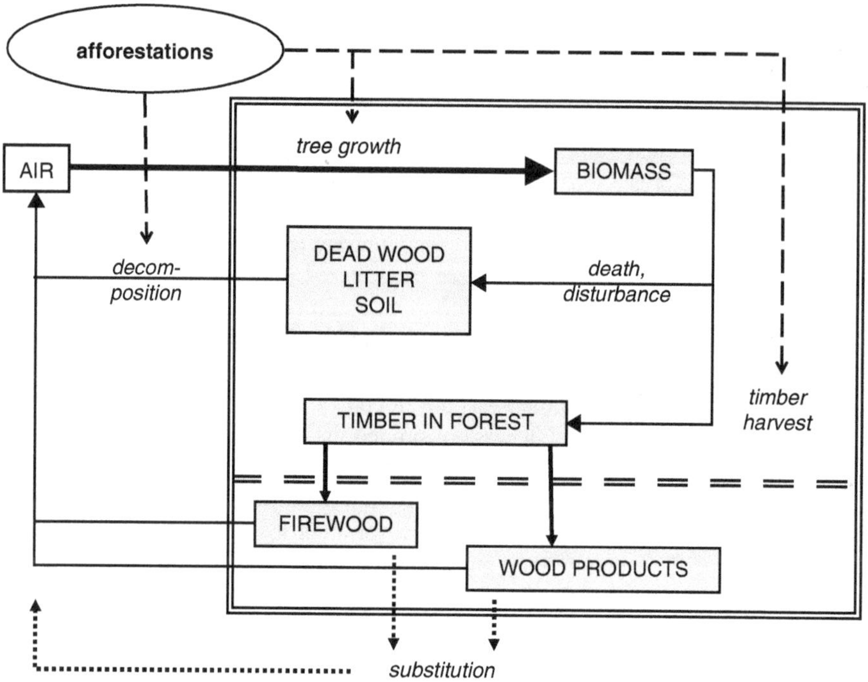

Fig. 9.1 Simplified systems diagram of CASMOFOR. Pools are indicated by *gray boxes*, whereas processes are shown by *full arrows*. External management factors, which are supposed to exert the largest direct human-induced effects, are indicated by the *ellipse* and the associated arrows. Note also that the main system boundary is defined by the *double line*, but because all outputs of the model include data on all pools and modeled processes, various system boundaries can be defined by the user. Such a user-defined boundary is shown by the *dashed double line*. Finally, the system could be even extended to include fossil fuels that could be substituted by wood of various life cycles

To fully describe CASMOFOR, which is an original Hungarian development, all accounting functions are included in the Appendix of this chapter.

The main structural elements of the model are the following:

- Aboveground tree volume growth is modeled using yield tables. Country-specific yield tables for all major tree species have been incorporated; thus, no parameterization is needed for the Hungarian conditions. These tables, however, can be replaced by any yield tables of similar structure. Note also that although tree growth may have changed in Hungary (Somogyi 2008), no adjustment has been made for that in the model due to lack of large-scale assessment of the changed rate of growth.
- Harvests and other disturbances are modeled by the updated versions of the so-called silvicultural models developed in the 1980s for the country. These models, which were developed by standard yield classes just like the yield tables, include both timing and intensity of thinnings. The original silvicultural models, which were developed and published in the 1980s, were revised to better reflect changes in forest management that took place since then.

- Leaf and belowground biomass (i.e., root) dynamics are modeled using constant factors from the literature, and factors developed based on IPCC (2003, 2006) default values and expert judgment.
- Deadwood, litter, dead roots, and harvested wood dynamics are modeled based on the aboveground production estimates of the yield tables using constant factors from the literature and using expert judgment, based on generalized species and site class specific mortality probabilities, and based on harvest regimes.
- Soil organic carbon dynamics are modeled supposing that a (constant) fraction of the deadwood, litter, and dead roots pools increases soil organic carbon, and supposing that emissions from soil occur. These emissions are estimated using generalized response curves from local case studies. They include studies to estimate emissions if site preparation occurs during afforestation, and if the afforestation is made on former grasslands (see Chapter 15).

All data needed to develop estimates for the Hungarian conditions are included in the model; thus, no further parameterization is needed, and the user can focus on developing and analyzing scenarios. In case country-specific data are not available, IPCC (2003, 2006) default values are used.

All parameters included in the carbon cycle model can be accessed through the help system of the model, and, together with references, can be found at http://www.scientia.hu/casmofor. All parameter values can be replaced; thus, in case compatible data sets are available, the model could be used outside Hungary. CASMOFOR has a full Monte Carlo analysis module to assess sensitivity of various parameters, and a detailed description of uncertainties.

Although not directly related to the description of the carbon cycle issue, the economic aspects of an afforestation program are highly relevant for decision makers. Therefore, an economic module was also added to CASMOFOR to analyze the most important revenues and costs associated with any afforestation scenarios. This module is based on a database of the mean costs and revenues of conducting the afforestation, and then managing and utilizing the established stands as any other stand in the country. Thus, the costs include those of soil preparation, planting or seeding, beating up, thinning and even final harvesting, and revenues include all revenues from marketing wood. As an option, revenues can include any income from marketing the amount of carbon dioxide sequestered by the forests under an emission trading scheme, e.g. that of the Kyoto Protocol. Figure 9.2 demonstrates an example of the extent and the timing of net costs/net revenues, which are always related to forest operations.

Finally, we note here that a more extensive description of the model, and how it is to be used can be found in the help system that is available at the website of the model. CASMOFOR is thus not only a research tool, it produces MS Excel graphs and data that can be copied to other applications and modified according to needs.

Acknowledgments Biome-BGC version 4.1.1 was provided by the NTSG at the University of Montana. NTSG assumes no responsibility for the proper use of Biome-BGC by others. We are very grateful to Maosheng Zhao (NTSG) for providing us the official MOD17 product, the version 5.1 model parameters, and the GMAO meteorological database. The DNDC simulations were

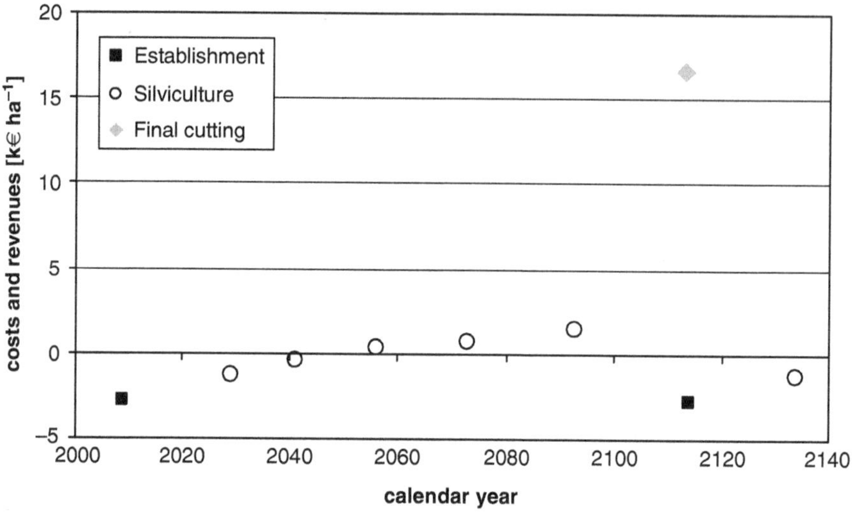

Fig. 9.2 Mean net costs and revenues of establishing and managing an oak stand at a price level of 2008

supported by NitroEurope Integrated Project of the European Commission's 6th R&D Framework Programme, and by GVOP AKF 3.1.1.2004–05 0358/3.0 research project. The development of CASMOFOR was supported by various Hungarian projects of the National Scientific Research Found, the Ministry of Environment and Water, the Ministry of Agriculture and Rural Development, as well as by the CarboInvent project (Multi-Source Inventory Methods For Quantifying Carbon Stocks And Stock Changes In European Forests; Contract number EVK2-CT-2002-00157), see www.joanneum.at/CarboInvent.

Appendix

I Equations and Variables Applied in the Accounting System of CASMOFOR to Model the Carbon Cycle of Forests and Forestry

Equations of CASMOFOR describing the carbon cycle are listed below by pool, and classified in various categories. Balance equations (BE) are ones that are used to calculate new values of the carbon pools at time t. FLUX IN refers to a process, i.e. a sink, by the yield of which the carbon content of the whole forestry system increases. Similarly, FLUX OUT refers to emissions by the amount of which the carbon content of the whole forestry system decreases. The emissions can come from carbon that was fixed by the FLUX IN processes, or from carbon that was already stored in the soil before the start of the simulation. Some carbon that flows through various pools eventually goes to the long-term SINK (in the soil), and

remains part of the forestry system for the entire period of the simulation. Equations that are not classified into any of the above categories are used to calculate variables in the above equations, and are noted by "other." By the forest and forestry system, all the pools and processes of the systems diagram in Fig. 9.1 are meant.
Variables related to carbon pools (in italics) and rates of fluxes are expressed as amount of carbon (tonnes of carbon – tC), and is calculated once for each calendar year of the simulation for each species and yield class.

In the equations below, index t refers to any year of the simulation, (t-1) to the previous year, whereas index (t – x) refers to year x year(s) before the current year. Index (i) means that the variable is a vector of values from year 1 to year x (the factor used in the formula varies with age), where x depends on the process.

To check that all carbon is accounted for, i.e. all incoming fluxes equal to all outgoing fluxes + changes of the carbon content of the system (including permanent sinks), an additional equation is tested to conserve mass:

$$\text{Total Flux IN}_t = \text{Total }(\text{POOLS}_t - \text{POOLS}_{t-1}) \\ -\text{Total Flux OUT}_t - \text{Total Flux to SINK}_t$$

Aboveground Biomass

Type of equation	Equations, variables and parameters (all pools and processes in terms of tC, and calculated once for year t)
BE	$AGWB_t = AGWB_{t-1} + CAIC - M - TH - FC$ *AGWB*: Aboveground woody biomass *CAIC*: current annual increment of aboveground woody biomass *M*: mortality *TH*: thinning *FC*: final cuttings
Flux IN	*CAIC* = CAI * d * cf *CAI:* current annual increment of tree volume; m^{-3} d: weight of oven-dry biomass/volume of fresh wood (at the stand level); tdm m^{-3} cf: carbon fraction of oven dry wood; tC tdm^{-1}
Other	$M = AGWB_{t-1} * (ddm + dim)$ ddm: density dependent mortality rate; species dependent; dimensionless dim: density independent mortality rate; randomly generated; dimensionless ddm + dim <= 0.4
Other	$TH = AGWB_{t-1}$ * thr thr: species specific, depends on age and yield class; dimensionless $FC = AGWB_{t-1}$ *of stands of rotation age* (final cutting is supposed to take place at the beginning of the year, and is immediately followed by regeneration) Rotation age: species specific, depends on yield class

Dead Wood

Type of equation	Equations, variables and parameter (all pools and processes in terms of tC ha^{-1})
BE	$DW = DW_{t-1} + DWI - EW$
	DW: dead wood
	DWI: deadwood increment
	EW: decomposition of (i.e., emission from) decomposable dead wood
Other	$DWI = [M + TH * (1 - \text{wpTH} - \text{fpTH}) + FC * (1 - \text{wpFC} - \text{fpFC})] * (1 - \text{ndf})$
	wpTH: wood product part of TH, dimensionless
	fpTH: fuelwood part of TH, dimensionless
	wpFC:wood product part of FC, dimentionless
	fpFC: fuelwood part of FC; dimensionless
	ndf: nondecomposable fraction; dimensionless
Flux OUT	$EW = Sum_i(DW_{t-1}/\text{DTw}_i)$ for i = 1 to x
	DTw: time needed for deadwood to decompose, years
Flux to SINK	$UWI = [M + TH * (1 - \text{wpTH} - \text{fpTH}) + FC * (1 - \text{wpFC} - \text{fpFC})] * \text{ndf}$
	UWI: Nondecomposable dead wood fraction

Living Leaves

Type of equation	Equations, variables, and parameters (all pools and processes in terms of tC ha^{-1})
BE	$LL = LL_{t-1} + LI - DLI - DLI * \text{ndf}/(1 - \text{ndf})$
	LL: amount of (living) leaves
	LI: increment due to tree growth
	DLI: decomposable leaf loss due to harvest and mortality at the end of year
Flux IN	$LI = CAIC$ * increment ratio
	increment ratio: leaf increment/aboveground woody biomass increment; dimensionless
other	$DLI = [(LL_{t-1} + LI) * \text{nll} + (LL_{t-1} + LI) * (1 - \text{nll}) * \text{fdlleaves}] * (1 - \text{ndf})$
	nll: ratio of nonliving/living biomass = $DWI/AGWB_{t-1}$; dimensionless
	fdlleaves: fraction of leaves dying and falling at the end of year; species specific (broadleaves: 1; conifers: <1); dimensionless
Flux to SINK	$ULI = DLI * \text{ndf}/(1 - \text{ndf})$
	ULI: Nondecomposable fraction; dimensionless

Dead Leaves

Type of equation	Equations, variables, and parameters (all pools and processes in terms of tC ha^{-1})
BE	$DL = DL_{t-1} + DLI - EL$ *DL*: dead decomposable leaves *DLI*: amount of decomposable leaves that die in year *EL*: decomposition of dead leaves
Flux OUT	$EL = Sum_i(DL_{t-i}/DTL_i)$ for i = 1 to x *EL:* Total emission due to decomposition of leaves DTL: time needed to decompose; year

Living Roots (Belowground Biomass)

Type of equation	Equations, variables and parameters (all pools and processes in terms of tC ha^{-1})
BE	$R = R_{t-1} + RI - DRI - DRI * \mathrm{ndf}/(1 - \mathrm{ndf})$ *R*: amount of roots *RI*: increase of root biomass *DRI*: amount of decomposable roots that die in year
Flux IN	$RI = CAIC * \mathrm{rts}$ rts*:* root-to-shoot ratio
Other	$DRI = [(R_{t-1} + RI) * nll + (LI_{t-1} + LRI) * (1 - \mathrm{nll}) * \mathrm{fdlroots}] * (1 - \mathrm{ndf})$ fdlroots*:* Fraction of roots of trees (relative to root increment) that dies at the end of year
Flux to SINK	$URI = DRI * \mathrm{ndf}/(1 - \mathrm{ndf})$ URI: Nondecomposable fraction; dimensionless

Dead Roots

Type of equation	Equations, variables, and parameters (all pools and processes in terms of tC ha^{-1})
BE	$DR = DR_{t-1} + DRI - ER$ *ER*: decomposition of (i.e., emission from) decomposable dead roots
Flux OUT	$ER = DR_{t-i}/\mathrm{DTR}_i$ DTR: time needed to decompose; year

(Industrial) Wood Products

Type of equation	Equations, variables, and parameters (all pools and processes in terms of tC ha^{-1})
BE	$WP = WP_{t-1} + WPI - WPU$ *WP*: wood products *WPI*: wood product increment *WPU*: wood products becoming unused
Other	$WPI = [FC$ * used part * $(1 -$ fuelwood part$)$ + TH * used part(t) * $(1 -$ fuelwood part$(t)]$ * $(1 -$ lost part$)$ used part: ratio of wood harvested that is used for wood products, dimensionless fuelwoodpart: ratio of wood harvested that is used for fuelwood, dimensionless lost part: ratio unused part of wood harvested, dimensionless
Other	$WPU = WPI_{t-i}$ Maximum of i: mean life time of wood product (species-specific); years
Flux OUT	$EUUWP = WPU_{t-1}$ * $(1 -$ unburnt$)$*WTD *EUUWP*: Emission from (decomposition of) unused and unburned wood product unburnt: unburned fraction; dimensionless WTD: (mean) time needed to decompose the wood products

Fuelwood

Type of equation	Equations, variables, and parameters (all pools and processes in terms of tC ha^{-1})
BE	$FW = FW_{t-1} + FWI + LWP + UWPF - EFW$ *FW*: fuelwood *FWI*: fuelwood increment (from harvest) *LWP*: loss in wood processing *UWPF*: unused wood products becoming fuelwood *EFW*: emission from firewood
Other	$FWI = FC$ * used part * fuelwood part + TH * used part$_t$ * fuelwood part$_t$
Other	$LWP = WPI$ * lost part/$(1 -$ lost part$)$
Other	$UWPF = WPU_{t-1}$ * unburnt
Flux OUT	$EFW = FW_{t-1}$ * $(1 -$ unburn$)$ unburn: unburnable fraction; dimensionless
Flux to Sink	$UFWI = FW_{t-1}$ * unburn *UFWI:* unburnable fuelwood

Soil

Type of equation	Equations, variables, and parameters (all pools and processes in terms of tC ha^{-1})
BE	$S = S_{t-1} + FS - CLO - GAL$
	S: soil
	FS: net flux to sink
	CLO: loss of C from soil due to afforestation and regeneration operations
	GAL: loss of C from soil due to afforesting grasslands
	(Net flux to sink, which includes soil respiration, and loss due to afforesting grasslands are only calculated for 75 years after afforestations. After that time, net flux to sink is supposed to be zero, i.e. transfer of carbon from other compartments to soil is equal to soil respiration, and no more losses are supposed to take place due to converting grassland to forest.)
Other	$CLO = A$ * asl
	A: afforestation area by tree species and yield class; ha (to be provided by the user before running of the model)
	asl: area specific carbon loss; to be specified by the user; tC ha^{-1}
Other	$GAL = A$ * glp * diff_CL_GL
	glp: percent of area afforested on former grassland; to be specified by the user; percent
	diff_CL_GL: time-dependent difference of carbon stock between cropland and grassland; tC ha^{-1}
Flux to SINK	$FS = ULI + URI + UWI + UFWI$

II Equations and Variables Applied to Model Carbon Economics in CASMOFOR

$$\text{Net annual forestry costs}_t = \text{Total of all annual costs}_t - \text{total of all annual revenues}_t$$

where *Total of all annual costs*$_t$ is the sum of all costs, at current prices, (in either EUR or Hungarian Forints, HUF) of all forestry operation on all area of the afforestation project that are due in year *t*; and *Total of all annual revenues*$_t$ is the sum of all revenues, at current prices, of all forestry operation on all area of the afforestation project that are due in year *t*.

$$\text{Total net forestry costs}_t = SUM_i\left(\text{Net annual forestry costs}_i\right)$$

where *Total net forestry costs*$_t$ is the total costs from the beginning of the afforestation project, i.e. year 1, summed up to year *t*; and i = 1 to t.

$$\text{Total carbon in forestry system}_t = AGWB_t + LL_t + R_t + DW_t + DL_t + DR_t + WP_t + FW_t + S_t$$

where *Total carbon in forestry system*$_t$ is the total amount of carbon accumulated in the afforestation project area up to year t in all carbon pools since year 1; all other symbols as in the table above.

$$Total\ net\ specific\ costs_t\ \left(in\ EUR\ /\ tCO_2\ or\ HUF\ /\ tCO\right)$$
$$= Total\ net\ forestry\ costs_t\ /\ Total\ carbon\ in\ forestry\ system_t$$

where *Total net specific costs*$_t$ is the total net forestry costs per tonne of CO_2 sequestered by the afforestation system by year t.

Total net specific costs$_t$ are also calculated with only the sum of $AGWB_t + R_t$ in the denominator to estimate total net specific costs of sequestering carbon in the tree biomass only.

References

Bürger G (1997) On the disaggregation of climatological means and anomalies. Climate Res 8:183–194

Büttner G, Bíró M, Maucha M, Petrik O (2000) Land Cover mapping at scale 1:50.000 in Hungary: Lessons learnt from the European CORINE programme, 20th EARSeL Symposium, 14–16 June 2000. In: A decade of Trans-European Remote Sensing Cooperation, pp 25–31

Farquhar GD, Caemmerer SV, Berry JA (1980) A biochemical-model of photosynthetic CO_2 assimilation in leaves of C-3 species. Planta 149:78–90

Franks SW, Beven KJ, Gash JHC (1999) Multi-objective conditioning of a simple SVAT model. Hydrol Earth Syst Sci 3:477–489

Herrmann A, Kelm M, Kornher A, Taube F (2005) Performance of grassland under different cutting regimes as affected by sward composition, nitrogen input, soil conditions and weather – a simulation study. Eur J Agron 22:141–158

Hollinger DY, Richardson AD (2005) Uncertainty in eddy covariance measurements and its application to physiological models. Tree Physiol 25:873–885

Horváth L, Asztalos M, Führer E, Mészáros R, Weidinger T (2005) Measurement of ammonia exchange over grassland in the Hungarian Great Plain. Agric For Meteorol 130:282–298

IPCC (2003) In: Penman J, Gytarsky M, Hiraishi T, Kruger D, Pipatti R, Buendia L, Miwa K, Ngara T, Tanabe K, Wagner F (eds) Good practice guidance for land use, land-use change and forestry. IPCC/IGES, Hayama, Japan

IPCC (2006) In: Eggleston HS, Miwa K, Ngara T, Tanabe K (eds) 2006 IPCC guidelines for national greenhouse gas inventories. IGES, Hayama, Japan

Jolly W, Nemani RR, Running SW (2005) A generalized, bioclimatic index to predict foliar phenology in response to climate. Global Change Biol 11:619–632

Kennedy M, O'Hagan A (2001) Bayesian calibration of computer models. J R Stat Soc 63B:425–463

Knyazikhin Y, Glassy J, Privette JL, Tian Y, Lotsch A, Zhang Y, Wang Y, Morisette JT, Votava P, Myneni RB, Nemani RR, Running SW (1999) MODIS Leaf Area Index (LAI) and Fraction of Photosynthetically Active Radiation Absorbed by Vegetation (FPAR) Product (MOD15) algorithm theoretical basis document. http://modis.gsfc.nasa.gov/data/atbd/atbd_mod15.pdf

Li CS (2000) Modeling trace gas emissions from agricultural ecosystems. Nutr Cycl Agroecosyst 58:259–276

Metropolis N, Rosenbluth AW, Rosenbluth MN, Teller AH, Teller EJ (1953) Equation of state calculation by fast computing machines. Chem Phys 21:1087–1092

Mo X, Beven K (2004) Multi-objective parameter conditioning of a three-source wheat canopy model. Agric For Meteorol 122:39–63

Monteith J (1972) Solar radiation and productivity in tropical ecosystems. J Appl Ecol 9:747–766

Monteith J (1977) Climate and efficiency of crop production in Britain. Phil Trans R Soc Lond Ser B 281:277–294

Mosegaard K, Tarantola A (1995) Monte Carlo sampling of solutions to inverse problems. J Geophys Res 100B:12431–12447

Nabuurs GJ, Garza-Caligaris JF, Kanninen M, Karjalainen T, Lapvetelainen T, Liski J, Masera O, Mohren GMJ, Pussinen A, Schelhaas MJ (2001) CO2FIX V2.0 – manual of a model for quantifying carbon sequestration in forest ecosystems and wood products. ALTERRA, Wageningen, The Netherlands

New M, Lister D, Hulme M, Makin I (2002) A high-resolution data set of surface climate over global land areas. Climate Res 21:1–25

Reichstein M (2001) Drought effects on carbon and water exchange in three Mediterranean ecosystems. PhD Thesis, Universitat Bayreuth, Max-Planck-Institute for Biogeochemistry, Hans-Knoll-Strasse 10, 07745, Jena, Germany

Running SW, Coughlan JC (1988) A general model of forest ecosystem processes for regional applications I. Hydrological balance, canopy gas exchange and primary production processes. Ecol Modell 42:125–154

Running SW, Gower ST (1991) A general model of forest ecosystem processes for regional applications II. Dynamic carbon allocation and nitrogen budgets. Tree Physiol 9:147–160

Running SW, Nemani RR, Glassy JM, Thornton PE (1999) MODIS Daily Photosynthesis (PSN) and Annual Net Primary Production (NPP) product (MOD17) algorithm theoretical basis document. www.ntsg.umt.edu/modis/ATBD/ATBD_MOD17_v21.pdf

Somogyi Z (2000) Possibilities for carbon mitigation in the forestry sector in Hungary. Biotechnol Agron Soc Environ 4:296–299

Somogyi Z (2008) Recent trends of tree growth in relation to climate change in Hungary. Acta Silvatica Lignaria Hungarica 4:17–27. http://aslh.nyme.hu/fileadmin/dokumentumok/fmk/acta_silvatica/cikkek/Vol04-2008/02_somogyi_p.pdf

Somogyi Z, Cienciala E, Mäkipää R, Muukkonen P, Lehtonen A, Weiss P (2007) Indirect methods of large-scale forest biomass estimation. Eur J For Res 126(2):197–207. doi:10.1007/s10342-006-0125-7

Strahler A, Muchoney, D., Borak, J., Friedl, M., Gopal, S., Lambin, E., Moody, A., 1999. MODIS Land cover product Algorithm Theoretical Basis Document (ATBD). http://modis.gsfc.nasa.gov/data/atbd/atbd_mod12.pdf

Tardieu F, Simonneau T (1998) Variability among species of stomatal control under fluctuating soil water status and evaporative demand: modelling isohydric and anisohydric behaviours. J Exp Bot 49:419–432

Thornton PE, Hasenauer H, White MA (2000) Simultaneous estimation of daily solar radiation and humidity from observed temperature and precipitation: application over complex terrain in Austria. Agric For Meteorol 104:255–271

Trusilova K, Trembath J, Churkina G (2009). Parameter estimation and validation of the terrestrial ecosystem model BIOME-BGC using eddy-covariance flux measurements. Technical Reports – Max-Planck-Institut für Biogeochemie 16:4–14

Tsubo M, Walker S (2005) Relationships between photosynthetically active radiation and clearness index at Bloemfontein, South Africa. Theor Appl Climatol 80:17–25

Van Oijen M, Rougier J, Smith R (2005) Bayesian calibration of process-based forest models: bridging the gap between models and data. Tree Physiol 25:915–927

Verbeeck H, Samson R, Verdonck F, Lemeur R (2006) Parameter sensitivity and uncertainty of the forest carbon flux model FORUG: a Monte Carlo analysis. Tree Physiol 26:807–817

Wang YP, Trudinger CM, Enting IG (2009) A review of applications of model–data fusion to studies of terrestrial carbon fluxes at different scales. Agric For Meteorol 149:1829–1842

White MA, Thornton PE, Running SW, Nemani RR (2000) Parameterization and sensitivity analysis of the Biome-BGC terrestrial ecosystem model: net primary production controls. Earth Interact 4:1–85

Wu X, Luo Y, Weng E, White L, Ma Y, Zhou X (2009) Conditional inversion to estimate parameters from eddy-flux observations. J Plant Ecol 2:1–14

Zhao M, Heinsch FA, Nemani RR, Running SW (2005) Improvements of the MODIS terrestrial gross and net primary production global data set. Remote Sens Environ 95:164–176. doi:10.1016/j.rse.2004.12.011

Zhao M, Running SW, Nemani RR (2006) Sensitivity of Moderate Resolution Imaging Spectroradiometer (MODIS) terrestrial primary production to the accuracy of meteorological reanalyses. J Geophys Res 111:G01002. doi:10.1029/2004JG000004

Chapter 10
Grasslands*

Dóra Hidy, Attila Machon, László Haszpra, Zoltán Nagy, Krisztina Pintér, Galina Churkina, Balázs Grosz, László Horváth, and Zoltán Barcza

Abstract In this chapter, we examine the effects of site-specific environmental conditions and management activities on water, carbon dioxide (CO_2), methane (CH_4), and nitrous oxide (N_2O) fluxes of Hungarian grasslands using two process-based ecosystem models. Biome-BGC is used to simulate the CO_2 budget, while DNDC is applied to simulate the CH_4 and N_2O budgets. Biome-BGC was originally developed to simulate the biogeochemical cycles of natural biomes; therefore,

*Citation: Hidy, D., Machon, A., Haszpra, L., Nagy, Z., Pintér, K., Churkina, G., Grosz, B., Horváth L., Barcza, Z., 2010: Modeling of biosphere–atmosphere exchange of greenhouse gases – Grasslands. In: Atmospheric Greenhouse Gases: The Hungarian Perspective (Ed.: Haszpra, L.), pp. 229–251.

D. Hidy (✉), A. Machon, L. Haszpra and L. Horváth
Hungarian Meteorological Service, H-1675, Budapest, P.O.Box 39, Hungary
e-mail: hidy.d@met.hu

A. Machon
Department of Botany and Plant Physiology, Szent István University, H-2103 Gödöllő, Páter K. u. 1., Hungary
and
Center for Environmental Science, Eötvös Loránd University, H-1117 Budapest, Pázmány P. s. 1/A, Hungary

Z. Nagy and K. Pintér
Plant Ecology Research Group of Hungarian Academy of Sciences, H-2103 Gödöllő, Páter K. u. 1., Hungary
and
Department of Botany and Ecophysiology, Szent István University, H-2103 Gödöllő, Páter K. u. 1., Hungary

G. Churkina
Leibniz-Centre for Agricultural Landscape Research, D-15374 Müncheberg, Eberswalder Strasse 84, Germany

B. Grosz
Department of General and Inorganic Chemistry, Eötvös Loránd University, H-1117 Budapest, Pázmány P. s. 1/A, Hungary

Z. Barcza
Department of Meteorology, Eötvös Loránd University, H-1117 Budapest, Pázmány P. s. 1/A, Hungary

L. Haszpra (ed.), *Atmospheric Greenhouse Gases: The Hungarian Perspective*,
DOI 10.1007/978-90-481-9950-1_10,

extension of the model source code was needed to simulate management practices. Besides implementing management modules, the soil water and phenology sub-models of Biome-BGC were developed. After the development and extension, the model was calibrated using a Bayesian approach by a novel *iterative* calibration method. DNDC model consists of two components: the first predicts soil temperature, moisture, alkalinity (pH), soil redox potential, and substrate concentration profiles based on ecological drivers; the other component calculates ammonia, nitrogen monoxide, nitrous oxide, and methane fluxes based on soil environmental conditions. DNDC handles several types of management activities. The simulation results of the models were validated using measured carbon dioxide and latent heat flux data from three Hungarian measurement sites, and also using N_2O and CH_4 fluxes measured over Hungarian grasslands. The calibrated Biome-BGC model was used to simulate the long-term effect of different management practices.

Keywords Process-based modeling • Bayesian calibration • Management activities • Grassland

10.1 Introduction

Human activity influences the climate of Earth by changing the chemical composition of the atmosphere (IPCC 2007). Power generation, industry, agriculture, and transportation release a huge amount of greenhouse gases into the atmosphere. Changes in the quantity of atmospheric greenhouse gases unbalance the energy budget of the Earth-atmosphere system and trigger global warming. The most important anthropogenic greenhouse gases are carbon dioxide (CO_2), methane (CH_4), and nitrous oxide (N_2O). Their concentrations have increased by about 40%, 150%, and 20% from the preindustrial values, respectively (IPCC 2007).

The biosphere is sensitive to environmental changes: the terrestrial carbon and nitrogen budgets can change rapidly and significantly in response to the changing environmental conditions (rainfall, temperature, radiation, nutrition status), and can interact with the climate through emission or uptake of greenhouse gases (Ciais et al. 1995). Since the terrestrial carbon and nitrogen budgets are influenced by a number of factors, the knowledge of the biogeochemical processes is essential in order to understand the possible causes of changes in the carbon and nitrogen budgets at ecosystem level. Numerical models are the most convenient instruments beyond the measurements to accomplish this task. Process-based models simulate mechanistically the ecosystem functioning. These models are able to incorporate the effect of environmental changes on ecosystem development.

Temperate grasslands contain about 10% of the global soil carbon pool; therefore, they have a significant role in the global carbon budget. It is important to examine their functioning in order to predict the possible changes in carbon fluxes due to the changing environmental conditions (Frank 2004). Methane has a global warming potential of 25 compared to CO_2 over a 100-year period (IPCC

2007). This means that methane emission will have 25 times higher impact on temperature than carbon dioxide emission of the same amount over the next 100 years. N_2O is derived from the soil as the result of the nitrification processes. The rate of nitrification is sensitive to the changing climate and increasing human nitrogen input to the ecosystem. N_2O is a greenhouse gas that is 298 times more effective per unit mass than CO_2 in absorbing infrared radiation (100-year time horizon, IPCC 2007). In addition, N_2O decomposed in the stratosphere in the presence of ultraviolet radiation catalyzes the destruction of stratospheric ozone.

Two process-based models, Biome-BGC and DNDC, are used to simulate the CO_2, CH_4, and N_2O budgets of the Hungarian grasslands and to examine their responses to the changing environmental conditions and human activity.

10.2 Simulation of Carbon Dioxide Exchange of Hungarian Grasslands with Biome-BGC Model

10.2.1 Implementation of the Calibration Procedure

We calibrated Biome-BGC for C3 grasslands using measured water and carbon dioxide exchange data from three Hungarian eddy covariance measurement sites with different site-specific conditions (local climate, soil type, species composition, management activity). One of the sites is a managed grassland on silt soil (Hegyhátsál, 46.95°N, 16.65°E, 248 m asl), the other site is a dry, managed grassland on sandy soil (Bugac, 46.69°N, 19.60°E, 111 m asl), while the third one is an unmanaged grassland on mountain heavy clay soil (Mátra, 47.85°N, 19.73°E, 300 m asl). The detailed description of the measurement sites can be found in Chapter 6. As it was mentioned in Chapter 9.1, there are certain ecophysiological and site-specific model parameters for which values are hard to obtain directly from experiments. We refer to them as *unknown model parameters* (they are listed in Table 9.1). Our calibration method is appropriate to estimate the optimal values of the *unknown model parameters* using measured and simulated flux data (see details in Chapter 9.1). Likelihood function is used to describe the accuracy of the model simulations. The likelihood function includes the relative error of the simulation and the square of the correlation between measured and simulated data. Relative error is the ratio of the mean absolute error of the simulation and the range of the given measured data (difference between maximum and minimum values). Multi-objective calibration means that different types of measured data are used in the calibration (carbon dioxide and water fluxes) in order to avoid one-sided optimization. Normalization is important in case of multi-objective calibration because the different misfits (biases from the different reference data) must be comparable (Mo and Beven 2004).

Our likelihood ($L(p)$) is expressed as the function of the relative error weighted by the square of the correlation coefficient:

$$L(p) = \frac{1}{C} \cdot \left(\left(1 - RE(p)\right) \cdot R^2(p) \right) \tag{10.1}$$

$$RE(p) = \frac{1}{n} \sum_{i=1}^{n} \frac{\left| sd_i(p) - md_i \right|}{\nu} \tag{10.2}$$

$$\nu = \max(md) - \min(md) \tag{10.3}$$

where C is a scaling constant that normalizes the likelihood to probability density function,[1] $RE(p)$ is the relative error of the simulation, $R^2(p)$ is the square of the correlation coefficient between measured and simulated data, and $sd(p)$ is the simulated data. These variables depend on the model parameter set (p). md refers to measured data, and ν is the range of measured data – here it is a weighting factor.

Correlation means a further weighting factor to avoid labeling a given run having small errors but poor correlation with the measured data with good likelihood value (Franks et al. 1999). The meaning of likelihood is as follows: if an unreal parameter set generates a large difference or incoherence (high relative error or low correlation) between the simulated and the measured data, the likelihood is close to zero. If a good parameter set generates a small difference and strong relationship (low relative error and high correlation) between the simulated and the measured data, the likelihood value will be close to 1.

We use gross primary production (GPP), total ecosystem respiration (R_{eco}) (calculated from the net ecosystem exchange (NEE) data through flux partitioning) and latent heat flux (LE) measured by eddy covariance technique as reference data. To achieve the combined likelihood, the likelihood of each variable are multiplied:

$$L(p) = L_{\mathrm{GPP}}(p) \cdot L_{\mathrm{Reco}}(p) \cdot L_{\mathrm{LE}}(p) \tag{10.4}$$

where L_{GPP}, L_{Reco}, and L_{LE} are the likelihood values calculated for GPP, R_{eco}, and LE, respectively.

An important and well-known problem of the global optimization methods is the parameter equifinality (Mo and Beven 2004). It means that different parameter combinations can produce equally good simulation results with respect to the reference data. To avoid parameter equifinality, we have developed a novel, n-step calibration method, which is used in a Bayesian framework. In our n-step approach, only the most "important" parameter is set to its optimal value (there is enough available information from the likelihood function for this parameter) at the first step in a chain of calibrations (it is referred as first *iteration section*). The word "important" is used here in the sense that the parameter has strong effect on model outputs (and through this, on goodness-of-fit). To choose the most important parameter, sensitivity analysis is performed (for the method, see Chapter 9.1). After

[1] Probability density function (PDF) of a variable is a function that describes the density of probability at each point in the sample space. The probability of a variable falling within a given set is estimated by the integral of its density function over the set (total integral of PDF is equal to 1).

10,000 iteration steps, the parameter with the highest sensitivity coefficient (if it is greater than 20%) is fixed on its optimal value; this means one iteration section in the chain. In the next *iteration section*, only $n-1$ parameters are varied randomly. Iteration sections are performed until each unknown model parameter is set to its optimal value (in our case, there are 21 iteration sections).

Based on the result of the least-square linearization (LSL) analysis, it is possible to rank the unknown model parameters for sensitivity using Monte Carlo method. We define a *sensitivity rank* to determine the importance of a given parameter as the function of the iteration section's number where the given parameter is fixed. As the most important parameter is fixed in the first iteration section using the *a posteriori* distribution described above, its *sensitivity rank* is maximal (in our case the number of the model parameters is 21, therefore the maximal sensitivity rank is 21). The less important parameter is fixed in the last iteration section, thus its *sensitivity rank* is 1. Calibration is performed using measured eddy covariance data from period of 2007 to 2008 for each measurement site. The calibrated model is validated using measured eddy covariance data from 2005 for Bugac and Mátra, and from 1999 for Hegyhátsál.

10.2.2 The Calibrated Model Parameters

If the given *unknown model parameter* (Table 9.1) has a high *sensitivity rank*, it means that the parameter has a strong effect on model output. In Fig. 10.1, we present the *sensitivity ranks* for different sites and the mean *sensitivity ranks*, which were calculated as the average of the *sensitivity ranks* for different sites.

There are seven parameters with mean *sensitivity ranks* greater than 15: transfer growth period as fraction of growing season (GP), annual whole-plant mortality fraction (WPM), ratio of new fine root carbon and new leaf carbon (FRC:LC), ratio of the carbon and nitrogen contents of leaves C:N(lv), canopy average specific leaf area (SLA), maximum stomatal conductance (MSC), and soil depth (SD). The *sensitivity rank* of nitrogen fixation (Nfix) is greater than 15 only at Mátra, where Nfix is the most important parameter.

We analyzed the site-specific dependence of these important model parameters (GP, WPM, FRC:LC, SLA, C:N(lv), MSC, SD, Nfix). The standard deviations of the optimal parameter values at different sites characterize the site-specific dependence of the given parameter. This standard deviation is normalized with the a priori interval of the given parameter, and this ratio is referred to as *spatial variance ratio*. *Spatial variance ratio* is considered a comparable metric to determine the site-specific and site-independent parameters.

Spatial variance ratio is less than 10% for SLA and MSC. It means that the values of these parameters are similar for different sites, therefore, they are practically independent from site-specific conditions (local climate, type of soil, etc.). SLA is a leaf morphological parameter, a measure of the leaf thickness: the leaf area per unit carbon content. Maximum stomatal conductance is the maximum rate

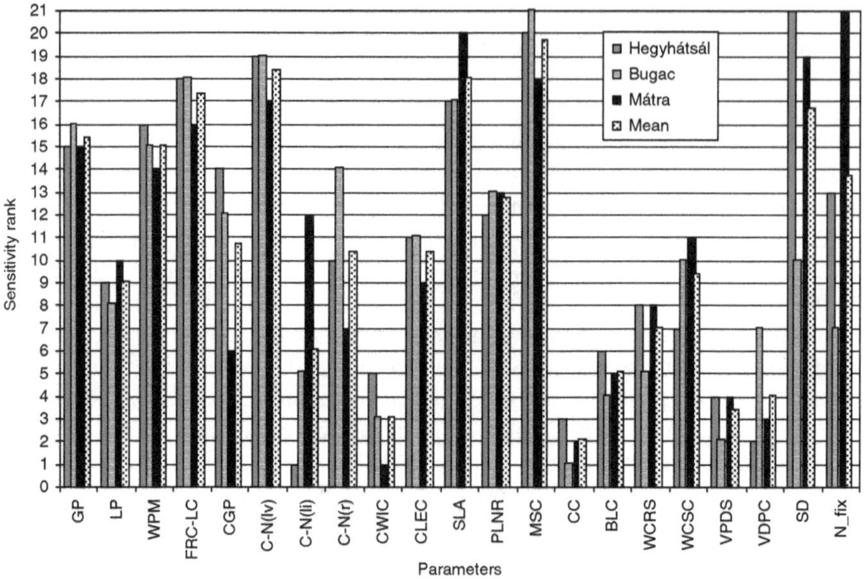

Fig. 10.1 The *sensitivity ranks* of the Biome-BGC model parameters for Hegyhátsál, Bugac, Mátra, and the mean *sensitivity ranks*. The parameter with the maximal sensitivity rank (21) has the strongest effect on the model output at the given site. For the explanation of the model parameters, see Table 9.1

of passage of both water vapor and carbon dioxide through the stomata. *Spatial variance ratios* are between 15% and 20% for GP and SD. It means that these parameters slightly depend on site-specific conditions. GP determines the transfer time during which plant material turns from the transfer pools to the actual carbon and nitrogen pools (leaf, root, etc.). This is affected by local climate. SD determines the depth of the bucket containing the simulated soil water available to the plant. It depends on the soil texture. In case of WPM, C:N(lv), FRC:LC, and Nfix *spatial variance ratios* are greater than 25%, which means that these parameters strongly depend on the site-specific conditions. WPM determines the mortality (fluxes from leaf or root pool to litter). C:N(lv) determines the nitrogen translocation processes. FRC:LC influences the allocation of nitrogen, while C:N(lv) and FRC:LC depend on species. Nfix (besides nitrogen deposition) is the natural external input of soil mineral nitrogen, which influences the nitrogen retranslocation processes.

10.2.3 Validation of the Calibrated Biome-BGC Model

As a result of the model developments, the incorporation of the management activities and the calibration of the model parameters (referred to as developed, extended, and calibrated model), and the simulations of the carbon dioxide and water fluxes of the three Hungarian grasslands have significantly improved. The success of calibration can be evaluated with the goodness-of-fit of the simulation and with the decrease in the uncertainty of the model parameters.

The confidence intervals of each parameter decreased (on average the *a posteriori* intervals are 50–60% of the a priori ones), i.e. the uncertainty of the *model parameters* decreased due to the calibration.

The model simulation is evaluated using the relative error (*RE* (%); mean difference between the measured and the simulated data relative to the difference between the maximum and the minimum measured) and the square of correlation coefficient (R^2). The results are presented in Table 10.1 for the three measurement sites using the original and the developed, extended, and calibrated model. Panel A of Table 10.1 refers to the calibration years (2007–2008), while Panel B to the validation year (Hegyhátsál 1999; Bugac and Mátra 2005). It can be seen that the relative errors decreased and the correlations got stronger as the result of the developments, extension, and calibration in both the calibration and validation years, which means that the modifications and the calibration procedure was successful.

In Fig. 10.2, we show the measured and simulated data using the original model with original model parameters (listed in Chapter 9.1) and the developed and extended model with calibrated model parameters for the validation year (Fig. 10.2a–c: Hegyhátsál; Bugac and Mátra). One can see that using the developed and extended model with calibrated parameters, the simulated data are much closer to the measured ones than using the original model with default parameter values.

Common error of the carbon dioxide flux simulations using the original model is the unrealistic start of the growing season (see Fig. 10.2). We developed the phenology module to calculate the start and end of the growing season in a more realistic way (see Chapter 9.1). Using this new method, the vegetation "awakes" earlier and the annual courses of GPP and R_{eco} become more realistic compared to the original model phenology. The model adjustments and the calibration have improved the outputs, which are characterized by lower relative error (defined above), higher correlation, and more realistic annual courses (Fig. 10.2 and Table 10.1).

Common error of the original evapotranspiration simulations is the occurrence of peaks in the seasonal course of the latent heat fluxes (Fig. 10.2). We have modified the calculation method of stomatal conductance to avoid this problem (see Chapter 9.1). In Fig. 10.3, we present a case study for Bugac to demonstrate the effect of the model developments: we examine the measured meteorological data (precipitation amount, vapor pressure deficit (VPD)), the simulated soil water status (soil water content (SWC) and soil water potential (ψ)), the simulated stomatal conductance, and the measured and simulated latent heat fluxes using the original and the modified model for the period between July 23, 2005, and August 7, 2005.

There was no precipitation between July 23 and August 3 (Fig. 10.3a). As a result, the air and the soil became drier and drier (VPD was increasing, soil water content was decreasing; Fig. 10.3b–c). Besides drought, the daytime mean temperature was between 20°C and 30°C with daily maximum between 25°C and 35°C during this period. Comparing the results of the original and the developed models, it can be seen that the stomatal conductance is higher in the original model, and

Table 10.1 Relative error (δ;%) and square of correlation (R^2) between measured and simulated CO_2 flux data at the different measurement sites. Panel A: calibration years (2007–2008); Panel B: validation year (Hegyhátsál: 1999, Bugac and Mátra: 2005)

Panel A	Hegyhátsál				Bugac				Mátra			
Calibration years	Original		Extended and developed		Original		Extended and developed		Original		Extended and developed	
	δ	R^2	δ	R^2	δ	R^2	δ	R^2	δ	R^2	δ	R^2
GPP	28.6	0.10	12.1	0.72	29.1	0.02	15.4	0.71	42.9	0.01	24.5	0.68
R_{eco}	27.3	0.14	7.20	0.82	17.8	0.10	13.7	0.60	22.7	0.13	20.6	0.72
LE	21.8	0.22	16.7	0.61	22.1	0.26	11.0	0.59	30.9	0.21	15.3	0.60
Panel B	Hegyhátsál				Bugac				Mátra			
Validation year	Original		Extended and developed		Original		Extended and developed		Original		Extended and developed	
	δ	R^2	δ	R^2	δ	R^2	δ	R^2	δ	R^2	δ	R^2
GPP	30.0	0.29	12.9	0.77	28.5	0.22	11.7	0.77	25.8	0.09	10.8	0.82
R_{eco}	31.1	0.52	12.7	0.93	15.5	0.53	11.5	0.75	16.3	0.35	18.6	0.81
LE	40.3	0.36	14.6	0.77	24.6	0.32	9.5	0.78	29.4	0.15	9.4	0.84

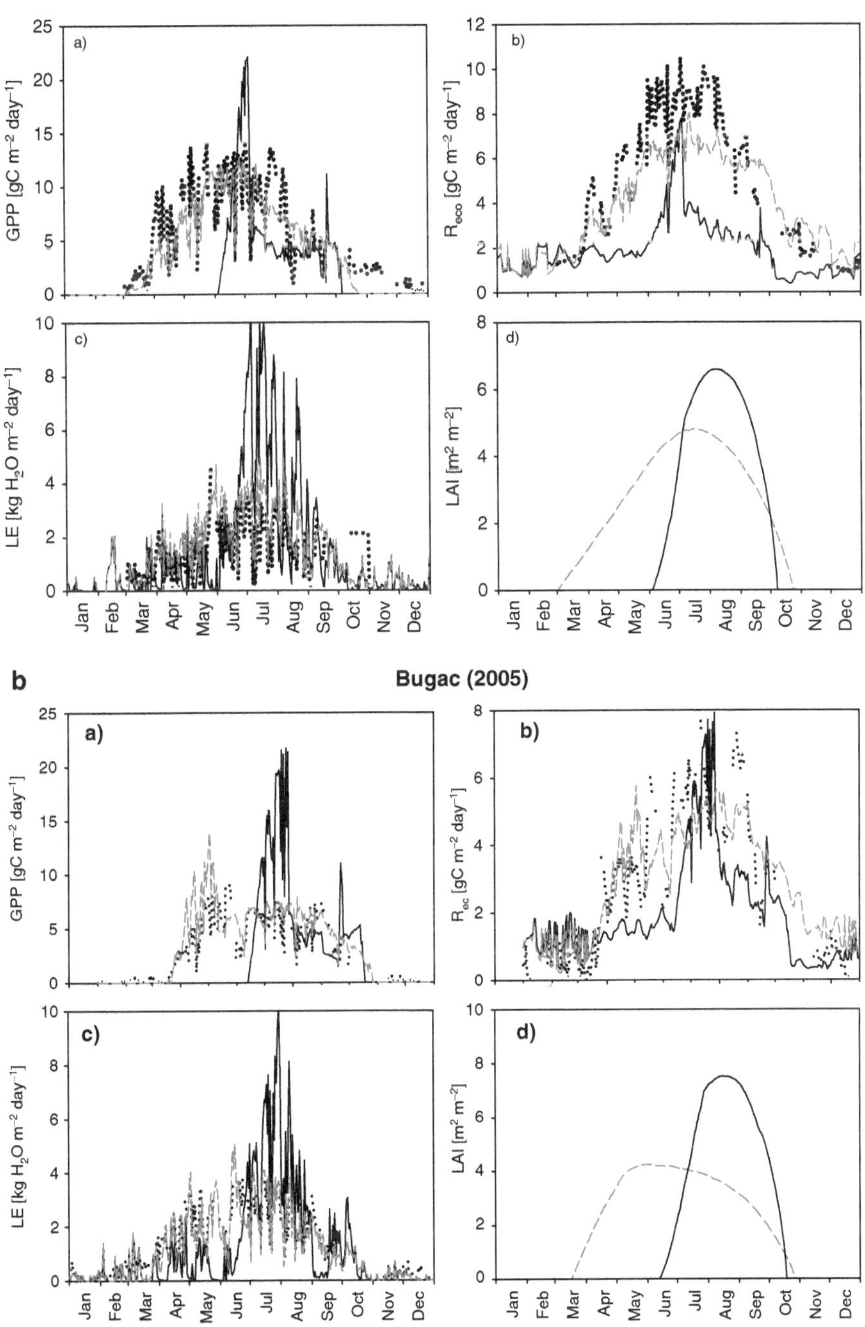

Fig. 10.2 The measured data (*black dots*), the simulated data using the original model with default parameter values (*black solid line*), and the results of the developed and extended model with calibrated model parameters (*gray dashed line*) for Hegyhátsál (A), Bugac (B), and Mátra (C) for the validation year: (**a**) gross primary production (GPP); (**b**) total ecosystem respiration (R_{eco}); (**c**) latent heat flux (LE); (**d**) leaf area index (LAI)

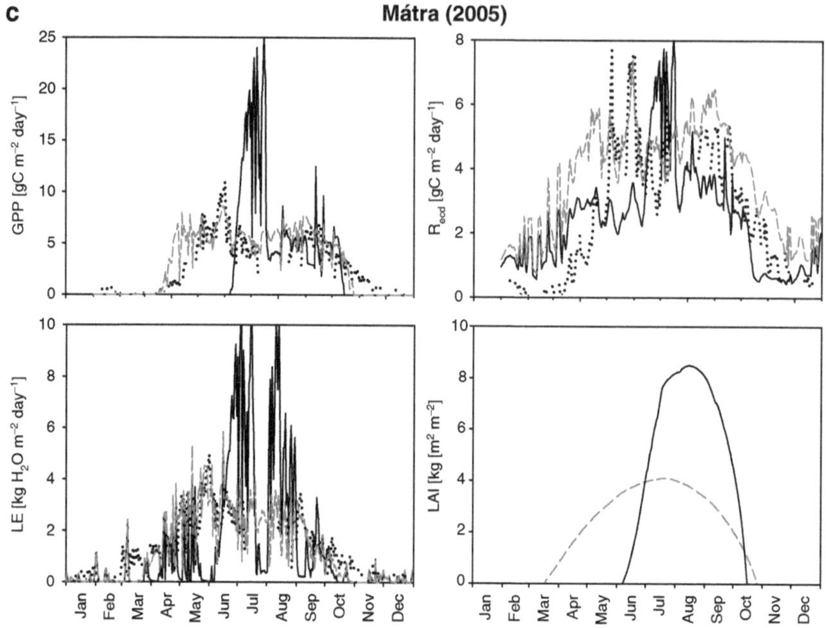

Fig. 10.2 (Continued)

therefore the simulated SWC is lower. Due to the high stomatal conductance, the latent heat flux is higher, and the simulated data overestimate the measured ones. Until July 31, the stomatal conductance does not decrease, the latent heat flux increases and reaches an unrealistic value of 10 kg H_2O m^{-2} day^{-1}. After July 31, the decrease of the stomatal opening is very sharp and it becomes zero on August 8. As a result, latent heat flux decreases to 2 kg H_2O m^{-2} day^{-1} in 2 days. Using the developed model, the decrease of stomatal conductance is smaller and more gradual (Fig. 10.3e). The simulated latent heat flux starts to decrease after July 27, and its course is closer to the measured one (Fig. 10.3f). On August 3, the daily sum of the precipitations was 15 mm, while on August 4, it was about 25 mm. Due to the rainfall, the simulated soil metric water potential became −0.2 MPa from −3 MPa (Fig. 10.3d), while stomatal conductance reached 1 m s^{-1} in 2 days using the original model. This is because soil metric water potential is close to the saturation, and not sensitive to the decrease in SWC. At lower SWC values, a small decrease in SWC can cause huge increase in the soil metric water potential (see also Chapter 9.1). In the developed model, the drought and rainfall affect stomatal conductance through SWC, the change of which with soil drying is more gradual than the change of soil metric potential. Therefore, the change in the latent heat flux is more gradual and more realistic.

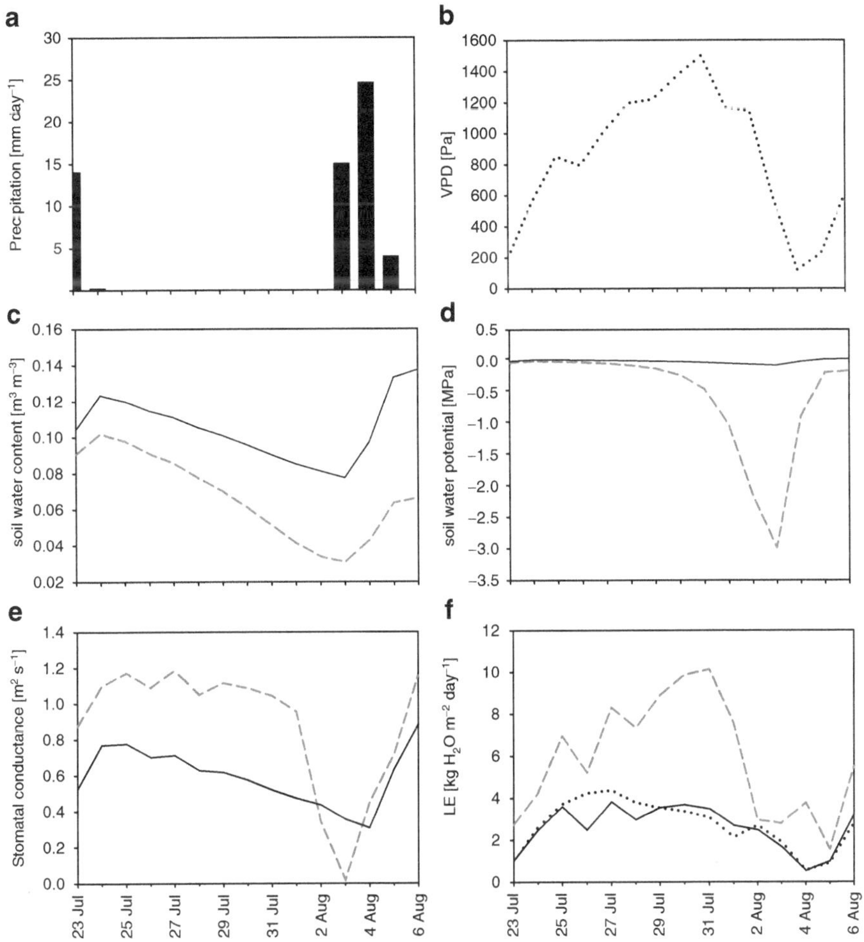

Fig. 10.3 A case study (Bugac, July 23–August 7, 2005) demonstrating the effect of the model development. Black dotted lines (in case of precipitation black columns): measured data; gray dashed lines: output of the original model; black solid lines: output of the developed model. (**a**) Precipitation; (**b**) vapor pressure deficit; (**c**) soil water content; (**d**) soil water potential; (**e**) stomatal conductance; (**f**) latent heat flux

10.2.4 Effect of the Site-Specific Conditions on Carbon Dioxide and Water Exchange

After the incorporation of management modules and the calibration, Biome-BGC has become an appropriate tool to simulate the carbon dioxide and water fluxes of the Hungarian grasslands. We applied the model to simulate the long-term dynamics of the carbon cycle components at three eddy covariance sites.

Because of the different ecophysiological parameters, local climate, and soil types, the carbon dioxide exchange is different at the three measurement sites. For the simulations, we used input meteorological data from 1997 to 2008 at Hegyhátsál, from 2002 to 2008 at Bugac and from 2003 to 2008 at Mátra. In Fig. 10.4, the mean seasonal variations of daily averaged simulation results based on long-term runs (Hegyhátsál: 12 years, Bugac: 7 years, Mátra: 6 years) are presented. We assumed mowing at Hegyhátsál in every year on day 150 and 234, and grazing at Bugac in every year from day 100 to 300.

The multi-annual average of gross carbon dioxide uptake (GPP) is the highest at the managed grassland on silt soil (Hegyhátsál) and the lowest at the grassland on mountain heavy clay soil (Mátra) (Fig. 10.4a). The differences among the sites are similar for total ecosystem respiration (R_{eco}) (Fig. 10.4b) and latent heat flux (Fig. 10.4c). Stomatal conductance is one of the most influential factors on carbon dioxide and water exchange fluxes. The amount of soil water necessary for the optimal carbon dioxide uptake processes depends on the soil type (available water for plant). The water-holding capacity of the soil is the highest in silt soil and the lowest in heavy clay soil (Tardieu and Simonneau 1998) as it is also detailed in Chapter 6. The mean annual temperature and mean annual sum of precipitation is 8.9°C and 750 mm at Hegyhátsál (silt soil), 10.4°C and 562 mm at Bugac (sandy soil) and 10.2°C and 622 mm at Mátra (clay soil), respectively (Barcza et al. 2003; Pintér et al. 2008). Therefore, the soil water limitation of carbon dioxide uptake is

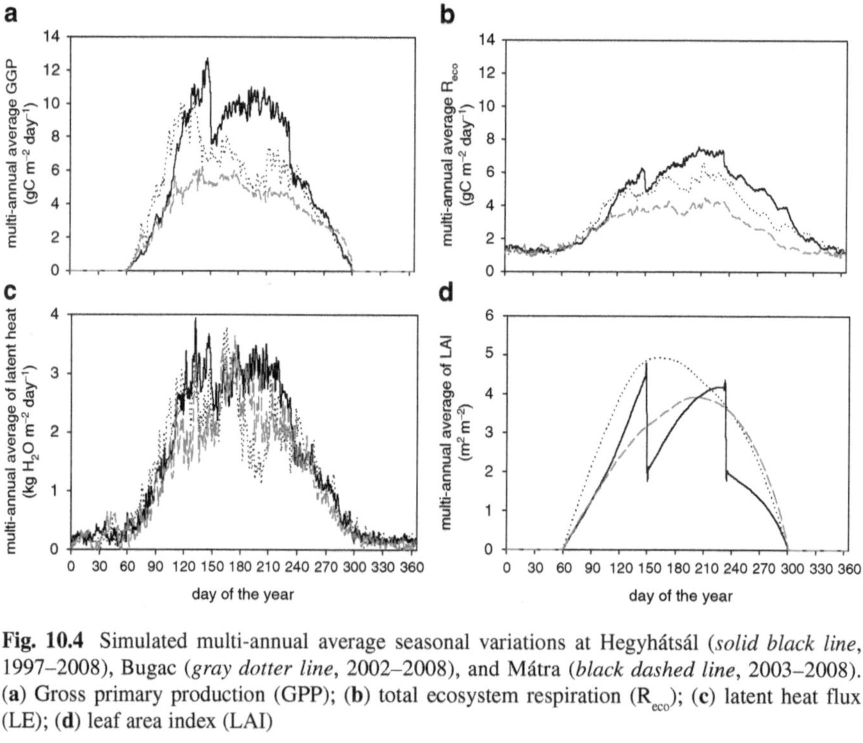

Fig. 10.4 Simulated multi-annual average seasonal variations at Hegyhátsál (*solid black line*, 1997–2008), Bugac (*gray dotter line*, 2002–2008), and Mátra (*black dashed line*, 2003–2008). (**a**) Gross primary production (GPP); (**b**) total ecosystem respiration (R_{eco}); (**c**) latent heat flux (LE); (**d**) leaf area index (LAI)

the lowest at Hegyhátsál (silt soil, low mean annual temperature, and high mean annual sum of precipitation) and the highest at Mátra (clay soil, high mean annual temperature, and moderate mean annual sum of precipitation; see also Chapter 6).

NEE is the difference between total ecosystem respiration (R_{eco}) and gross carbon dioxide uptake (GPP), where carbon loss/surplus caused by human activity (mowing, grazing, fertilizing, etc.) is not included. When NEE is positive, the ecosystem is a net carbon dioxide source to the atmosphere, while when it is negative, the ecosystem is a net carbon dioxide sink. At Hegyhátsál and Mátra, the multi-annual average GPP is higher than that of R_{eco} in the growing season, which is not surprising, because grasslands are usually net carbon dioxide sinks during that period of the year. In contrast, at the dry, managed grassland on sandy soil (Bugac), the multi-annual average of GPP is lower than that of R_{eco} for circa a month in the growing season (from day 180 to 220; July and August). This is because the soil usually dries out and the soil water content decreases below the wilting point during that period. Usually after the rainy days in late August, the carbon dioxide uptake increases again. This second growing period (from late August until early September) is an important and well-known feature of the grassland at Bugac (Pintér et al. 2008; see also Chapter 6).

In Fig. 10.4d, the effect of the grass management on the multi-annual average leaf area index (LAI) can be seen for Hegyhátsál. The variation is caused by the prescribed mowing on day 150 and 240. Usually, LAI is the lowest at Mátra.

The annual carbon dioxide uptake is the highest at Hegyhátsál and the lowest at Mátra, but usually the ecosystems are net carbon sinks in the growing season. At Bugac, there are two growing periods because of soil drying-out in July–August. The agreement between the simulated and measured data is good for all measurement sites.

10.2.5 *Effects of Management on Carbon Dioxide and Water Exchange*

Several management modules have been implemented in Biome-BGC (see Chapter 9.1). At Hegyhátsál, grass is mowed twice a year, while cattle grazing is occurring at Bugac. The results of the simulation of grazing and mowing can be verified, and the performance of the model can be evaluated using the measurements. The main effect of grazing and mowing is the removal of leaves. In case of grazing, a given proportion of the aboveground biomass flows to the litter compartment as the result of trampling and excretal returns.

To analyze the effect of these management practices, Biome-BGC was run for Hegyhátsál for a 38-year long period (1971–2008). We examined the effect of mowing and grazing, although grazing does not occur at Hegyhátsál in the reality. Mowing is assumed on day 150 and 235 in every year, while grazing is assumed from day 100 to 300. Fig. 10.5 presents the results of this 38-year run using the NO MANAGEMENT scenario (all management modules are turned off in the extended and calibrated model) and the MOWING scenario (mowing module is enabled; mowing occurs twice a year on the two prescribed days).

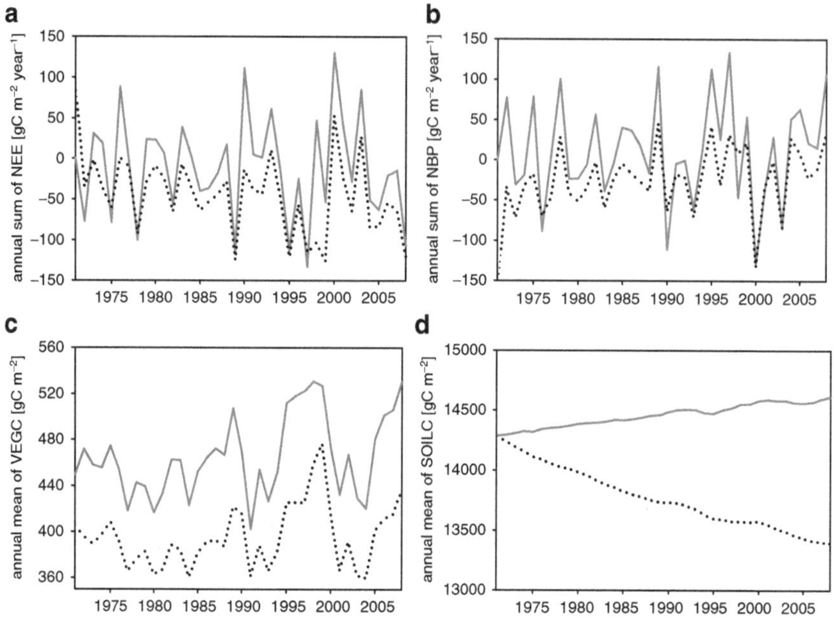

Fig. 10.5 Results of the 38-year long simulation with NO MANAGEMENT scenario (*solid lines*) and MOWING scenario (*dotted lines*) for Hegyhátsál. (**a**) Annual sums of net ecosystem exchange (NEE); (**b**) annual sums of net biome production (NBP); (**c**) annual mean of vegetation carbon content (VEGC); (**d**) annual mean of soil carbon content (SOILC)

Net biome production (NBP) gives the long-term net gain or loss of carbon of an ecosystem or region (Chapin et al. 2005). It is equal to the net ecosystem production (NEP; equals -NEE) minus the carbon removed by disturbance (mowing, grazing, etc.). NBP is positive when the ecosystem (taken into account the effect of disturbances) is a net sink of carbon and negative if it is a net source. In Fig. 10.5a, it can be seen that the annual sums of NEE are smaller in the MOWING run than in the NO MANAGEMENT run. The leaf area, which intercepts and evaporates a certain part of the precipitation, decreases to a fixed value twice a year; therefore, more precipitation penetrates into the soil. Its result is the increase in the soil water content, which is an important limiting factor of the stomatal conductance. If more soil water is available, the stomatal conductance is less limited, and higher carbon dioxide uptake is possible. The average NBP is 11 g C year^{-1} in the NO MANAGEMENT simulation and −26.3 g C year^{-1} in the MOWING run, which means that the ecosystem is a net carbon sink on the average if it is undisturbed (i.e. soil carbon is increasing), while it becomes a net carbon source if it is mowed (Fig. 10.5b). This result is consistent with the NBP estimation presented in Chapter 6.

Vegetation carbon content (VEGC) in Biome-BGC is the sum of the carbon content of actual leaf and fine root pools, leaf and fine root storage pools, and leaf and fine root transfer pools (Fig. 10.5c) (for definitions see Chapter 9.1). In every simulation year, VEGC is greater in case of the NO MANAGMENT run and the mean difference is 32.5 g C m^{-2}. In the NO MANAGEMENT run, the soil carbon content (SOILC) slowly increases (Fig. 10.5d), while in the MOWING run, it decreases continuously. It corroborates that in the MOWING run, the ecosystem is a net carbon source (NBP is negative) due the harvested and removed biomass, in contrast to the NO MANAGEMENT scenario.

Figure 10.6 presents the results of the 38-year NO MANAGEMENT run and GRAZING run (grazing module is enabled; grazing is occurring every day from day 100 to 300 of the year).

The situation is similar to the MOWING run. Annual sum of NEE is lower in the GRAZING run (higher net carbon dioxide uptake). Annual means of VEGC are also lower than in the reference scenario (NO MANAGEMENT) because of animal intake (except in 1999). We have already mentioned that as a result of decreasing leaf area, the soil water content can increase, which enables higher carbon dioxide uptake and a parallel higher carbon storage. Therefore, it is possible that VEGC is higher in certain years in the GRAZING run (Fig. 10.6c). Annual sums of NBP are less (but to a lesser extent than in case of MOWING) in the GRAZING scenario than in the reference one (Fig. 10.6b). SOILC are lower in the GRAZING run (similar to MOWING run) but SOILC decreases slower during the simulation period (Fig. 10.6d). This is because the carbon loss (decrease of plant material) by grazing (animal intake) is less than carbon loss by mowing. Carbon loss by animal intake is

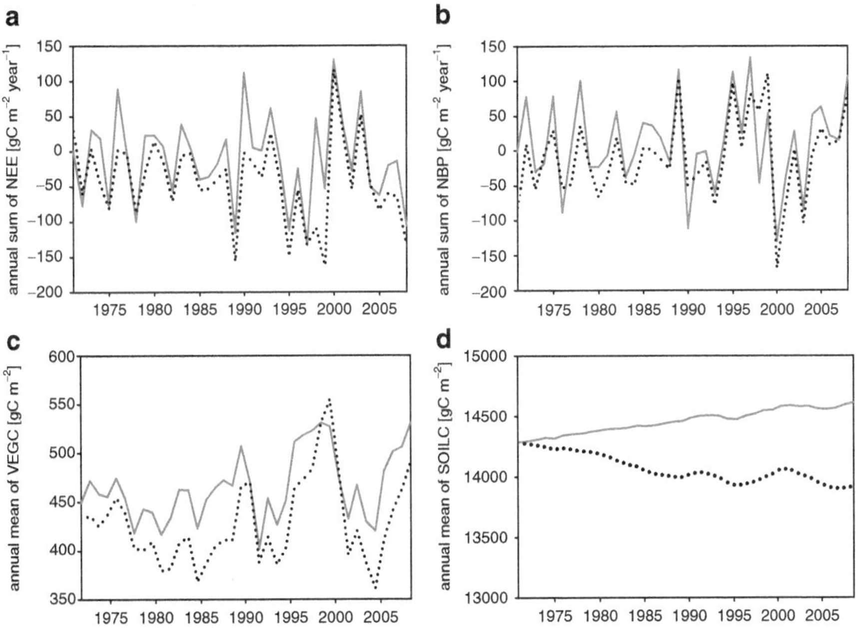

Fig. 10.6 Results of the 38-year long simulation with NO MANAGEMENT scenario (*solid lines*) and GRAZING scenario (*dotted lines*) for Hegyhátsál. (**a**) Annual sums of net ecosystem exchange (NEE); (**b**) annual sums of net biome production (NBP); (**c**) annual mean of vegetation carbon content (VEGC); (**d**) annual mean of soil carbon content (SOILC)

partly recovered by carbon and nitrogen surplus from excretal returns in contrast to mowing (cut-down fraction is removed, which means net loss for the system).

Further long-term simulation experiments are needed together with reliable measurement data (including e.g. soil carbon stock changes) to corroborate the model results and to verify the applicability of the model for grassland ecosystems on larger spatial scales.

10.3 Simulation of Nitrous Oxide and Methane Exchange of Hungarian Grasslands with the DNDC Model

10.3.1 Validation of the Model

Nitrous oxide and methane exchanges of Hungarian grasslands were simulated by the DNDC model. DNDC is a public domain software, but it is not an open source one. Therefore, we have no possibility to modify the source code. All we can do to achieve the best simulation results is specifying the best input data for the model. The model can be used for small areas in the so-called "spot mode" when direct local measurements are used as input data. In "regional mode," the model runs on spatially aggregated input data representing the average conditions of the region modeled. The main difference between these two modes is that DNDC in regional mode uses fewer input parameters because certain parameters are internally generated via submodels instead of incorporating direct measurements.

For the validation of the model in spot mode, we compared the simulated N_2O (and NO – not discussed here) soil flux data to measured fluxes at the semi-arid, seminatural grassland site, Bugac, for a selected 1-year-long period (August, 2006–July, 2007) (Table 10.2).

Square of the correlation between the modeled and static chamber based measured monthly N_2O fluxes shows a relatively good agreement ($R^2=0.5$). The emission peaks in the simulation correlate with temperature, and reflect the rain events. After rain, the denitrification processes can produce an N_2O emission peak due to the anaerobic period. In contrast, soil conditions at low water-filled pore space (WFPS), when soil is well aerated, do not favor the denitrifier bacteria. The soil

Table 10.2 Comparison of measured and simulated minimum, maximum, daily mean, and cumulative N_2O fluxes for the year of testing (August 2006–July 2007)

	N_2O flux	
Compared parameters	Measured	Modeled
Minimum value (mg N m^{-2} s^{-1})	−1.31	0.0
Maximum value (mg N m^{-2} s^{-1})	2.85	7.99
Daily mean value (mg N m^{-2} s^{-1})	0.43	0.44
Cumulative emission (mg N m^{-2} $year^{-1}$)	13	14

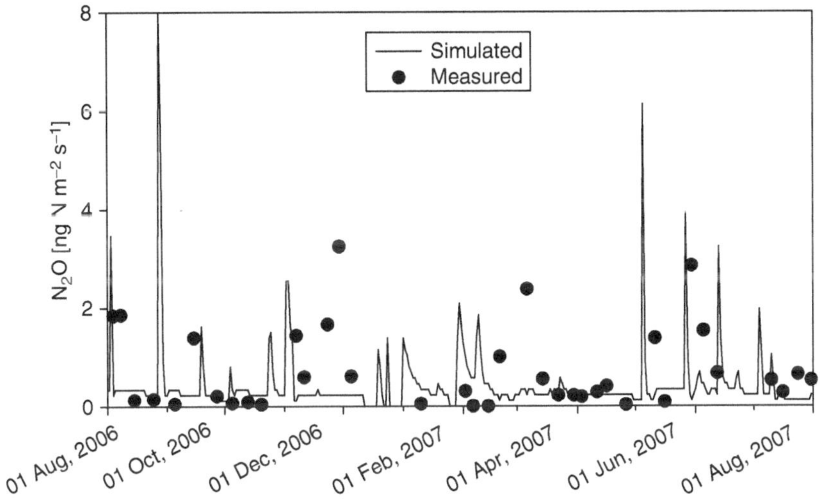

Fig. 10.7 Simulated and measured soil N_2O fluxes at Bugac (August 1, 2006–August 1, 2007)

N_2O flux measured indicates weak emission as a consequence of the low precipitation amount and the dry soil during the period examined (Fig. 10.7).

The soil surface is generally frozen during the cold periods and trace gases produced in the deeper soil layers are stored in the unfrozen subsoil. In spring, when thawing normally results in high N_2O emission peaks (Müller et al. 2002), we probably missed some of the N_2O emission peaks because of the weekly or fortnightly sampling. However, the magnitude of episodic emissions after thawing is low and has negligible effect on the yearly total flux. We could not validate the simulated CH_4 flux, because there was no CH_4 measurement during this period.

For the validation of the model in regional mode, the territory of the country was parceled by a $1/6 \times 1/6°$ grid resulting in 466 grid cells of approximately 200 km^2 each. Meteorological and soil property information were compiled for this grid. In this case, the input data are derived from spatial averaging of direct measurements. For more details, see Chapter 9.3.

For the validation of the model, we choose the closest cells to Bugac (46.69°N, 19.60°E, 111 m asl), Isaszeg (47.53°N, 19.32°E, 252 m asl), and Gödöllő (47.58°N, 19.37°E, 249 m asl.) monitoring sites, and compared the measured fluxes with the

Table 10.3 Comparison of the measured and the simulated N_2O emissions at three different grassland sites

	Measured (kg N_2O ha^{-1})	Modeled (kg N_2O ha^{-1})
Bugac 2003	1.29	1.01
Bugac 2004	0.91	1.59
Isaszeg 2004	1.41	1.48
Gödöllő 2004	1.18	1.45

estimations of DNDC for the relevant grid cells. The results are displayed in Table 10.3. One can see that the difference between the measured and modeled data is 20.2% (regarding the measured data) on average. The main reason for the difference may be that the point measurements are not representative enough for the grid cell simulated by the model in regional mode.

Results of comparisons suggest that the model in regional mode is likely applicable on the whole territory of Hungary in order to estimate the methane and nitrous oxide fluxes over grasslands.

10.3.2 *Analyzing the Relationships Between the Microclimate, Soil Processes, and the Soil Emission of Nitrous Oxide and Methane*

It is well known that meteorological factors influence the soil conditions and the soil conditions have impacts on soil GHG emission through soil processes. For this reason, effectiveness of soil microbial processes (e.g. denitrification) strongly depends on the influence of extreme weather conditions (e.g. water or heat stress). Trace gas emissions vary year-by-year, but the soil processes more or less explain the variability through the meteorological parameters (like air or soil temperature, precipitation amount, and soil moisture, etc.). Some uncertainties may arise from the complexity of soil and atmospheric processes and/or accuracy of modeling.

The magnitude of the annual sum of nitrous oxide flux (74–86 mg N m^{-2} $year^{-1}$) measured over a sandy, semiarid grass field (Bugac; for more details see: Chapter 6) is in a good agreement with the results of the simulation (48–100 mg N m^{-2} $year^{-1}$; see Table 10.4). Simulations were run in regional mode; therefore, the modeled fluxes represent a larger area than that of the measured values (plot scale vs. regional scale). The regional mode of the DNDC model calculates with uniform values of soil and meteorological parameters in a given grid cell not taking into account their spatial variability.

Based on global warming potential of N_2O, which is 298 times higher than the global warming potential of CO_2 on a 100-year time horizon (IPCC 2007), both the

Table 10.4 Comparison of the measured and modeled (DNDC regional mode) N_2O fluxes in dry and wet years at Bugac

Measurement period	Measured mean N_2O flux (± stand. dev.) (mg N m^{-2} $year^{-1}$)	Simulated mean N_2O flux (mg N m^{-2} $year^{-1}$)	Annual mean temperature (°C)	Precip. amount (mm)
Dry year 2002 (Sept.–Dec.)	86±140	48	11.5[a]	413[a]
Dry year 2003	80±100	64	9.8	456
Wet year 2004	74±90	100	10.0	728

[a] whole year

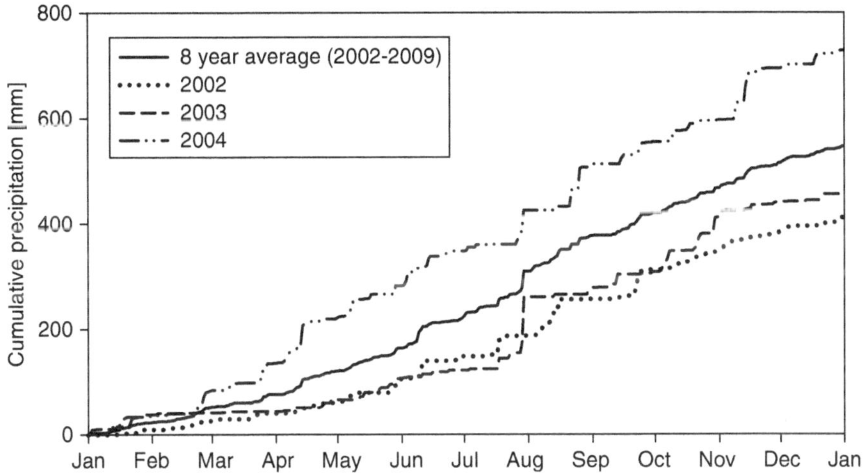

Fig. 10.8 Cumulative precipitation over the semiarid grassland at Bugac in different years, and the 8-year average (2002–2009)

measured and simulated N_2O emissions ranged between 144 and 300 kg CO_2-eq ha^{-1} $year^{-1}$ during the period examined.

In the years 2002–2003, the yearly precipitation amounts were much lower than the multiyear average (546 mm) (Fig. 10.8). The deficit in the precipitation mainly occurred in summer. In 2002, the annual mean temperature was 1°C higher than the multi-annual average (10.4°C). Simulated daily nitrous oxide fluxes are positively correlated with soil temperature as soil microbial activity is increasing with soil temperature (Meixner and Yang 2006). The more intensive microbial activity at higher soil temperature can be suppressed by soil moisture level not favoring the production of N_2O. In 2003, due to the extreme low precipitation (Fig. 10.8), the denitrification process was still limited since soil wetness was usually lower than optimum for nitrous oxide production and following emission (WFPS 50–70%; Davidson 1991).

In 2004, the annual mean temperature was the same as in 2003, but the sum of the annual precipitation was higher. This is why the model gave a significantly increased emission. In contrast, substantial increase was not observed in the measured flux. It can be explained partly by the fact that the sandy soil dried out within a short time after the rain events resulted that the effective time for denitrifier bacterial community preferring anaerobic condition was shorter than in soils with higher water-retaining capacity. On the other hand, in extreme wet soils, loss of nitrous oxide is suppressed, reduction leads to molecular nitrogen (N_2). The weather conditions (heat or water stress) prevented keeping the optimum soil condition for nitrous oxide production in extended periods of the observation.

During 2002–2004, the simulated annual total methane fluxes were zero. In the DNDC model, a submodel simulates the CH_4 production/consumption (Li 2007) through aeration status determined by a kinetic scheme. Based on the predicted redox potential (calculated by Nerst equation), the soil layers are divided into relatively aerobic and anaerobic volumetric fractions (driven by precipitation, irrigation, etc.). Based on the ratio of these two microsites, the substrates (e.g., NO_3^-, NH_4^+) are allocated into the oxic and anoxic parts in the soil. Within these two volumetric fractions, the substrates can only be involved in the oxidation reactions (e.g., nitrification, methanotrophy) or in the reductive processes (e.g., denitrification, methanogenesis) (Li 2007). Both processes can occur in the soil simultaneously. With decreasing soil redox potential, the CH_4 production is increasing rapidly. Based on the observations, DNDC model predicts CH_4 production rate as a function of dissolved organic content (DOC) and temperature, when the estimated soil redox potential reaches −150 mV. CH_4 diffusion between soil layers is simulated as a function of CH_4 concentration gradients, temperature, and porosity in the soil. Methane oxidation rate depends on the estimated soil CH_4 concentration and soil redox potential.

The zero methane flux resulted by the model can be explained by the dry and hot summers and extremely dry soil in the examined period (especially in 2003). It did not favor either the catalytic surface decomposition or the anaerobic formation (emission) of methane. According to the field observations, CH_4 emission is strongly controlled by soil carbon available, soil redox potential (Eh), temperature, and moisture (Machon et al. 2008). The reduction of available carbon to CH_4 is mediated by anaerobic microbes (e.g., methanogens) that are only active when the soil redox potential is low enough. At Bugac grassland, the sandy soil was usually well aerated (also in the relatively wet 2004) and the soil redox potential was high due to the low water-filled pore space (see Fig. 10.9). Therefore, CH_4 production was inhibited both years, especially in 2003.

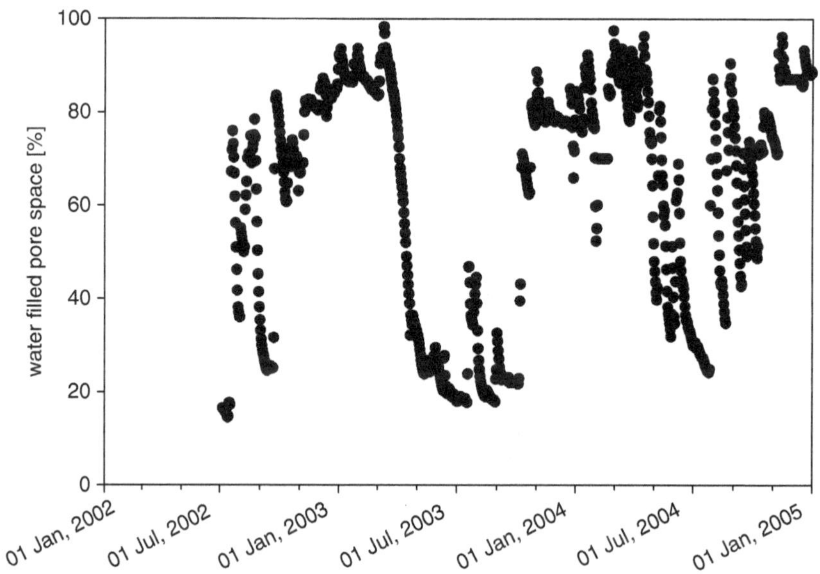

Fig. 10.9 Variation of the daily average water-filled pore space at Bugac (2002–2004)

On the other hand, methane can be oxidized by aerobic methanotrophs in the soil. It is assumed that CH_4 produced in deeper layers with low soil redox potential by anaerobic microcommunity diffuses up to aerated, topsoil layer where aerobic bacterial community dominates and the higher redox circumstances accelerate the oxidation of methane (Li 2000). In this way, a significant part of CH_4 produced in the deeper layers can be oxidized before leaving the soil (Schipper and Reddy 1996). Oxidation rate of CH_4 is calculated by the model as a function of CH_4 concentration and soil redox potential.

Machon et al. (2008) reported a 1.8 $\mu gC\ m^{-2}\ h^{-1}$ mean methane flux from Bugac under meteorological condition similar to that during the modeled period. This methane flux could be considered negligible because the yearly emission summed up only 5.2 kg CO_2-eq ha^{-1} taking into account the global warming potential of methane on a 100-year time horizon (25; IPCC 2007).

10.4 Conclusion

Using the Biome-BGC and DNDC models, we simulated the carbon and nitrogen budgets of Hungarian grasslands. Calibration of the Biome-BGC model was performed using Bayesian approach with measurement data from three Hungarian measurement sites (Bugac, Hegyhátsál, and Mátra). It was found that after calibration, the model is an appropriate tool for the estimation of the carbon dioxide exchange and the latent heat flux. We examined the site-specific dependence of the model parameters. Using the extended, calibrated, and validated model, we compared the carbon dioxide budgets of the three Hungarian measurement sites with different site-specific conditions. We have found that carbon dioxide fluxes are strongly affected by the soil type: carbon dioxide uptake is the highest at the managed grassland on silt soil (Hegyhátsál) and the lowest at the grassland on mountain heavy clay soil (Mátra). Thanks to the implementation of management modules, we could test the short-term and long-term effects of management activities (mowing and grazing). We pointed out that the mowed or grazed ecosystem can turn into a net carbon dioxide source from a net carbon dioxide sink (NBP is negative assuming mowing or grazing)

Soil N_2O emission simulation of DNDC in spot and regional mode was validated using data measured at Bugac. The differences between the measured and modeled data were 20.2% on average. The main reason of the difference is the lack of local measurements regarding to the input data.

The most important factors influencing the fluxes are denitrification and nitrification for N_2O, and decomposition for CH_4. These processes are strongly depending on the actual weather conditions as they control the state of the soil. We analyzed the N_2O and CH_4 fluxes for the period of 2002–2004 regarding Bugac, and we examined the effect of the different weather conditions on the decomposition processes. We pointed out that the extreme hot weather and the drying-out of the soil limited the N_2O fluxes.

The results of the validations suggest that the Biome-BGC and DNDC models can be used to estimate the biospheric fluxes of CO_2, N_2O, and CH_4 accurately in Hungary.

Acknowledgments Biome-BGC version 4.1.1 was provided by the Numerical Terradynamic Simulation Group (NTSG) at the University of Montana. NTSG assumes no responsibility for the proper use of Biome-BGC by others. We acknowledge the support of the Hungarian National Scientific Research Fund (OTKA T23811, T42941, OTKA-NKTH CK77550), the Hungarian Ministry of Economy and Transport (GVOP-3.2.1.-2004-04-0107/3.0, GVOP-AKF-3.1.1.-2004-05-0358/3.0), the INTERREG IIIB CADSES program (5D038), and the European Commission's 5th and 6th R&D Framework Programmes (EVK2-CT-1999-00013, EVK2-CT-2002-00105, EVK2-CT-2002-00163, GOCE-CT-2003-505572, GOCE-CT-17841, GOCE-037005, IMECC – Project No. 026188). The authors are grateful for the permission of measurements on the territory of Kiskunság National Park.

References

Barcza Z, Haszpra L, Kondo H, Saigusa N, Yamamoto S, Bartholy J (2003) Carbon exchange of grass in Hungary. Tellus B 55:187–196

Chapin FS, Woodwell GM, Randerson JT, Rastetter EB, Lovett GM, Baldocchi DD, Clark DA, Harmon ME, Schimel DS, Valentini R, Wirth C, Aber JD, Cole JJ, Goulden ML, Harden JW, Heimann M, Howarth RW, Matson PA, McGuire AD, Melillo JM, Mooney HA, Neff JC, Houghton RA, Pace ML, Ryan MG, Running SW, Sala OE, Schlesinger WH, Schulze E-D (2005) Reconciling carbon-cycle concepts, terminology and methods. Ecosystems 9:1041–1050. doi:10.1007/s10021-005-0105-7

Ciais P, Tans PP, Trolier MJ, White WC, Francey RJ (1995) A large Northern Hemisphere terrestrial CO_2 sink indicated by the $^{13}C/^{12}C$ ratio of atmospheric CO_2. Science 269:1098–1102

Davidson EA (1991) Fluxes of nitrous oxide and nitric oxide from terrestrial ecosystems. In: Roger JE, Whitman WB (eds) Microbial production and consumption of greenhouse gases: methane, nitrogen oxides and halomethanes. Am Soc Microbiol, Washington DC, pp 219–235

Frank AB (2004) Six years of CO_2 flux measurements for a moderately grazed mixed-grass prairie. Environ Manage 33:S426–S431

Franks SW, Beven KJ, Gash JHC (1999) Multi-objective conditioning of a simple SVAT model. Hydrol Earth Syst Sci 3:477–489

IPCC (2007) The physical science basis – contribution of working group I to the fourth assessment report of the IPCC. Cambridge University Press, Cambridge, New York

Li CS (2000) Modeling trace gas emissions from agricultural ecosystems. Nutr Cycl Agroecosyst 58:259–276

Li CS (2007) Quantifying greenhouse gas emissions from soils: Scientific basis and modeling approach. Soil Sci Plant Nutr 53:344–352

Machon A, Grosz B, Horváth L, Pintér K, Tuba Z (2008) Non-CO_2 greenhouse gas flux measurement above a nature reserve grassland in Kiskunság in an unusual year. Cereal Res Commun 36:203–206

Meixner FX, Yang WX (2006) Biogenic emissions of nitric oxide and nitrous oxide from arid and semi-arid land. In: D'Odorico P, Porporato A (eds) Dryland ecohydrology. Springer, Dordrecht, Berlin, Heidelberg, New York, pp 233–255. ISBN 1-4020-4261-2

Mo X, Beven K (2004) Multi-objective parameter conditioning of a three-source wheat canopy model. Agric For Meteorol 122:39–63

Müller C, Martin M, Stevens RJ, Laughlin RJ, Kammann C, Ottow JCG (2002) Processes leading to N_2O emissions in grassland soil during freezing and thawing. Soil Biol Biochem 34:1325–1331
Pintér K, Barcza Z, Balogh J, Czóbel S, Csintalan Z, Tuba Z, Nagy Z (2008) Interannual variability of grasslands' carbon balance depends on soil type. Community Ecol 9:43–48
Schipper LA, Reddy KR (1996) Determination of methane oxidation in the rhizosphere of Sagittaria lancifolia using methyl fluoride. Soil Sci Soc Am J 60:611–616
Tardieu F, Simonneau T (1998) Variability among species of stomatal control under fluctuating soil water status and evaporative demand: modelling isohydric and anisohydric behaviours. J Exp Bot 49:419–432

Chapter 11
Forests*

Zoltán Somogyi

Abstract For modeling carbon stock changes in forests, currently a dedicated Hungarian model, CASMOFOR can be used. CASMOFOR was originally developed mainly to allow one to estimate the carbon stock changes of various afforestation projects. This section presents modeling results for afforestations that have taken place in Hungary since 1990, as well as that could be implemented in the coming decades. For modeling future afforestations, several scenarios were developed with regard to the intensity, species structure, and site distribution of possible projects. The amount of carbon stock changes estimated by the model varied quite substantially by these factors, but also by carbon pools and calendar year. In addition, the specific costs of carbon sequestration were assessed, and they showed a sharply declining value over time. Thus, the results not only demonstrate the dynamics of carbon sequestration of forests, but also the competitiveness of removing carbon from the air through afforestations.

Keywords CASMOFOR • Carbon stock changes • Afforestation • Kyoto Protocol • UNFCCC • Specific costs

---

*Citation: Somogyi, Z., 2010: Modeling of biosphere–atmosphere exchange of greenhouse gases – Forests. In: Atmospheric Greenhouse Gases: The Hungarian Perspective (Ed.: Haszpra, L.), pp. 253–261.

Z. Somogyi (✉)
Hungarian Forest Research Institute, H-1142 Budapest, Stefánia út. 14., Hungary
e-mail: som9013@helka.iif.hu

L. Haszpra (ed.), *Atmospheric Greenhouse Gases: The Hungarian Perspective*,
DOI 10.1007/978-90-481-9950-1_11,

11.1 Introduction

For forests, there is currently one model in Hungary, i.e. CASMOFOR (see Chapter 9), that can be used to develop estimates of emissions and removals at country level for forests. This model has been used to estimate the carbon stock changes of both past afforestations and future options for further conversions of land to forests. What follows is a summary of recent modeling exercises to demonstrate how the estimates were developed, how much sequestration was estimated for past afforestations, how much the further afforestation sink potential might be if specific afforestation programs are implemented, and how CASMOFOR could be used in decision making related to afforestation programs.

11.2 Estimation of the Amount of Carbon Sequestered by Afforestation Programs Since 1990

Under the Kyoto Protocol (KP) of the United Nations Framework Convention on Climate Change (UNFCCC), the net removals of all forest areas that were afforested after January 1, 1990 (so-called Article 3.3 AR[1] land), have to be estimated for each calendar year of 2008–2012. In order to facilitate this estimation, and to assess the dynamics of carbon sequestration, the carbon sequestration of the Article 3.3 AR land for 1990–2012 was estimated in 2006. Presenting the estimates also provides an example of how estimates for various future afforestation programs can be developed.

Projections can only be made using models as no stand level carbon-related information on the afforested lands, which could have been used for extrapolation, was collected until 2006, therefore, using CASMOFOR was the only way to develop estimates with respect to the carbon sequestration of the afforestation program. In 2006, only the area of the above afforestations was known for each year by species; however, expert judgment was necessary to allocate the area by site class. In order to be conservative, below-than-average tree growth rates were supposed.

The area of land afforested between 1990 and 2005 was over 170,000 ha (Somogyi and Horváth 2006). No afforestations were assumed after 2006, again in order to be conservative. The results of the estimation for 1990–2012 as produced by CASMOFOR (Fig. 11.1) indicate that the afforested land is a strong net sink of carbon. About half of all carbon is fixed in the aboveground biomass pool, but additional carbon is fixed in the belowground biomass carbon pool, as well as in the litter, which mainly acts as sink until it is fully built up. The annual rate of the (net) sequestration in all pools as well as their ratio are the net resultant of a complex chain of fluxes. Both the annual rates and their ratios change over time.

[1] Article 3.3 is an essential article of the KP, which states that carbon stock changes in land that is afforested or reforested (AR), as well as land that is deforested, since January 1, 1990, must be accounted for to comply with the commitments made in the KP itself.

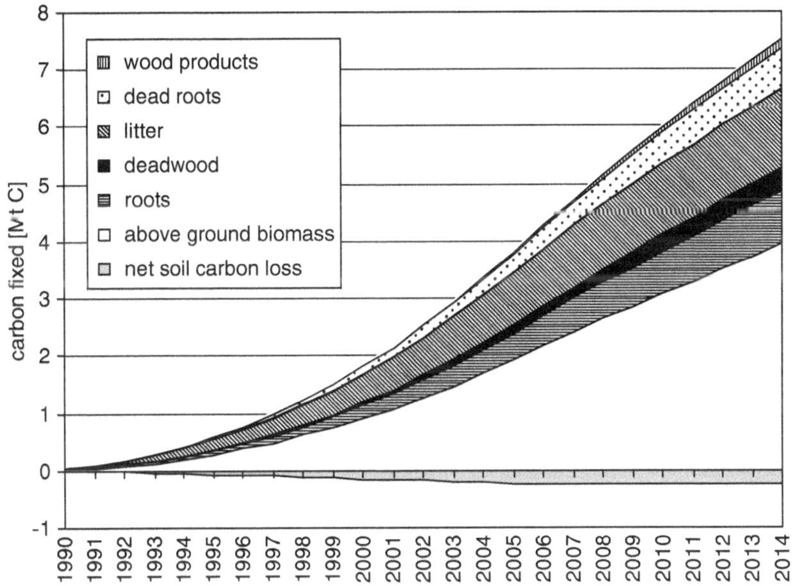

Fig. 11.1 Carbon stocks over time (1990–2014) in the various pools of all forests that were established between 1990 and 2005, supposing that half of all lands had been cropland, whereas the other half had been grassland. Note that the figure includes the pool of wood products for demonstration purposes only as no robust assumption can be made on how long carbon remains sequestered in the various wood products in the future

The annual rate of change in each pool depends not only on the age and species composition of the stands, but also on the site patterns.

Note that land was only afforested between 1990 and 2005, and adding land to the system was one major factor in the accumulation of carbon in the system. After 2006, as mentioned above, no more afforestations are assumed, and the increase in the carbon of the aboveground biomass pool becomes less steep. However, the amount of carbon in the system keeps growing for decades after 2005.

Note also how CASMOFOR models carbon dynamics for all major forest carbon pools. Concerning soil carbon, the chronosequence estimates by Horváth (2006) are used that predict carbon losses for decades if grassland is converted to forest, but no losses if cropland is converted (see also Chapter 15). However, due to lack of proper data, assumptions had to be made with respect to the area of the pre-afforestation land uses, i.e. cropland and grassland. By conservatively supposing that half of the afforested area was cropland, and the other half was grassland before the afforestation, the results indicate a small net loss of soil carbon (0.23 MtC in 2014, the "net soil carbon loss" values in Fig. 11.1). On the other hand, when supposing in another simulation that two-thirds of the area was cropland, the initial net losses are offset by net gains, and the soil starts to sequester carbon after 2010. Based on recently analyzed historical afforestation data (Central Agricultural Authority, personal communication, 2010), it is supposed that more than four-fifth of the afforested land was cropland before the afforestations, thus, little, if any, carbon stock losses occurred in the afforested land; therefore, the estimated emissions in Fig. 11.1 are overestimations.

Note that CASMOFOR separately models the deadwood pool and the dead roots pool, with the deadwood only containing carbon that is transferred to it from the aboveground biomass pool. This pool definition differs from that of IPCC (2003, 2006), according to which the dead roots are also included in the deadwood pool. In fact, some carbon of both pools is transferred to the soil pool through a process that takes years. During this time, the carbon is not in the soil pool. Therefore, CASMOFOR separately accounts for the carbon in the two dead organic matter pools; however, these amounts could be added up as one dead wood pool for specific accounting purposes (such as accounting under the KP).

As an alternative to the IPCC definitions, it could also be assumed that most of the carbon in these pools is immediately emitted as CO_2, just like with the harvested wood in the IPCC Tier 1 assumption (IPCC 2003, 2006), and all of the nondegradable carbon in these pools is immediately transferred into the soil pool. In this case, the carbon stock changes in the soil would include the nondegradable carbon from both the deadwood and the dead roots pool, and these latter two pools would be considered as empty ones.

11.3 Estimating the Carbon Sequestration Potential of Hungary

CASMOFOR could also be used to assess the development of the carbon stocks in future afforestation programs (Somogyi 2000). In order to estimate the carbon sequestration potentials in the country for the next two decades, the following storylines were assumed:

- The forest area will continue to increase. The estimated future rate will vary between 5,000 and 20,000 ha annually.
- The level of harvests is going to stay at the current level. In other words, a sustainable forest management is supposed, and it is assumed that the current silvicultural models can be used in the coming two decades. This is supported by current timber market trends. However, these markets may become rather volatile, and the two decades of the projection period is too long to see what may happen in case of possible wood shortages later.
- The rate of volume increment will not change. This assumption may not hold true, because Somogyi (2008) found that tree growth rate has increased in the past few decades, and, if the underlying factors of this increase continue to change, they may further increase tree growth at least in the next two decades. Thus, the assumption of no tree volume increment change is a conservative one.
- This rate does not depend much on any activity that is within the limits of sustainable forestry, rather, it depends on previous interventions, and possible future changes of site conditions.

When developing scenarios to study the carbon sequestration potentials, several factors must be considered; the first of which being the area of land available for conversion. Based on the historical rates, it can be predicted that, depending on national resources, the annual afforestation rate can change between about 5,000 and 25,000 ha in the future.

The site fertility of the area available for the afforestations, termed in forestry as the "yield potential" of the area, is modeled through selecting different so-called yield classes. In Hungary, six yield classes are used, yield class 1 indicating the best sites, and yield class 6 indicating the poorest ones.

Yet another factor that considerably affects carbon sequestration is the tree species selected. Tree growth, or yield, is species dependent in both terms of the amount of wood produced until any reference time and the pattern of growth, i.e. the growth rate over time. There are "fast growing" species, like Black locust and poplars, which grow fast in the first 1–2 decades after establishment, however, they slow down or stop growing later, whereas the "slow growing species" like oak and beech grow more slowly in the first decades but keep growing for many decades. Mainly because of this, the amount of carbon sequestered highly depends on time (i.e., calendar year), and any evaluation of results must strictly follow this dependence.

Finally, the amount of carbon sequestered also highly depends on the timing of the afforestations themselves. Therefore, it is also important to consider whether more afforestation is done in the first years of a scenario, and less is done later (which is more favorable as it would yield more sequestration), or less is done in the first years and the afforestation rate is increased later.

Based on the above, the following area scenarios have been developed (Somogyi 2009):

- Four scenarios of constant rate of afforestations, i.e. 5, 10, 15, and 20 thousand ha a year until 2020
- Additionally, four scenarios varying rates of the afforestations, i.e. one with an increase from the current level (i.e. 10,000 ha a year) to 15,000 ha; one with an increase from current level to 20,000 ha a year; one with a continuous decrease from current level to 5,000 ha; and one with a continuous decrease from 15,000 to 5 ha by 2020.

For all of the above eight area scenarios, two scenarios were developed by species composition: one with predominantly fast growing species and the other with predominantly slow growing species (the ratio of the species being the same over time; see Table 11.1). Finally, for all the above 16 scenarios, two scenarios were defined by yield class: one with assuming better sites (i.e. yield class 3) and the other with assuming poorer sites (i.e. yield class 4).

Thus, altogether $(4 + 4) \times 2 \times 2 = 32$ scenarios were defined, of which one can be considered as baseline (i.e. constant 10,000 ha annual afforestation of predominantly slow growing species and assuming poorer sites) and the other 31 scenarios can be considered as mitigation scenarios to analyze.

All scenarios were simulated by CASMOFOR using the above specifications. The results of the calculations to estimate the amount of carbon fixed can be found in Table 11.2.

Table 11.1 The species composition of the "slow" and "fast" species groups used for the simulations

"Slow"	Percent of area	"Fast"	Percent of area
Sessile oak	20	Sessile oak	10
Pedunculate oak	20	Pedunculate oak	10
Turkey oak	30	Turkey oak	10
Black locust	15	Black locust	30
Poplar	8	Poplar	30
Scotch pine	7	Scotch pine	10

Table 11.2 The results of the simulations to estimate the amount of carbon fixed by afforestation scenarios by 2100

Scenario	Total area (10^3 ha year^{-1})	Species structure type	Yield class	Amount of carbon dioxide fixed by calendar year (Mt CO_2)			
				2015	2020	2025	2100
1	5, constant	Slow	3	0.72	2.85	6.45	48.08
2	10, constant	Slow	3	1.44	5.70	12.91	96.16
3	15, constant	Slow	3	2.16	8.55	19.36	144.25
4	20, constant	Slow	3	2.87	11.40	25.81	192.33
5	10 → 15	Slow	3	1.50	6.21	14.73	120.35
6	10 → 20	Slow	3	1.56	6.73	16.55	144.53
7	10 → 5	Slow	3	1.38	5.18	11.09	71.98
8	15 → 5	Slow	3	2.03	7.52	15.72	95.88
9	5, constant	Fast	3	0.95	3.63	7.94	37.04
10	10, constant	Fast	3	1.90	7.25	15.88	74.09
11	15, constant	Fast	3	2.85	10.88	23.81	111.13
12	20, constant	Fast	3	3.81	14.51	31.75	148.18
13	10 → 15	Fast	3	1.99	7.93	18.18	92.81
14	10 → 20	Fast	3	2.07	8.60	20.49	111.53
15	10 → 5	Fast	3	1.82	6.58	13.57	55.37
16	15 → 5	Fast	3	2.69	9.54	19.20	73.69
17	5, constant	Slow	4	0.58	2.36	5.34	38.90
18	10, constant	Slow	4	1.17	4.72	10.68	77.81
19	15, constant	Slow	4	1.75	7.08	16.02	116.71
20	20, constant	Slow	4	2.34	9.43	21.36	155.62
21	10 → 15	Slow	4	1.22	5.14	12.18	97.24
22	10 → 20	Slow	4	1.26	5.56	13.69	116.67
23	10 → 5	Slow	4	1.12	4.30	9.18	58.38
24	15 → 5	Slow	4	1.66	6.23	13.01	77.85
25	5, constant	Fast	4	0.75	2.89	6.33	29.71
26	10, constant	Fast	4	1.50	5.77	12.65	59.43
27	15, constant	Fast	4	2.25	8.66	18.98	89.14
28	20, constant	Fast	4	3.00	11.54	25.30	118.85
29	10 → 15	Fast	4	1.56	6.30	14.49	74.35
30	10 → 20	Fast	4	1.63	6.83	16.32	89.28
31	10 → 5	Fast	4	1.43	5.24	10.82	44.50
32	15 → 5	Fast	4	2.12	7.60	15.31	59.29

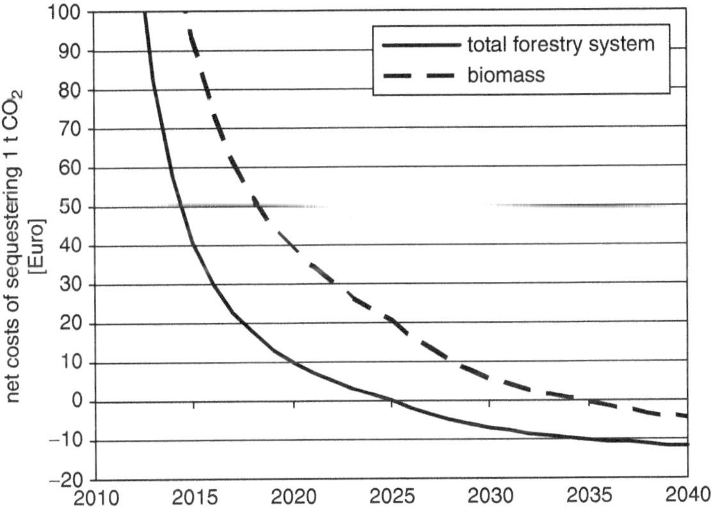

Fig. 11.2 Cost curves for the scenario of 10,000 ha of afforestation each year until 2020 with predominantly "fast" growing species on "good" sites (yield class 3) for both biomass (including belowground biomass) and the total forestry system (Scenario 18 in Table 11.2). Net costs exclude net revenues from selling the sequestered carbon for

Some general conclusions that can be drawn from Table 11.1 include:

- As expected, the amount of carbon sequestered is strongly related to the area afforested.
- Also as expected, the rate of afforestation over time has a strong effect, too.
- The quality of the sites, modeled here by the respective yield classes, is also important; the amount sequestered can be increased by some 25% if trees are planted on better sites.
- The effect of the species composition depends on the length of the projection. In the short run, the "fast" system seems to be superior; however, in the long run, the "slow" system sequesters more carbon.

That projection length has an effect is, in general, true for all major factors that affect the amount of carbon sequestered. But it also determines the unit cost of sequestering carbon, i.e. how much it costs to sequester 1 t of CO_2. The general tendency of this specific cost as predicted by CASMOFOR is demonstrated in Fig. 11.2. The specific cost relative to either the biomass or the total forestry system dramatically decreases over time, and is reduced to less than 20€ by 2025 and 2018, respectively, demonstrating how cost-effective afforestations can be.

The amount of carbon sequestered in combination with the specific cost information is also very useful to optimize where and how afforestation programs could be conducted to achieve the double objectives of climate change mitigation, i.e. carbon sequestration and cost-effectiveness. As Fig. 11.3 demonstrates, there are relatively large differences between the various programs, and, given the appropriate

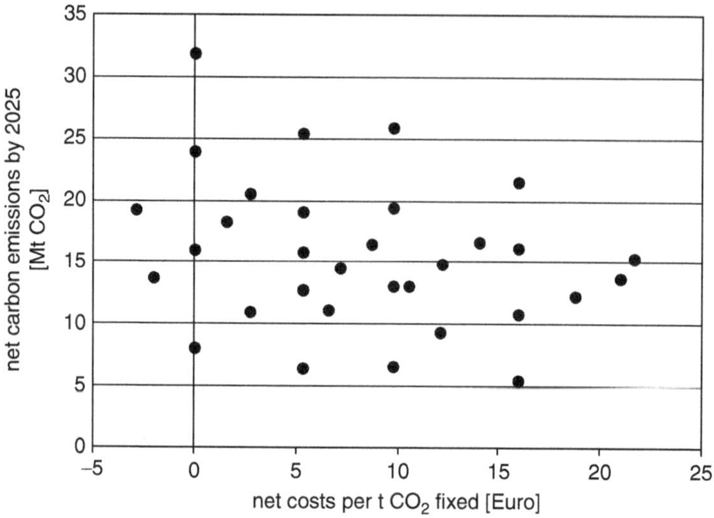

Fig. 11.3 Total net carbon removals by 2025, and net specific costs of sequestering 1 t CO_2 as estimated for the 32 afforestation scenarios

information, there are large potentials in taking decisions to maximize carbon benefits, minimize costs, as well as optimize all other benefits of afforestation programs.

Acknowledgments This research was funded by the European Commission as part of the CarboInvent project (Multi-Source Inventory Methods For Quantifying Carbon Stocks And Stock Changes In European Forests; Contract number EVK2-CT-2002-00157), see www.joanneum.at/CarboInvent. COST E21 also provided opportunities for the author to obtain valuable information at both whole action meetings, as well as task force meetings. The author is also grateful to the following organizations for providing assistance for the preparation of this paper: DG Joint Research Centre, Ispra; Erdészeti Tudományos Intézet (Forest Research Institute, ERTI), Budapest; Nyugat-Magyarországi Egyetem (University of West Hungary), Sopron; Kiskunsági Erdészeti és Faipari Rt. (Kiskunsag Forestry and Wood Company, KEFAG Rt.), Kecskemét; Állami Erdészeti Szolgálat Kecskeméti Igazgatósága (Kecskemet Directorate of the State Forest Service), Kecskemét, and the Ministry of Environment and Water.

References

Horváth B (2006) C-Accumulation in the soil after afforestation: contribution to C-mitigation in Hungary? Forstarchiv 77:63–68

IPCC (2003) In: Penman J, Gytarsky M, Hiraishi T, Kruger D, Pipatti R, Buendia L, Miwa K, Ngara T, Tanabe K, Wagner F (eds) Good practice guidance for land use, land-use change and forestry. IPCC/IGES, Hayama, Japan

IPCC (2006) In: Eggleston HS, Miwa K, Ngara T, Tanabe K (eds) 2006 IPCC guidelines for national greenhouse gas inventories. IGES, Hayama, Japan

Somogyi Z (2000) Possibilities for carbon mitigation in the forestry sector in Hungary. Biotechnol Agron Soc Environ 4:296–299

Somogyi Z (2008) Recent trends of tree growth in relation to climate change in Hungary. Acta Silvatica Lignaria Hungarica 4:17–27. http://aslh.nyme.hu/fileadmin/dokumentumok/fmk/acta_silvatica/cikkek/Vol04-2008/02_somogyi_p.pdf

Somogyi Z (2009) CASMOFOR: a decision-making tool for analysing projections of afforestations. Technical workshop on projections of GHG emissions and removals in the LULUCF sector. Joint Research Centre, Ispra, Varese, Italy, pp 27–28, January 2009. http://afoludata.jrc.ec.europa.eu/v2007/events/Kyoto_projections_workshop/Somogyi.pdf

Somogyi Z, Horváth B (2006) Az 1930 óta telepített erdők szénlekötéséről (On the carbon sequestration of forests established after 1930). Erdészeti Lapok 116:257–259

Chapter 12
Arable Lands*

Balázs Grosz, Györgyi Gelybó, Galina Churkina, László Haszpra, Dóra Hidy, László Horváth, Anikó Kern, Natascha Kljun, Attila Machon, László Pásztor, and Zoltán Barcza

*Citation: Grosz B, Gelybó Gy, Churkina G, Haszpra L, Hidy D, Horváth L, Kern A, Kljun N, Machon A, Pásztor L, Barcza Z (2010) Modeling of biosphere–atmosphere exchange of greenhouse gases – Arable lands. In: Haszpra L (ed) Atmospheric greenhouse gases: the Hungarian perspective, pp. 263–293

B. Grosz (✉)
Department of General and Inorganic Chemistry, Eötvös Loránd University, H-1117 Budapest, Pázmány P. s. 1/A, Hungary
e-mail: grosz@mail.datanet.hu

L. Pásztor
Research Institute for Soil Science and Agricultural Chemistry of the Hungarian Academy of Sciences, Herman O. út 15, H-1022 Budapest, Hungary

A. Machon
Department of Botany and Plant Physiology, Szent István University, Páter K. u. 1., H-2103 Gödöllő, Hungary

L. Haszpra, D. Hidy, L. Horváth and A. Machon
Hungarian Meteorological Service, H-1675 Budapest, P.O. Box 39, Hungary

G. Churkina
Leibniz-Centre for Agricultural Landscape Research, D-15374, Müncheberg, Eberswalder Strasse 84, Germany

Gy. Gelybó, A. Kern and Z. Barcza
Department of Meteorology, Eötvös Loránd University, H-1117 Budapest, Pázmány P. s. 1/A, Hungary

N. Kljun
Department of Geography, Swansea University, Singleton Park, SA2 8PP, Swansea, United Kingdom

A. Machon
Hungarian Meteorological Service, H-1675 Budapest, P.O. Box 38, Hungary
and
Center for Environmental Science, Eötvös Loránd University, Pázmány P. s. 1/A, H-1117, Budapest, Hungary
and
Department of Botany and Plant Physiology, Szent István University, H-2103 Gödöllő, Páter K. u. 1., Hungary

A. Kern
Adaptation to Climate Change Research Group, Hungarian Academy of Sciences, Pázmány P. s. 1/A, H-1117, Budapest, Hungary

L. Haszpra (ed.), *Atmospheric Greenhouse Gases: The Hungarian Perspective*,
DOI 10.1007/978-90-481-9950-1_12,

Abstract Three different, state-of-the-art biogeochemical models are adapted in order to quantify the carbon dioxide (CO_2), nitrous oxide (N_2O), and methane (CH_4) exchanges between arable lands and the atmosphere. Biome-BGC and MOD17 are validated using the long-term measurement results from Hegyhátsál tall tower site (Western Hungary). After a simple bias correction, it was found that Biome-BGC is capable to describe the net ecosystem exchange (NEE) of Hungarian arable lands. The model is used to estimate the net primary production (NPP) and the net biome production (NBP) in the vicinity of the tall tower for the period of 1997–2008. The validation of MOD17 shows that the model is suitable to describe the gross primary production (GPP) of arable lands, although the model has limitations in years with higher precipitation amount. Introducing a new method MOD17 is downscaled to 250 m resolution, which is close to the size of agricultural parcels in the region. The downscaling helps us to better understand the biogeochemical processes of individual parcels with different crop cultivars. The DNDC model is parameterized using country-specific fertilizer input and climate data to provide estimates of the N_2O and CH_4 emissions for different crop cultivars. Sensitivity analysis was performed to find the most important parameters needed for accurate emission estimation. The results show that N_2O emission is highly variable for the different crop types, and it has significant contribution to the GHG balance of arable lands. Methane has bidirectional fluxes and arable lands were generally net sinks of CH_4 in the study period (2002–2004).

Keywords Biogeochemical models • greenhouse gases • net ecosystem exchange • net primary production • nitrification/denitrification

12.1 Introduction

Arable lands occupy about half of the total territory of Hungary. Due to their large extent arable lands are expected to dominate the biospheric greenhouse gas (GHG) balance of Hungary. In order to constrain the Hungarian biospheric GHG balance, we need to provide accurate and possibly bias-free, estimates on plot or regional scale. This can be accomplished through the adaptation of state-of-the-art biogeochemical models and with the full utilization of the existing measurements.

In this chapter, we present the adaptation of three different models (Biome-BGC, MOD17, and DNDC) that are capable to simulate GHG fluxes of arable lands. Biome-BGC is calibrated using the Hungarian tall tower (Hegyhátsál) eddy covariance measurements to provide reliable CO_2 flux estimates at plot scale and possibly at country level. Net biome production (NBP) of arable lands is estimated using ancillary crop yield data and information about residue management in the Hegyhátsál region. The MOD17 remote-sensing-based light use efficiency model is also used for the Hegyhátsál region. As the footprint of the Hungarian tall tower eddy covariance measurement is comparable to the spatial resolution of the MOD17 model, we can critically estimate the accuracy of the model product for arable lands. The third model is the DNDC (denitrification–decomposition) model. The fluxes of non-CO_2 greenhouse gases were simulated for the whole agriculture area (5.12 million hectare) of Hungary by DNDC.

12.2 Simulation of Biospheric Carbon Dioxide Exchange of Arable Lands with Biome-BGC

12.2.1 Introduction

Biome-BGC and other process oriented biogeochemical models simulate the carbon dioxide and water exchange between the biosphere and the atmosphere. Traditionally, the models were developed to simulate biogeochemical processes of undisturbed ecosystems (Running and Hunt 1993; Vetter et al. 2008). The research focus has switched to managed ecosystems after the recognition that the human alteration of the world has a superior role in the formation of the land surface processes through agricultural practices, forest management, land use conversion, erosion, peatland drainage, fire suppression, irrigation, grazing, etc. (Vitousek et al. 1997). The modeling community has faced the problems that biogeochemical models designed for natural ecosystems are unable to quantify the carbon dioxide exchange of managed ecosystems adequately.

The biogeochemical processes of arable lands are inherently affected by management practices (deep plowing, sowing, harvest, application of fertilizers, crop rotation, intercropping, etc.) that strongly influence the carbon, water, and nitrogen cycles of the ecosystem. Harvest and plowing cause a sudden drop in the standing biomass and alters the carbon pools (aboveground biomass, litter), which in turn affects the carbon balance and the physiological response of the ecosystem to the environmental conditions. The current version of Biome-BGC (version 4.1.1) is unable to simulate disturbance and cropland management; therefore, in many cases, croplands are handled as natural or fertilized grasses (Vetter et al. 2008). Using the grassland parameterization for crops inevitably causes problems as the growing season of agricultural crops generally does not follow the natural growing period for grasslands (Vetter et al. 2008). For example, usually the growing season of winter crops ends in July due to harvest but for natural grass it ends much later. Consequently, substantial biases may occur in the annual carbon balance components if we compare the model results with measurements at plot level (Gervois et al. 2004).

However, considering larger regions where both winter and summer crops are present, the annual cycle of net ecosystem exchange (NEE) becomes more or less balanced in time (i.e., there is no sudden decrease in NEE caused by harvest), as it is obvious from the large scale eddy covariance data measured over a mixed agricultural area in Hungary (Haszpra et al. 2005; Figs. 5 and 8 therein). The main cause of the balanced behavior is the joint presence of winter crops (e.g., winter wheat, harvested around June–July) and summer crops (e.g., maize, harvested around October–November; Barcza et al. 2009a). The measurement data suggest that the overall carbon dioxide balance of arable lands can be approximated with the carbon dynamics of a perennial herbaceous ecosystem. In this sense, croplands can be handled as seminatural grasslands without any sharp decrease in the standing biomass (thus no modification in the model logic is needed for the simulation). This is the only reason why Biome-BGC can be used to simulate agricultural NEE with reasonable accuracy and why crops can be handled as a kind of "super grass" (i.e., fertilized grass, see Vetter et al. 2008) using the existing C3 grass parameterization. The success or failure of the present modeling effort strongly depends on the quality and proper use of the dataset used for the training of the model.

The only monitoring project collecting CO_2 net ecosystem exchange data for mixed agricultural fields in Hungary is carried out at Hegyhátsál (Haszpra et al. 2001, 2005; Barcza et al. 2009a). The eddy covariance system installed at 82 m elevation above the ground on a tall tower has been providing regional-scale NEE data since 1997 (see Chapter 8 of this book). Because mixed agricultural activity including winter and summer crops is typical for a large part of Hungary (~50% of total area), the data measured here can be used for the calibration of Biome-BGC for the general Hungarian conditions.

12.2.2 Modeling Design

For the spin-up phase of the model simulation (see Chapter 9 of this book for details), the climate variables from CRU TS 1.2 database were used (New et al. 2002). In the normal model simulation, local meteorological measurements together with data from a nearby meteorological station were used.

Soil depth (0.4 m) and soil physical properties (sand: 30%, silt: 50%, clay: 20%) were estimated based on data provided by the Research Institute for Soil Science and Agricultural Chemistry of the Hungarian Academy of Sciences (AGROTOPO database, Tóth et al. 2007). Atmospheric CO_2 mixing ratio was set to 290 ppm during the spin-up phase reflecting preindustrial conditions and was set to actual value for the normal phase of the simulations. Preindustrial and present day atmospheric nitrogen depositions were estimated based on data presented in Churkina et al. (2003). According to the "super grass" approach we have to take into account the application of fertilizers. The amount of nitrogen added as fertilizer was estimated from the data of the Hungarian Central Statistical Office (http://www.ksh.hu). Country average fertilizer input was calculated based on data from the 2000–2003 period, taking into account the area of arable lands. As part of the fertilizer input is removed from the ecosystem by runoff, erosion, leaching, and denitrification, the nitrogen amount (i.e., nitrogen available for plants) was decreased by 50% according to Peoples et al. (1995). The sum of nitrogen from fertilizers and from atmospheric deposition (3.4 gN m^{-2} $year^{-1}$) was used in the model simulations.

12.2.3 Calibration of the Biome-BGC Model

The different ecological systems need different ecophysiological parameters; thus the model should be adapted to the systems studied (White et al. 2000). Using the default, generalized parameter set of Biome-BGC the model approximates the carbon dioxide exchange of the different ecological systems with moderate accuracy. Substantially, better results can be achieved if the model is adapted to the real Hungarian conditions, which means that the model is "calibrated."

There are only very few publications that deal with the ecophysiological parameterization of Biome-BGC for crops. Wang et al. (2005) tested the applicability of the Biome-BGC model for agricultural crops. As the parameters presented in their study are specific for China, we cannot use them in the present framework.

In case of the "super grass" simulation we have 21 ecophysiological parameters that can be adjusted to the local conditions (see Chapter 9). This means that the model has high degree of freedom, which is a disadvantage from the point of view of model calibration. In order to decrease the degree of freedom we used the least squares linearization method (LSL; Verbeeck et al. 2006). Using the LSL analysis, it is possible to rank the model parameters for sensitivity, so that the number of model parameters to be constrained can be reasonably decreased (see Chapter 9 for details).

Taking into account the results of the sensitivity analysis, calibration of the Biome-BGC model can take place. In our case, calibration was performed based on the daily gross primary production (GPP) and total ecosystem respiration (R_{eco}) data observed using a simple Monte Carlo Maximum Likelihood (MCML) approach (Hollinger and Richardson 2005). The calibration was accomplished using 9 years of measured eddy covariance data (1997–1999, 2001–2006; Haszpra et al. 2005). After the calibration, data from 2 years of measurements (2007–2008) were used to validate the model results.

In order to quantify the relationship between the model results and the measurement data, and to estimate the misfit, we used the following quantities:

$$\mathrm{RMSE} = \sqrt{\frac{1}{n}\sum_i \left(y^i(p) - y^i_{obs}\right)^2} \qquad 12.1$$

$$L(p) = \exp\left(-\frac{1}{2}\frac{\mathrm{RMSE}}{\sigma^2}\right) \qquad 12.2$$

where RMSE is the root mean square error, $y^i(p)$ is the model output that depends on the parameter set p, y^i_{obs} is the measurement data, n is the number of observation-model data pairs used for the analysis, σ is the uncertainty of the measurement data estimated based on Chevallier et al. (2006), and $L(p)$ is the likelihood function. In maximum likelihood paradigm, the main aim is to maximize $L(p)$, which is a function of RMSE.

12.2.4 Results of the Sensitivity Analysis

Sensitivity coefficient of each parameter was estimated based on the methodology described above using 10,000 model runs with Monte Carlo technique (Verbeeck et al. 2006). The possible parameter intervals were estimated based on White et al. (2000). To quantify the dependence of the model results on the ecophysiological parameters, we used four different measures: the likelihood ($L(p)$), the RMSE, the sum of R_{eco}, and the sum of GPP for the entire simulation time interval (1997–1999, 2001–2006). Variability of the sums of GPP and R_{eco} is related to model bias, thus all four variables provide information about the "goodness" of the model results.

Table 12.1 shows the sensitivity coefficients for all 21 model parameters. It can be seen that, using different quantities to calculate model output variability (likelihood, RMSE, GPP, and R_{eco}), the sensitivity coefficients change significantly. Sensitivity coefficients tends to covary for GPP and R_{eco} ($R^2 = 0.96$). This may be the consequence of the close relationship between the two fluxes. The correlation between RMSE and the fluxes is lower ($R^2 = 0.72$ for GPP and $R^2 = 0.64$ for R_{eco}),

Table 12.1 Results of the sensitivity analysis of the Biome-BGC model. Sensitivity is given as percent of total sensitivity (Verbeeck et al. 2006). L(p) is likelihood, RMSE is root mean square error, modeled GPP and R_{eco} are the sums of the daily modeled GPP and R_{eco} for the entire simulation period. Bold font shows the parameters the sensitivity of which is greater than 5%

Ecophysiological parameter	L(p)	RMSE	GPP modeled	R_{eco} modeled
Yearday to start new growth	0.0	1.9	**5.7**	**5.4**
Yearday to end litterfall	1.6	0.5	**5.5**	**10.6**
Transfer growth period as fraction of growing season	0.5	3.8	3.3	3.1
Litterfall as fraction of growing season	0.4	1.0	1.2	1.8
Annual whole-plant mortality fraction	4.0	**5.8**	**5.1**	**5.7**
(Allocation) New fine root C : new leaf C	**15.7**	**6.0**	3.9	3.6
(Allocation) Current growth proportion	1.7	1.6	3.6	2.5
C:N of leaves	**6.3**	**21.8**	**25.0**	**25.2**
C:N of leaf litter, after retranslocation	0.1	0.5	0.9	0.7
C:N of fine roots	0.4	0.0	0.1	0.1
Canopy water interception coefficient	0.2	0.1	3.8	4.5
Canopy light extinction coefficient	**7.6**	**22.5**	**5.8**	**5.6**
Canopy average specific leaf area	**28.3**	0.6	**7.5**	**6.6**
Fraction of leaf N in Rubisco	**25.8**	**32.0**	**24.7**	**21.7**
Maximum stomatal conductance	2.8	1.6	0.4	0.2
Cuticular conductance	0.7	0.4	0.5	0.5
Boundary layer conductance	2.3	0.1	2.0	1.4
Leaf water potential: start of reduction	1.4	0.1	0.8	0.7
Leaf water potential: complete reduction	0.2	0.0	0.0	0.0
Vapor pressure deficit: start of reduction	0.0	0.0	0.1	0.0
Vapor pressure deficit: complete reduction	0.1	0.0	0.1	0.1

and there is an even weaker correlation between *L*(*p*) and the other measures (R^2~0.3). One of the reasons for the different behavior might be the fact that likelihood values are strongly biased to very low values (close to zero) due to the exponential function used, while the others are not.

Although there are differences among the sensitivity results, we see a robust result for parameter ranking. There are a number of ecophysiological parameters that have a low effect on the overall variability of the results. We selected those parameters for calibration for which the sensitivity exceeded 5%, at least for one measure. Based on this criteria eight parameters were chosen (Table 12.1). The key parameters are in good accordance with those published by White et al. (2000). The eight most important parameters contribute to 80–90% of the variability of the misfit measures (for likelihood and RMSE the contribution is higher).

12.2.5 Results of the Calibration and the Validation

MCML calibration was performed using the results of the sensitivity analysis. Our aim was to estimate the values of the eight parameters that were selected as the most sensitive ones. In order to perform the MCML calibration, we assigned intervals defined by the possible minimum and maximum values of the ecophysiological parameters (Table 12.2). These intervals are the same that were used for the sensitivity analysis.

The MCML calibration was first performed with 10,000 model simulations using the GPP and R_{eco} data for the period 1997–2006 from the tall tower eddy covariance system located at Hegyhátsál (for detailed site description see Chapter 5). At each step, the eight parameters were varied randomly inside their possible intervals. The likelihood function was calculated at each step. After the first calibration, another MCML simulation was performed using only three parameters, while the other five parameters were fixed using the results of the first MCML experiment. The three parameters were C:N of leaves, canopy light extinction coefficient, and fraction of leaf nitrogen in Rubisco. These three parameters belong to the most sensitive parameters (Table 12.1), and they also tend to covary, which

Table 12.2 Parameter intervals used for the calibration of the model. The results of the calibration are also presented

Parameter	Minimum	Maximum	Calibrated
Yearday to start new growth	30	100	31
Yearday to end litterfall	260	360	330
Annual whole-plant mortality fraction ($year^{-1}$)	0.01	1	0.46
(Allocation) New fine root C : new leaf C	1.2	2.3	1.74
C:N of leaves ($kgC\ kgN^{-1}$)	10	80	60.06
Canopy light extinction coefficient	0.3	0.8	0.76
Canopy average specific leaf area ($m^2\ kgC^{-1}$)	3	20	6.814
Fraction of leaf N in Rubisco	0.1	0.3	0.253

means that they should be calibrated simultaneously (Thornton 2000). The final model parameters were estimated based on the second MCML experiment using 100,000 steps. The final results are presented in Table 12.2. The full ecophysiological parameterization is not presented here, as the other parameters were not modified substantially. Only maximum stomatal conductance was set to 0.005 m s^{-1} based on a previous analysis (Barcza et al. 2009b). We refer here to White et al. (2000) where the default grass parameterization is presented in detail for the Biome-BGC model.

As the result of the calibration, modeled GPP explained about 75% of the measured GPP variance (i.e. $R^2 = 0.75$), while modeled total ecosystem respiration (R_{eco}) explained about 50% of the total variance. Modeled NEE explained 54% of the measured NEE variance. After the spin-up phase, soil carbon stock was 7.78 kg m^{-2}, while soil nitrogen stock was 0.78 kg m^{-2}. Those estimates are close to the data provided by the inventory based Kreybig database, where estimated soil carbon pool is 8.3 kg m^{-2}, while soil nitrogen pool is 1.24 kg m^{-2} in the vicinity of the monitoring site (Pásztor et al. 2006).

Model validation was performed independently from calibration (Aber 1997). We used measured NEE data from 2007 and 2008 for the validation. The variability of the measured data is considerably higher than the variability of the modeled data (Fig. 12.1). The results are very similar for GPP and R_{eco}. There is a better agreement between the time series if the measured dataset is smoothed. The large variability might be the consequence of the heterogeneous agricultural landscape around the tower (Barcza et al. 2009a), as the model works over homogeneous virtual vegetation. The model captured the onset and offset of the growing seasons well in both years. We can see a possible overestimation of NEE by the model during the dormant season relative to the measurement data. This overestimation will be discussed later. During the 2 years, square of the correlation coefficient between the measured and modeled NEE was 0.56, while

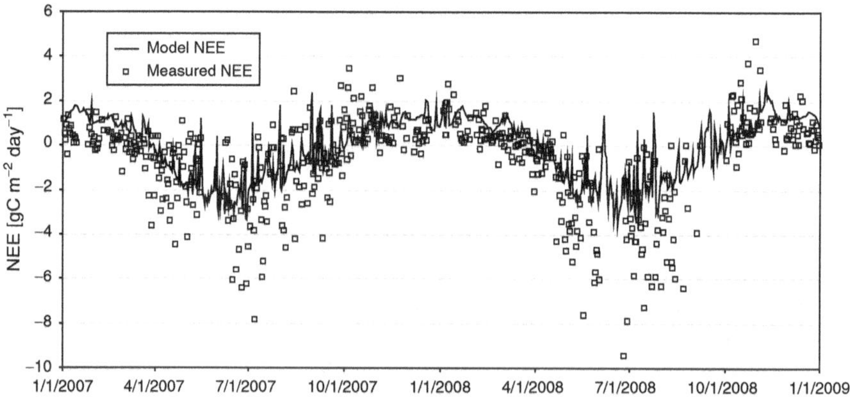

Fig. 12.1 Result of the model validation for NEE for 2007–2008. Continuous line shows the modeled NEE, while squares show the measurement data

it was 0.72 for GPP and 0.48 for R_{eco}, respectively. There was no significant deviation from the 1:1 line for the three variables. Based on the validation, we conclude that the "super grass" method is applicable for the simulation of mixed agricultural croplands, but we have to put emphasis on thc quantification of possible model bias.

12.2.6 Simulation of Cropland Carbon Cycle at Hegyhátsál

One of the most important parameters of cropland carbon cycle is the net primary production (NPP). NPP is the difference between GPP and autotrophic respiration (R_a). NPP is closely related to crop yield, so the magnitude and possible variability of NPP is in the center of interest for sustainable development. Unfortunately, it is not possible to estimate NPP directly with the eddy covariance technique (only by means of ancillary measurements like soil respiration to estimate heterotrophic respiration, and so to estimate autotrophic respiration as the difference between total ecosystem respiration and heterotrophic respiration). The calibrated Biome-BGC model can be used to estimate NPP, as it simulates both autotrophic and heterotrophic respirations.

Figure 12.2 shows the mean monthly sums of the simulated NPP for the 1997–2008 time period and their variabilities. In order to be consistent with the temporal coverage of the eddy covariance measurement, we excluded year 2000 from the analysis.

According to the figure, we see significant carbon dioxide uptake as early in the year as in March (contribution of winter crops), while growth ends in November

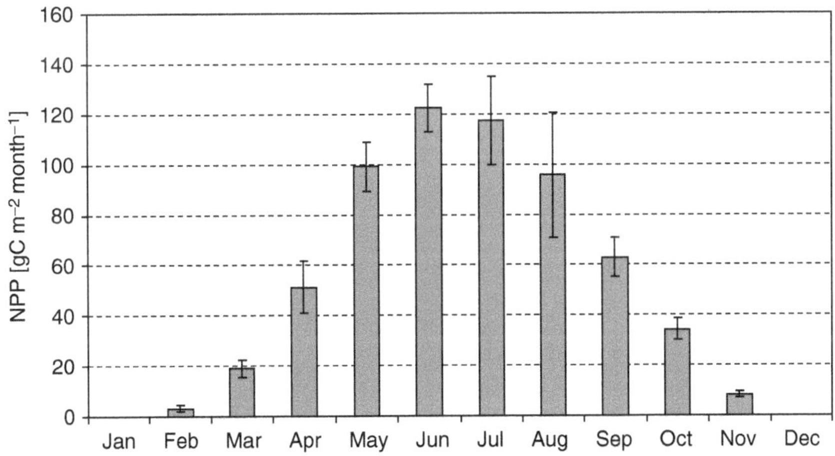

Fig. 12.2 Mean monthly simulated NPP (1997–2008; year 2000 is excluded – see text) and its variability (± one standard deviation is plotted) for the mixed agricultural landscape at Hegyhátsál

(summer crops or secondary greening of winter crops). The simulations suggest that NPP peaks around June with relatively small interannual variability. NPP in July and August has a larger variability most probably due to the variability and timing of precipitation events. Monthly maximum NPP sum is around 120 gC m^{-2} $month^{-1}$ in June and July, while it is close to 100 gC m^{-2} $month^{-1}$ in May and August. The simulations show zero NPP in January and December, which is not realistic according to the experience. Because of mild winters, wheat can grow even this season in some years.

NPP can be estimated from local yield data or national yield statistics based on the harvest index (HI) and other conversion ratios (Goudriaan et al. 2001; Ciais et al. 2005, 2007, 2010). We used country-specific HI and conversion data for the estimation of NPP. Dry matter content of biomass and carbon content of the different plant pools were estimated from Goudriaan et al. (2001). As mainly winter crops (winter wheat) and summers crops (maize) are grown in the region, we used the county level (Vas County, NUTS level 3 region, code HU222) harvest statistics of wheat and maize published by the Hungarian Central Statistical Office. The estimated HI for winter wheat was 0.543, while it was 0.443 for maize. For the estimation of root production as percentage of the total aboveground biomass, we used 19.4% for wheat and 32.5% for maize (M. Birkás, personal communication).

Figure 12.3 shows the simulated and HI based estimated annual NPP data for the period of 1997–2008. We calculated HI-based NPP for wheat and maize, and we also compared the average NPP of the two main crops with the simulation results. It can be seen in Fig. 12.3 that simulated NPP has a low interannual variability relative to the HI-based estimates. The lower NPP in the dry years (2000–2003) was

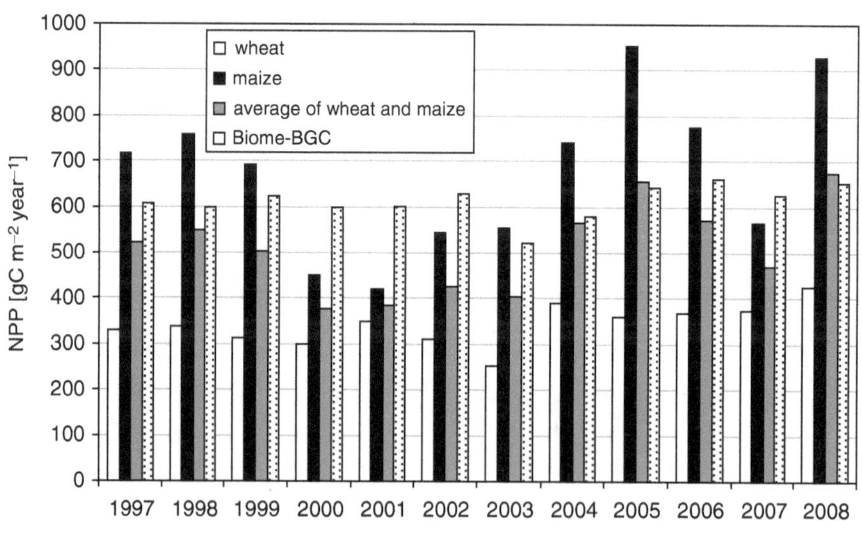

Fig. 12.3 Agricultural NPP estimates using the harvest index method and the Biome-BGC results for the 1997–2008 time period

not simulated well. Wheat NPP is always lower than the Biome-BGC based NPP because of the shorter growing period (wheat is usually harvested around July). Maize has a longer growing period (typically from April until October–November), and it can have a relatively large NPP (more than 900 gC m^{-2} $year^{-1}$). The model was unable to capture the high NPP during the most productive years.

During the simulation period, the mean HI based NPP of wheat was estimated to be 342 gC m^{-2} $year^{-1}$, while the mean NPP of maize was 675 gC m^{-2} $year^{-1}$. The mean NPP of wheat and maize is 509 gC m^{-2} $year^{-1}$. The "super grass" average NPP was 612 gC m^{-2} $year^{-1}$ during 1997–2008, which overestimates the HI-based results. The simulated NPP/GPP ratio was 0.62, which is slightly higher than the estimate presented in Ciais et al. (2010).

It is interesting to note that if we use the HI and root production ratio parameters of Goudriaan et al. (2001) the mean NPP of wheat would be 427 gC m^{-2} $year^{-1}$, and the mean NPP of maize would be 579 gC m^{-2} $year^{-1}$. Note that the HI-based method still has uncertainties arising from the spatial and temporal variability of HI and allocation parameters (above- and belowground partitioning of NPP), so the results should be interpreted with caution (Ciais et al. 2010). We neglected other cultivars (e.g., barley, alfalfa) also grown in the region due to the lack of information about the spatial pattern of crop types (Barcza et al. 2009a). Clearly, more research is needed to constrain the NPP estimates for the Hegyhátsál region.

NEE is the difference betweeen GPP and R_{eco}, and it is interesting to check the behavior of the "super grass" GPP and R_{eco}, and their relationship with the measurement-based best estimates. Figure 12.4 shows the mean monthly sums of simulated GPP and R_{eco} together with the gap filled, measurement based GPP and R_{eco} results (see Chapter 8). We excluded the simulation data from year 2000 for consistency with the measurements. While the model captured the annual dynamics of R_{eco} well, the simulated respiration is generally higher than the measured one. The variability of observed R_{eco} is higher than that of the simulated one especially during the growing season. This finding is in accordance with the results obtained for NEE during the validation (Fig. 12.1). Simulated GPP also follows the measurement-based estimates quite well. In the first 5 months of the year, Biome-BGC underestimates the magnitude of GPP, but it overestimates GPP afterwards until November. The variability of the measurements is higher than that for the model simulation, quite similarly to R_{eco}. Mean annual simulated GPP is –986 gC m^{-2} $year^{-1}$, while the measurement-based estimate is –1,014 gC m^{-2} $year^{-1}$ (for uncertainty intervals of GPP and R_{eco} measurements see Haszpra et al. 2005). Mean simulated R_{eco} is 944 gC m^{-2} $year^{-1}$, while its measurement-based estimate is 909 gC m^{-2} $year^{-1}$.

In order to use the model at country level, we have to eliminate model bias as the upscaling process magnifies even a small systematic difference which could distort the results and could lead to inaccurate estimates. As NEE is the small residual of the two large fluxes of GPP and R_{eco}, we expect a biased model NEE based on the results presented above. Figure 12.4 also shows the mean monthly NEE sums for the entire simulation period (leaving out year 2000). According to our expectations modeled NEE is generally biased toward more positive values

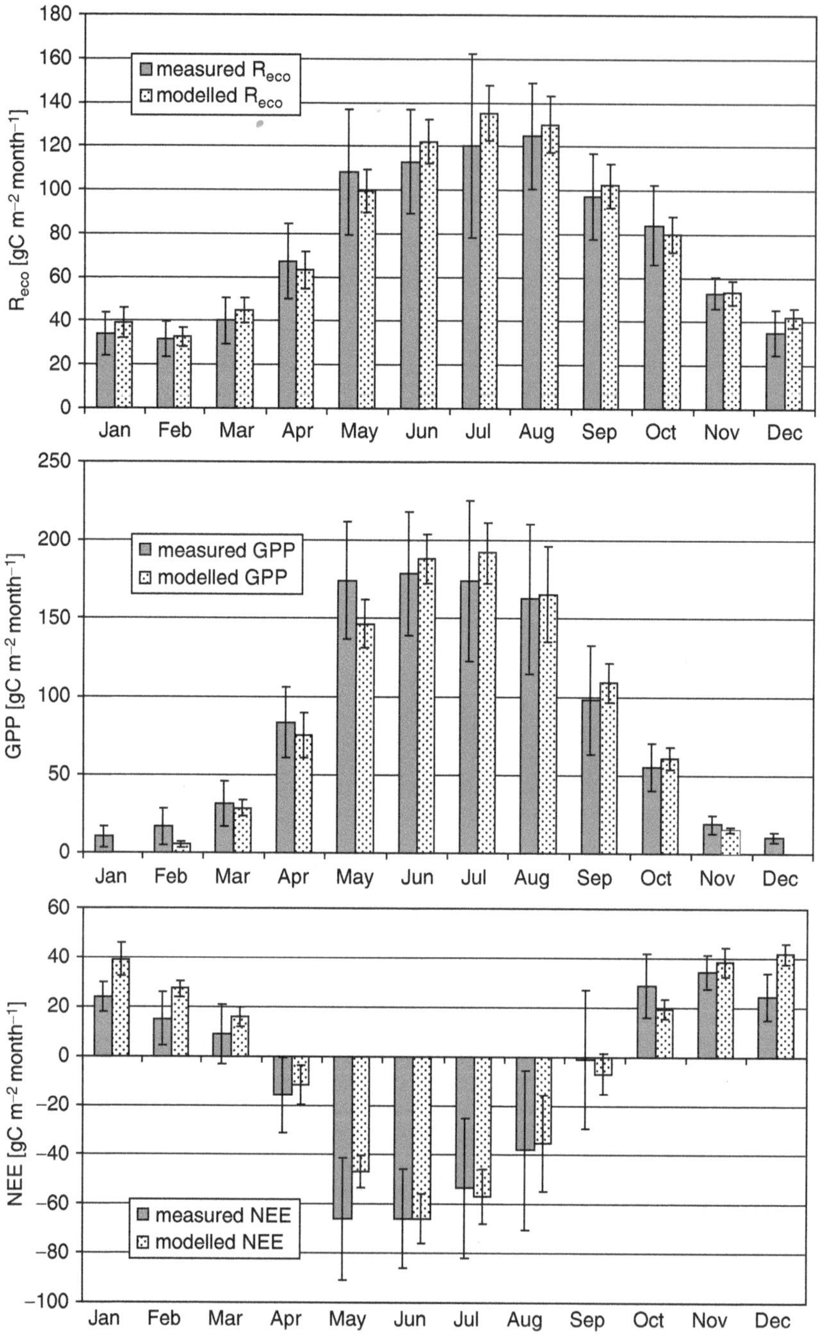

Fig. 12.4 Mean monthly model simulated and measured R_{eco}, GPP and NEE (1997–2008; year 2000 is excluded – see text) and their variability (± one standard deviation is plotted) for the mixed agricultural landscape at Hegyhátsál

(i.e., toward lower uptake). This is especially pronounced during the dormant season. Multiyear mean modeled NEE is -43 gC m^{-2} $year^{-1}$, while the measured mean NEE is -105 gC m^{-2} $year^{-1}$ (uncertainties of NEE measurements are presented in Fig. 12.6; for details on uncertainty estimations see Haszpra et al. 2005).

As the above comparison is performed with gap filled data, which contains modeled NEE that may cause additional uncertainty in the comparison, it is important to scrutinize the bias using exclusively measurement data. Taking into account the gap-filled time periods, we also filtered the model data so that the resulting model and measurement databases have the same temporal coverage in the 1997–2008 period. The filtered datasets were used to calculate multiyear annual courses of NEE for the investigated period (Fig. 12.5). The figure corroborates the abovementioned finding about the biased NEE throughout the year. Simulated NEE is usually biased toward more positive values. Bias is the highest during the dormant season.

The most likely cause of the bias is cropland management: as large part of the biomass is removed via harvest, it cannot decompose in the field and respire back to the atmosphere. Using the current version of Biome-BGC harvest and biomass removals cannot be simulated, thus respiration is overestimated. This is why NEE is generally biased toward more positive values. The removed biomass is generally consumed by humans or animals, and after digestion its carbon content returns to the atmosphere typically within a year (Ciais et al. 2007). It means that respiration manifests elsewhere, not in the arable fields where carbon dioxide is taken up from the atmosphere. If we would like to calculate changes in the ecosystem carbon content, we have to take into account the removed carbon (Chapin et al. 2005).

In order to provide the most reliable NEE estimate, we need to find a way to eliminate the bias as much as possible. As the model is planned to be used at country level, we have to find a parameter for the correction that is (1) independent from the eddy covariance measurements, and (2) available in explicit spatial context.

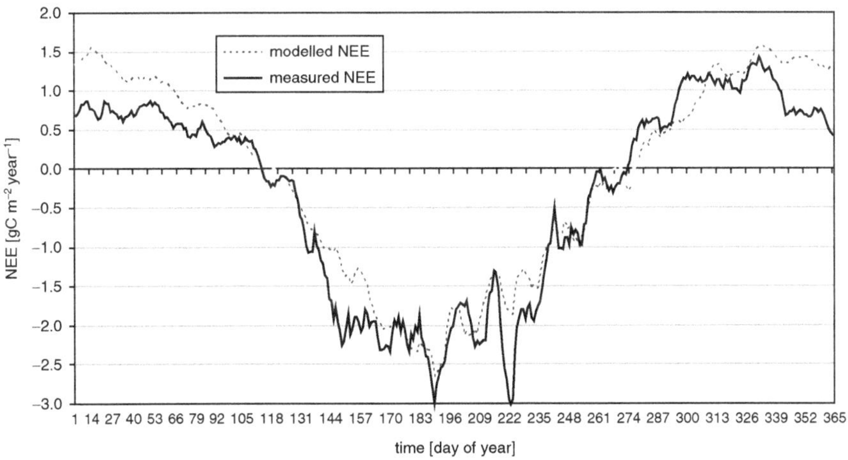

Fig. 12.5 Mean annual courses of the simulated and measured NEE based on the entire measurement period. Gap filled days were excluded from the analysis

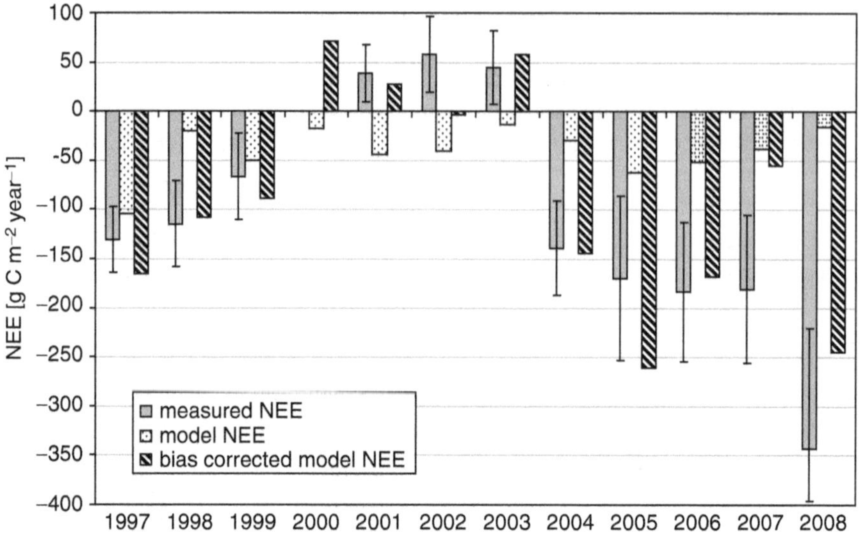

Fig. 12.6 Annual sums of measured NEE, modeled NEE and bias corrected model NEE for 1997–2008. Uncertainty for the measured NEE is also plotted using the methodology published in Haszpra et al. (2005). There were no measurements in 2000 for technical reasons

As the bias is related to harvest, we sought a relationship between yield (or any other yield based productivity estimates; see above) and annual NEE bias. We found strong linear relationship between NEE bias and the average yields of wheat and maize for Vas county ($R^2 = 0.75$, $P < 0.001$). For other counties R^2 is lower (varying between 0.04 and 0.57). We used the established linear relationship and yield data to correct NEE modeled by Biome-BGC. Figure 12.6 shows the annual sums of the measured NEE, the original simulated NEE, and the bias corrected NEE. The multiyear mean of the corrected NEE sum is very close to the measured one, though there are discrepancies in some years. Nevertheless, we conclude that our simple bias correction is an appropriate tool for the estimation of the regional-scale NEE around Hegyhátsál, so we may try to generalize the results to country scale.

The ultimate aim of the carbon cycle-related studies is the estimation of the net biome production (NBP), which is the long-term integral of larger scale carbon accumulation or loss of the ecosystem (the latter is called net ecosystem carbon balance, NECB; Chapin et al. 2005). Although generally it is not possible to estimate all components of the carbon balance, in arable lands, NBP is heavily affected by the management practices. As harvested plant material is generally digested and thus returns to the atmosphere outside the fields where crops are grown, this horizontally displaced carbon has to be taken into account when calculating the total carbon balance of an ecosystem. Net biome production can be approximated as the difference between by definition, NEP equals NEE times minus 1-NEE (=net ecosystem production, NEP) and the removed carbon. The removed carbon has positive sign as it is associated with respiration (Ciais et al. 2010). NBP has positive sign if carbon is accumulating at the ecosystem.

Table 12.3 Measured and corrected simulated NEE, horizontally removed carbon and NBP for the Hegyhátsál agricultural region for the 2003–2008 time period. Numbers are given in gC m^{-2} year^{-1}. Negative NEE means carbon uptake by the ecosystem, while negative NBP means net carbon loss from the ecosystem (Chapin et al. 2005). Removed carbon is a net loss from the ecosystem

	2003	2004	2005	2006	2007	2008
Measured NEE	44	−139	−170	−184	−181	−309
Bias corrected simulated NEE	49	−142	−253	−165	−36	−250
Removed carbon (footprint weighted)	144	188	222	197	164	225
NBP based on measured NEE	−189	−49	−53	−12	11	88
NBP based on simulated NEE	−202	−44	38	−29	−114	24

Using the country-specific allocation factors and the results of the footprint analysis for the Hegyhátsál tall tower eddy covariance measurements (Barcza et al. 2009a; also see Chapter 8), it is possible to estimate the horizontally displaced carbon due to harvest and residue management. Yield was assumed to be removed from the fields in all cases. Additionally, based on information provided by a local farmer, we assumed that 2/3 of the wheat residues (straw) was removed from the field, while the rest was plowed, and thus returned to the soil. For maize, we assumed that all crop residues returned to the soil. The contribution of the different crop types was estimated from the results of the Barcza et al. (2009a) study for the period of 2003–2008 (see their Table 1 and 2). NBP was estimated as the difference between the measured and simulated NEP and the removed carbon.

The estimated NBP is presented in Table 12.3. According to the results, during the period of 2003–2008, the Hegyhátsál region was a net source of carbon to the atmosphere, which means that the soil organic matter content generally decreased. However, the region acted as a net sink in some years, which means that soil carbon can increase due to climatic conditions and/or management practices. Mean removed carbon during the 2003–2008 period was 190 gC m^{-2} year^{-1}. Mean NBP was −34 gC m^{-2} year^{-1} based on the measured NEE, while it was −54 gC m^{-2} year^{-1} when it was derived from the bias corrected simulated NEE. These NBP estimates are close to the European mean estimates published by Ciais et al. (2010) though our results show higher loss than the European average. Obviously, more research and longer measurement time series are needed to support this relatively high soil carbon loss estimate.

12.2.7 Discussion

We presented a modeling approach to simulate the carbon cycle of mixed agricultural landscapes in Hungary. We used the grass submodel of Biome-BGC for this purpose. Calibration of the model was performed by means of the measurement-based GPP and R_{eco} estimates. The calibrated and validated model was used to estimate the carbon balance components of the region near the Hegyhátsál tall tower monitoring site with ancillary information gathered from local yield statistics.

It was found that the model is appropriate for the estimation of NPP, GPP, and R_{eco}, but there is a considerable uncertainty in the estimation of NEE. NEE is a small residual of the two large components (GPP and R_{eco}); thus, it is sensitive to the biases in GPP and R_{eco}. We pointed out that the main drawback of the "super grass" approach is the overestimation of NEE (higher CO_2 release) during the dormant season. As harvested and removed biomass changes the overall carbon balance, the calibrated model has deficiencies. We found that with a simple bias correction based on the county's yield statistics NEE can be estimated more accurately.

The cropland-related simulations could be improved with information about the crop types and harvest data. There is a relatively large variability in yield due to the actual meteorological conditions, method and timing of soil disturbance, cultivars sown, fertilizer, and weed/pest killers' amounts applied. As the fields belong to individuals, it is hard to get information about them. Uncertainties also arise from HI-based estimates since we used mean HI, which neglects the interannual and spatial variability of HI, and root allocation ratio. Our study provides the first comprehensive estimate of the carbon balance of mixed agricultural landscape in Hungary. Future work is needed to constrain these estimates and to describe the cropland carbon cycle for longer time periods. This is especially important as the utilization of crop residues as biofuels is becoming popular in the European Union. As there can be a change in residue management, this could sensitively affect the soil carbon pools. Our results suggest that carbon could accumulate in the soil in 2007 and 2008, which demonstrates carbon sequestration potential for Hungarian soils. If residues are burned as biofuels, this could negatively affect the soil quality.

As management is not simulated in the current version of Biome-BGC, developments are needed in the model logic in order to provide more realistic carbon balance estimates for croplands (see, e.g., Gervois et al. 2004, 2008). In Chapter 10 of the present book, we introduced the current developments that are related to grass mowing and grazing. Plowing, sowing, fertilization, and harvest routines have to be implemented and tested in the future to help us to provide more accurate cropland carbon balance estimates for Hungary.

12.3 Simulation of Biospheric Carbon Dioxide Exchange of Arable Lands with MOD17

12.3.1 Introduction

Efforts to constrain the greenhouse gas balance of agricultural lands have raised the need for the synthesis of several independent methods and datasets to get defensible and accurate results (Ciais et al. 2010). Estimations relying purely on modeling results are not defensible and are highly uncertain in some cases (Vetter et al. 2008; Barcza et al. 2009b). The available plot-level measurements can help to constrain estimations, e.g., through calibration of ecosystem models. However, due to their small spatial representativeness, these measurements are unable to influence regional or country scale estimations directly, which means that the results cannot be simply upscaled (Barcza et al. 2009a). Satellite-based radiance measurements provide invaluable information about the properties of the land and ocean surface and about the atmosphere with high accuracy. The radiances can be used to construct sophisticated models that provide insight into several processes related to the environment (Turner et al. 2004).

Being more artificial than forests or other natural biomes, agricultural lands are among the most heterogeneous land use types in Central Europe. Land use changes in the last 20 years further shifted the typical Hungarian farms toward smaller fields and more complex cultivation patterns. As a consequence, the need for spatially representative estimations is the most urgent in case of arable lands, especially considering that a significant area of the country is covered by arable lands. Due to their global spatial coverage, remote-sensing-based estimations can help to constrain the greenhouse gas balance estimations on regional or country scale. However, it is essential to evaluate and – if necessary – to improve the performance of the remote-sensing-based models on plot level before applying them on larger scale.

Here we focus on remote-sensing-based gross primary production (GPP) estimations. The MOD17 light use efficiency (LUE) GPP model can distinguish 11 plant functional types (PFTs, for list of PFTs, see Chapter 9) that include only one agricultural category. This seems to be a huge oversimplification even if only the high diversity of worldwide crop types is considered. This property of the model automatically leads to the consequence that characterizing all agricultural ecosystems with one set of parameters globally generally causes significant biases compared to individual eddy covariance (EC) flux tower measurements as ground truth (Zhang et al. 2008).

It also needs to be emphasized that the discrepancies between measurements and model results (MOD17) can also be caused by the mismatch in spatial representativeness of remote sensing and flux tower measurements. This statement is undoubtedly true in case of short flux towers, where the footprint is definitely smaller than the 1 km spatial resolution of the MOD17 model (at least for daytime unstable stratification when the satellite measurements take place). However, the situation is somewhat different in case of tall towers. Data from a tall flux tower near Hegyhátsál, Hungary, are used in this study to evaluate remote-sensing-based GPP estimations. The footprint of the tower is characterized in Chapter 8 of this book. The daytime footprint is shown to extend further from the tower by more than 1 km as it can be seen in Fig. 8.1 in Chapter 8. Ideally, tall tower measurements should be in better agreement with estimations from remote sensing data, as their spatial representativeness is rather similar. However, this can only become reality if the vegetation (and land cover in general) around the tower is homogeneous, and/or the footprint is identically representative to the whole area. As it can be expected, these conditions generally do not meet in reality (Wang et al. 2006; Barcza et al. 2009a).

Due to the heterogeneity of land use around the tall tower, the 1-km resolution of the GPP model might smooth out small-scale effects that influence the EC measurements. In order to find a better match between the spatial scales of the EC measurements and remote sensing here we introduce a new approach, where the MOD17 model is downscaled to a 250 m grid (corresponding to the MODIS Normalized Difference Vegetation Index (NDVI) product) for consistency with previous studies. This downscaling is also necessary because of the size of the individual fields surrounding the tower. The downscaled GPP model is used to simulate crop-specific CO_2 exchange that is comparable with the measurement results (see Chapter 8).

12.3.2 Model Design

12.3.2.1 Estimation of GPP Using Different Approaches

Three different estimations of GPP, all of them based on the MOD17 model are used in this study. The official MOD17 GPP product (UMT NTSG v5.1, see Chapter 9) was downloaded and validated using the data of the tall EC tower at Hegyhátsál for years 2001–2006 (these estimations are referred as "GPP-official" later on). In this case, the discrepancies between GPP based on EC tower measurements and modeled GPP can be caused by errors in the input data or originated from the structure of the model. To eliminate the effect of possible errors in the input data, besides the validation of the official product, we adapted the model and performed model calculations using two different approaches. In the first step, we replaced the global meteorological reanalysis data with local (on-site) meteorological measurements (referred as "GPP-met"). As a second step toward more accurate estimations, the ~1 km spatial resolution fraction of absorbed photosynthetically active radiation (FPAR) data of the MOD15 product (Collection 5) were replaced by alternative FPAR data calculated from a ~250 m resolution NDVI time series described in Chapter 8 (referred as "GPP-NDVI"). Note that in this latter case, meteorological input for model runs also comes from local measurements.

Calculations are performed on a daily basis followed by the determination of 8-day averages in order to compare the results with the official product. The datasets used for validation are daily sums of GPP described in Chapter 8.

12.3.2.2 Calculations Using Local Meteorological Measurements

The MOD17 model has been shown to be sensitive to meteorological input dataset (Zhao et al. 2006). To evaluate the real performance of the model that is not affected by errors in meteorological datasets, we implemented the MOD17 model and performed calculations using local meteorological measurements. We followed the description in Zhao et al. (2005) regarding FPAR quality control and filling the resulting gaps in the dataset. To check whether our implementation is successful,

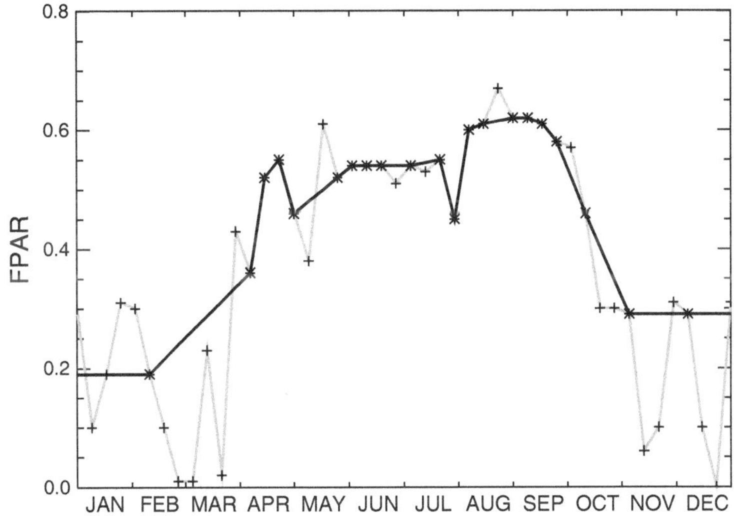

Fig. 12.7 FPAR Collection 5 data for Hegyhátsál before (*gray line, plus symbols*) and after (*black line, asterisks*) the quality control and filling the resulting data gaps (Data from 2004 is shown)

we simulated the GPP-official estimations using the GMAO (Global Modeling and Assimilation Office) meteorological dataset. Due to the original coarse resolution of the GMAO data that also causes inherited errors in GPP, interpolation of meteorological data is necessary (interpolation routine is based on Zhao et al. 2005). Although quality control is subjective for a certain degree that causes minor discrepancies, the results of the test runs were in a good accordance with the official product. FPAR time series, before and after the quality control and gap filling for year 2004, are presented in Fig. 12.7.

In GPP calculations, local meteorological measurements were used (vapor pressure deficit and global radiation) together with the data from a nearby meteorological station (daily minimum and maximum temperature, precipitation). The MTCLIM model was used here to gap fill the missing meteorological measurements (Thornton et al. 2000). We used the measured data (excluding gap-filled values) to validate the interpolated GMAO dataset used in GPP-official.

12.3.2.3 Downscaling Using NDVI Data of 250 m Spatial Resolution

Downscaling of model calculations is supposed to answer the question whether the coarse spatial resolution of the FPAR product is responsible for model inaccuracies for a certain degree. It has been shown in Chapter 8 that different locations (agricultural parcels) over the heterogeneous landscape surrounding the tower do not have identical contributions to the measured signal; therefore, a mismatch between representativeness of the MODIS pixel and that of the tower occurs. In order to synchronize the representativeness of the model to that of the tower, instead of using the FPAR of the MOD15 product, we applied a downscaling of FPAR data using the 250 m NDVI time series described in Sect. 8.2. The FPAR-NDVI relationship used was set according to Sims et al. (2005). The use of the 250-m spatial resolution NDVI dataset offers the possibility of taking into account both the hourly footprint of the tower measurements and the phenology of the individual fields with finer resolution in GPP calculations. This can lead to an improvement in differentiation of winter and summer crop productivities instead of modeling them as a fictional "average crop." On the other hand, the simple parameterization can have disadvantages compared to the more sophisticated retrieval of the official 1-km resolution FPAR product.

These alternative 250-m resolution modeling efforts are substantially based on results of the footprint analysis and the crop-specific information retrieved from remote sensing measurements (see Chapter 8) using the same model setup described above. The discretized footprint climatology and the CORINE2000 land cover information re-gridded to the same grid are used in the calculations. NDVI signals of pixels in footprint locations are used in FPAR and GPP retrieval, and the occurring gaps in GPP (when the footprint analysis had no result) are filled by linear interpolation. Only two of the eleven PFTs occurred in the calculations, mostly croplands PFT, but in some cases, the neighboring forests (deciduous broadleaf forest) also had slight contributions. We used the BPLUT (biome properties lookup table, which describes the model coefficients) version 5.1 in the calculations (M. Zhao, personal communication).

Crop-specific (summer and winter crops) GPP calculations were also carried out using the same input data and methodology, but only NDVI signals of pixels covered by the specific crop type were used. These results can be compared to crop-specific NEE time series (Chapter 8).

12.3.3 Evaluation of the Results

Validation of the GMAO reanalysis data for the entire study period (2000–2006) can be seen in Fig. 12.8 for vapor pressure deficit (VPD), global radiation (GR), and minimum temperature. The regression lines are also presented in the figures. Although, generally, there is a good agreement between the datasets, VPD is overestimated by GMAO. This fact can affect model results especially given that the site is located in a relatively wet part of the country.

The GPP-official estimations and the GPP calculated from tall tower measurements (GPP-tower) can be seen in Fig. 12.9 (8 day average daily sums). It is clear that, in most of the years, the model results are in a good agreement with tower GPP, although in the last 3 years (2004–2006), the measured GPP increased drastically (for GPP annual sums at Hegyhátsál, see Chapter 8) that is not reflected by the GPP-official. The accuracy of global radiation and minimum temperature do not vary among years, which could explain the underestimation of GPP in these

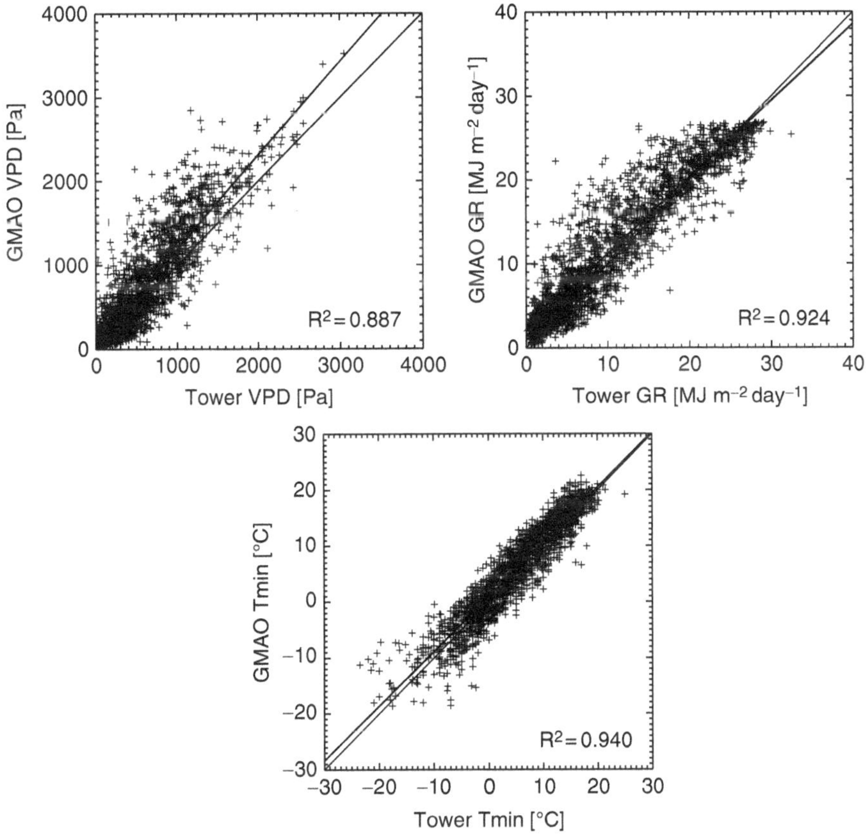

Fig. 12.8 Evaluation of GMAO meteorological database against local measurements of vapor pressure deficit (VPD), global radiation (GR) and daily minimum temperature (T_{min}). The 1:1 line and the fitted linear regression line are also plotted

years. However, as we pointed out, VPD is overestimated by GMAO, and it was found that the error is larger in the last 3 years (Fig. 12.10). It is interesting in Fig. 12.9 that, in the beginning of 2004, the model results are closer to the measured GPP. Based on this, it can be presumed that besides errors in VPD, there might be other factors interfering with this effect.

If VPD is responsible for the discrepancies between EC-based GPP and GPP-official in 2004–2006, the results are supposed to match better if we use local meteorology (Fig. 12.9). The results show that it is not only the sometimes significant error in the meteorological input data that is responsible for the underestimation of the measured GPP, as the use of local meteorology did not improve the estimations significantly. Effect of meteorological data on the MOD17 model has been shown to be site-specific, and sometimes, like in our case, the use of local measurements tends to decrease the estimations of annual GPP sums (Heinsch et al. 2006). Annual sums are plotted for the six years in Fig. 12.11, calculated from

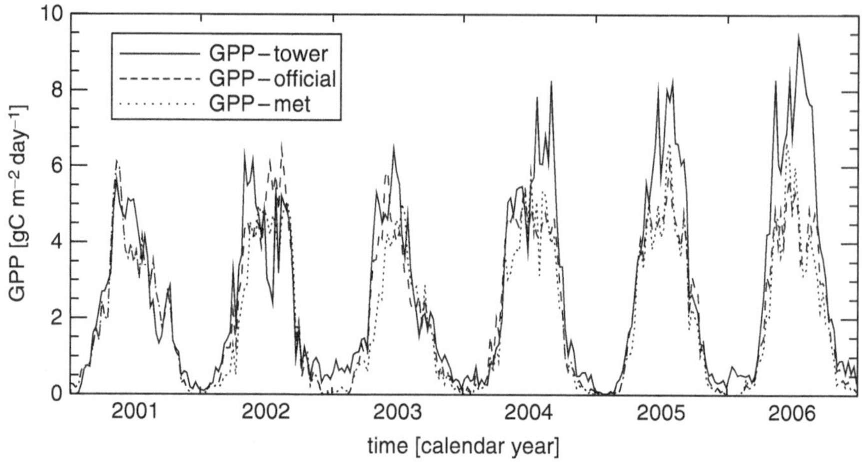

Fig. 12.9 GPP based on EC measurements (GPP-tower; *solid line*), GPP-official (*dashed line*) and GPP-met (*dotted line*) for the period of 2001–2006 (see text for definitions)

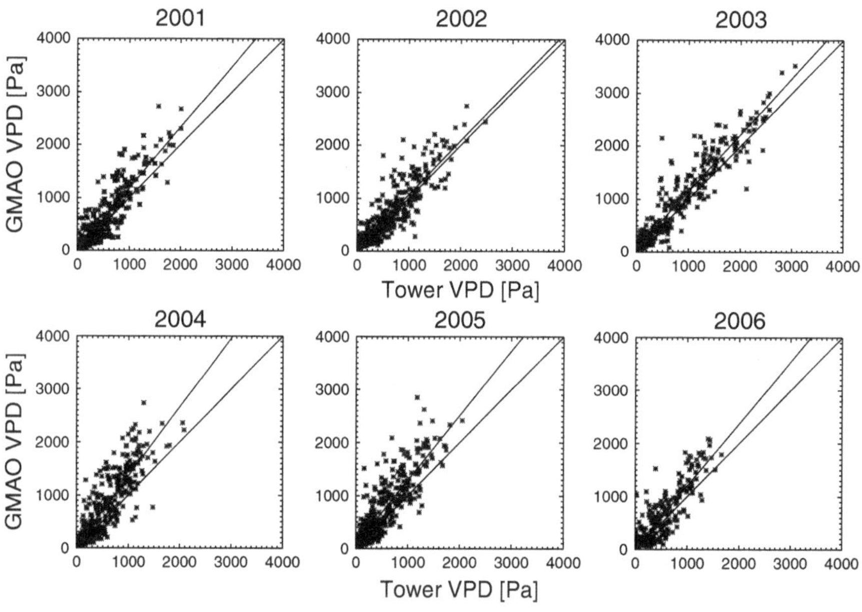

Fig. 12.10 Annual relationship between VPD from the GMAO database (GMAO VPD) and local VPD measurements (tower VPD) for 2001–2006

GPP-official, GPP-met and EC measurements (GPP-tower). It is evident that GPP-met runs provide very similar or slightly lower estimates compared to GPP-official. The significant underestimation of maximum light use efficiency (ε_{max}) together with the lack of soil water availability stress parameterization (Hwang et al. 2008)

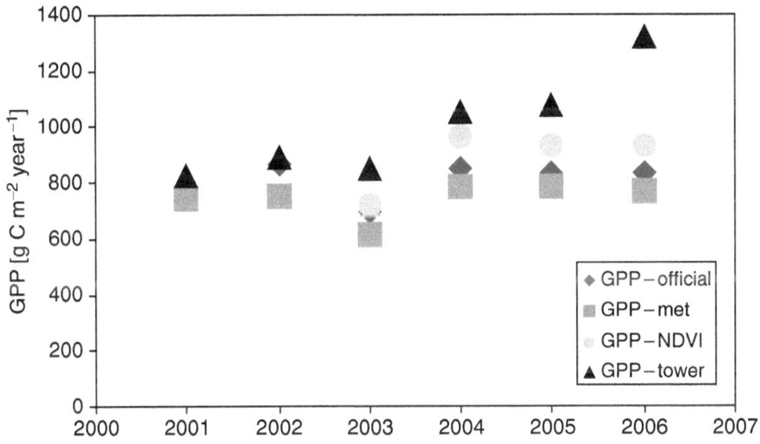

Fig. 12.11 Estimated annual sums of GPP for Hegyhátsál for 2001–2006. Plots show the results of the tower measurements and three different modeling results (see text for details)

can be a possible explanation to this behavior of the model. While the first condition results in an underestimation, the latter causes overestimation. The trade-off between these effects results in spatially varying accuracy among sites and ecosystems, and occasionally, like in this case, temporarily varying results from year to year. Apparently, in case of agroecosystems, the underestimation of light use efficiency is the more influencing factor. However, this is not necessarily the case with other biomes (Turner et al. 2006; Heinsch et al. 2006).

These findings are also supported by Zhang et al. (2008), where they evaluated the performance of the MOD17 model using both GMAO and local meteorology over a cropland (winter wheat–maize) and an alpine meadow. They found significant underestimation by MOD17 results particularly in case of the cropland site regardless of the meteorology used in the model runs, and that the underestimation of ε_{max} is mostly responsible for the performance of the model.

Another interesting feature occurred in 2004, when the contributions of winter and summer crops to the measured flux were nearly equal (Sect. 8.2). This ended up in a two-peaked GPP curve that can be considered to be typical at our site. GPP-official estimation follows the measured GPP time series quite well early in the year, but the second peak (contribution of summer crops, mostly maize) cannot be observed at all.

Although the use of local meteorology does not have significant effect on the resulting GPP curve, as it can be expected, GPP-met estimations show greater variability than those based on interpolated reanalysis data. In 2004, the underestimation after the beginning of the growing season is probably caused by a drop in FPAR and not by meteorological factors. It is clear from Fig. 12.9 that differences in meteorological data affect the short-term variation of the model results, but are not responsible for the substantial difference between model results and measurements. The second peak of the measured GPP cannot be discovered in the model results.

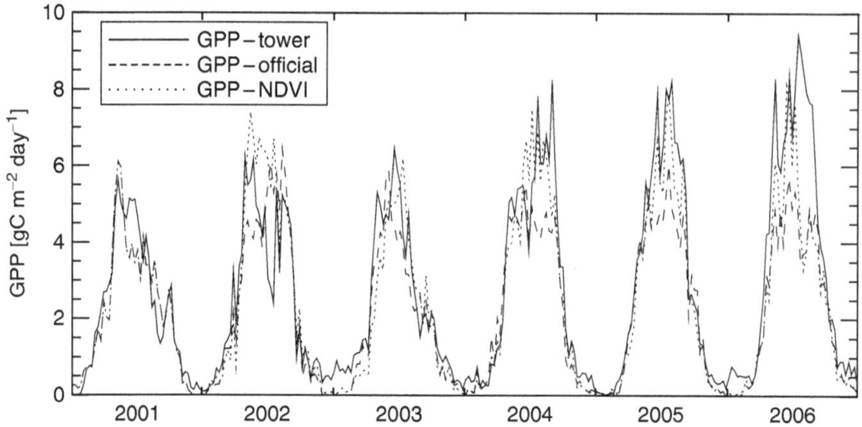

Fig. 12.12 GPP based on EC measurements (GPP-tower; *solid line*), GPP-official (*dashed line*) and GPP-NDVI (*dotted line*) (see text for definitions)

In order to see whether this two-peaked signal can be captured by a simple downscaling and additional use of footprint information, we performed alternative calculations. Note that this procedure can only eliminate errors in FPAR resulting from spatial resolution and representativeness of the remote sensing measurements. Structural problems of the model, i.e., the too general characterization of crop types cannot be resolved by this algorithm. Figure 12.12 presents the comparison of EC tower–based GPP (GPP-tower), the GPP-official and results based on the downscaling method (GPP-NDVI). Although GPP increased using the alternative method, the second peak in the time series is still not very pronounced. The increase in annual sums can also be seen in Fig. 12.11 (GPP-NDVI symbol). The tower is mainly surrounded by winter wheat (C3) and maize (C4; for land cover information see Chapter 8); however, the important difference between the properties of C3 and C4 crops is not reflected in this case, although as it can be seen on the measured curve, C4 plants usually have higher CO_2 uptake.

The algorithm that uses footprint information and is based on finer resolution NDVI data provides the methodological background for crop-specific calculations. Using the actual footprint information and the NDVI signals of footprint locations, we can estimate crop-specific GPP time series. Comparison of these results to the crop-specific tower GPP (for details see Sect. 8.2) calculations shows that the two independent estimations are in a good agreement. The comparison of the two datasets for 2004 is shown in Fig. 12.13. This is the only year when crop-specific GPP data from tower measurements are available for both crop types. It can be seen that in the second part of the year, the observed GPP is underestimated by the model calculations in case of winter crops. Considering the human intervention, i.e., harvest at this time of the year that causes a sudden drop in biomass, the modeled curve can be realistic. In this part of the country, a secondary greening of winter crops is common after harvest (see Barcza et al. 2009a), but since the NDVI time series reflect this feature, it cannot explain the underestimation. One possible explanation is that the tall tower data can be subject to contaminations by signals of other crop

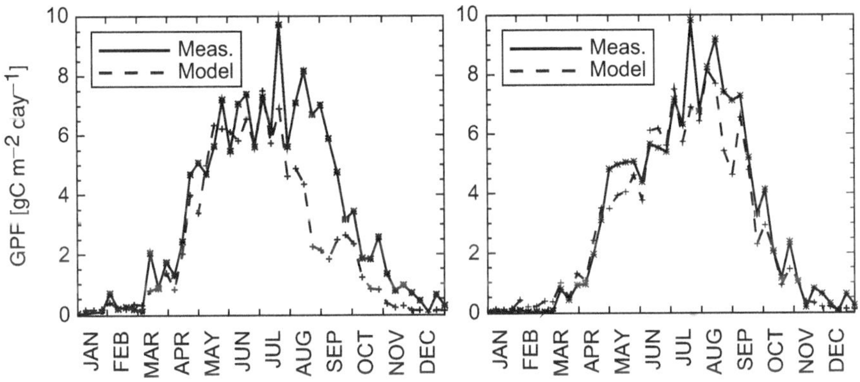

Fig. 12.13 Crop specific GPP signals in 2004. GPP based on EC measurements (*solid line*) and crop specific model calculations (*dashed line*) are shown. Left plot: winter crops; right plot: summer crops

types more than remote sensing data, especially in the transition period when both summer and winter crops are present around the tower.

It needs to be emphasized that only a single remotely sensed measurement is used as input in our crop-specific (NDVI-based) model, in conjunction with advanced modeling techniques. For example, in Yan et al. (2009), authors presented a GPP model that is based on the combination of three vegetation indices to estimate GPP in case of winter whcat–maize double cropping system. According to our results the modified MOD17 model (with the original model parameters, no calibration has taken place) using only a single remote-sensing-based input (NDVI) showed good results in case of crop-specific GPP estimations. The higher CO_2 uptake of C4 plants (maize in our case) is also reflected by the results without the calibration of the model.

12.3.4 Conclusions

It has been demonstrated that our new downscaling technique significantly improves the MOD17 model results, although as it can be seen in case of crop-specific calculations, GPP of both major crop types are underestimated. Hence, calibration of the parameter set might be the next important step toward more reliable estimations.

The presented methodological advances toward improving the accuracy of the model pointed out the need of more sophisticated methods in remote-sensing-based modeling. These methods can establish a more flexible applicability of remote sensing for simulation of crop productivity on local or regional scale over typically heterogeneous landscape. Although the proper choice of input datasets can help improving model accuracy, errors caused by the poor characterization of the ecosystem cannot be eliminated without appropriate calibration of the model. The physiological differences in properties of winter (at our site typically winter wheat, C3) and summer (typically maize, C4) crops should be included in model parameters through calibration.

The evaluation of the official MOD17 product showed that, in most cases, the product provides useful and realistic estimations for our tall tower site, and it is feasible to use the product on regional or larger scales. Therefore, we use the official product in the following chapters, although we have to be aware of its properties and possible errors.

12.4 Modeling of Biosphere–Atmosphere Exchange of Nitrous Oxide and Methane of the Hungarian Agricultural Lands by DNDC

In the following section, we discuss the methane and nitrous oxide fluxes between the Hungarian arable lands and the atmosphere. Although the definition of arable lands does not include grapes, fruits, berries, and fallow, these agriculture fields were also simulated by the DNDC model. The land cover data applied in the simulations are based on the CLC-50 land cover dataset (Büttner et al. 2000).

For the simulations DNDC was used in regional mode. For the application of this mode the territory of the country was parceled by a 1/6° × 1/6° grid yielding 466 grid cells of approximately 200 km^2 each. Meteorological and soil property information were compiled for this grid. DNDC is a software in the public domain, but it is not an open source one, therefore, its internal parameters cannot be adjusted to the local conditions. We can evaluate the performance of the model through the comparison of its output with measurements at selected monitoring sites. The comparison of the emission/absorption range given by the simulations with the range of the measured values concluded that the minimum values are closer to the measurements. Nevertheless, the model gives reasonable estimates of the real-world fluxes (see also Chapter 10).

The DNDC model can simulate all types of Hungarian agricultural lands. The total area of these lands is 5.12 million hectare. Note that the forests and pastures are not included. The DNDC model can be adjusted to the characteristic of farming management. In the lack of other information, we specified an average management for the whole country. The parameterization of the average crop management type can be adjusted in the model but the default one proved to be adequate. The default farming and crop management parameterization is available online at the website of the DNDC model (http://www.dndc.sr.unh.edu/).

The required soil and land cover data were available from the Research Institute for Soil Science and Agricultural Chemistry of the Hungarian Academy of Sciences and the CLC-50 database mentioned above. Quantity and timing of fertilization are among the most important factors for the emission of the greenhouse gases studied. Table 12.4 contains the information used in the simulations.

Methane and nitrous oxide fluxes of Hungarian agricultural lands were simulated by means of the DNDC model for the period of 2002–2004. In the simulation, actual meteorological data were used, while the management practice was assumed to be the same in each year. Table 12.5 gives the total area of the different kinds of

Table 12.4 Quantity and timing of fertilization applied in the simulations

	Amount of fertilizer (kg N ha^{-1})	Time of fertilization
Corn	140	1 April
Winter wheat	40	1 November
	80	1 February
Sunflower	140	1 May
Barley	100	1 June

Table 12.5 Specific and country scale emissions of nitrous oxide and methane calculated by the DNDC model. Negative sign indicates uptake by the soil

		N_2O		CH_4	
		Spec.emission	Country	Spec.emission	Country
	Area(1,000 ha)	(g m^{-2} $year^{-1}$)	(Gg $year^{-1}$)	(g m^{-2} $year^{-1}$)	(Gg $year^{-1}$)
Alfalfa	181	0.36	0.65		
Barley	373	0.71	2.64	−0.07	−0.26
Beans	4	0.54	0.02		
Beet	78	0.31	0.24		
Berries	2	1.21	0.03		
Corn	1645	1.50	24.72	−0.29	−4.77
Fallow	401	0.54	2.16	−0.15	−0.60
Fruits	67	0.16	0.11		
Grapes	139	0.28	0.39	−0.03	−0.04
Hay	81	1.18	0.95		
Millet	9	0.68	0.06		
Nursery	3	1.05	0.03		
Oats	66	0.42	0.28		
Onion	7	0.25	0.02		
Potato	59	0.34	0.20		
Rapes	132	1.24	1.63		
Rice	12	0.28	0.03	10.75	1.25
Rye	49	0.50	0.24		
Silo corn	168	2.77	4.66		
Sorghum	5	0.67	0.03		
Soybean	24	0.71	0.17		
Sunflower	342	0.77	2.62	−0.01	−0.03
Tobacco	6	0.37	0.02		
Tomato	7	0.95	0.06		
Vegetables	89	0.29	0.26		
Winter wheat	1165	0.35	4.06	−0.27	−3.19
Total	**5115**	–	**46.30**	–	**−7.64**
Country average	–	**0.91**	–	**−0.15**	–

land cover categories, their simulated specific nitrous oxide and methane emissions, and the total emission of the given land cover category. The table also gives the calculated mean annual nitrous oxide and methane fluxes for the Hungarian

Table 12.6 Interannual variations in the annual mean fluxes as simulated by the DNDC model

	2002	2003	2004
N_2O (g N_2O m^{-2} $year^{-1}$)	0.86	1.13	0.65
CH_4 (g CH_4 m^{-2} $year^{-1}$)	−0.16	−0.15	−0.15

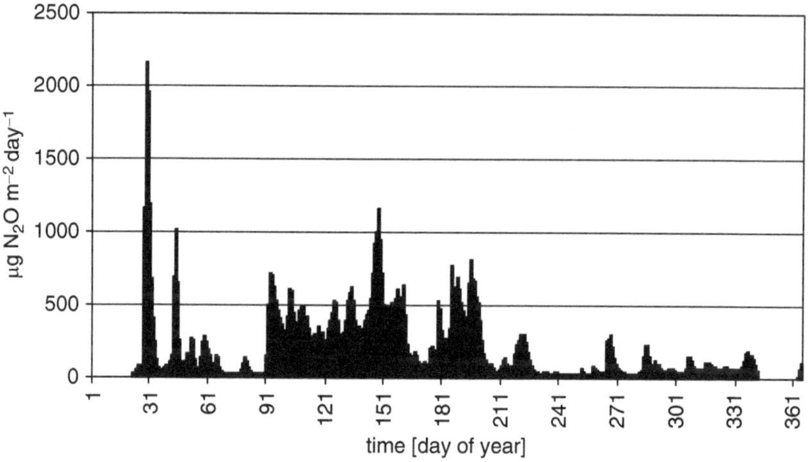

Fig. 12.14 Country scale daily N_2O flux over agricultural lands in 2002 as calculated by the DNDC model

agricultural lands for the period of the simulation. We note that in the case of several crop types methane flux is practically zero according to the simulation. For these crop types, no methane fluxes are given in Table 12.5. As we can see in Table 12.5, for the various crops, the specific N_2O emissions vary in a range of roughly three magnitudes. Additionally, our investigations have shown that fertilization of croplands enhances the emission significantly. It is also worth noting that the Hungarian croplands, except for rice plantations, are net methane sinks on average. Emission of rice fields cannot compensate the methane uptake of the other croplands; thus, the overall balance of the Hungarian croplands is negative.

According to the data in Table 12.5, corn and winter wheat emissions dominate the methane and nitrous oxide budgets of the Hungarian agricultural lands due to their high share (55%).

Considering the interannual variation in the nitrous oxide and methane emission, the data in Table 12.6 gives an impression. While the methane emission shows little variation, nitrous oxide release ranges between 0.65 and 1.13 g N_2O m^{-2} $year^{-1}$.

Figures 12.14 and 12.15 show examples for the daily N_2O and CH_4 emissions of the country's agricultural lands for the year 2002. Since the topsoil is frozen during winter in Hungary, in most cases, we can see a prominent N_2O emission peak at the beginning of the spring (Fig. 12.14). This emission is caused by the thawing soil. The deeper soil layers are warmer and the gases are aroused by the microbial soil

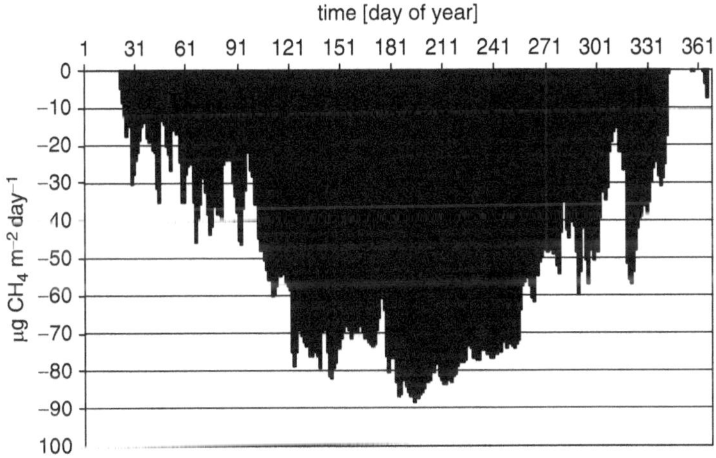

Fig. 12.15 Country scale daily CH_4 flux over agricultural lands in 2002 as calculated by the DNDC model. The negative values indicate uptake by the soil

processes. The frozen soil holds nascent gases in the soil during the winter, but the thaw causes the rapid disengage of these them. Rainfalls (i.e., increment in the soil water content) caused the other prominent peaks, since comparing the meteorological data with the simulation results the correlation is rather obvious.

Acknowledgments Biome-BGC version 4.1.1 was provided by the Numerical Terradynamic Simulation Group (NTSG) at the University of Montana. NTSG assumes no responsibility for the proper use of Biome-BGC by others. We are very grateful to Maosheng Zhao (NTSG) for providing us the official MOD17 product, the version 5.1 model parameters and the GMAO meteorological database. We are also very grateful to Attila Bakler (owner of a field close to the Hegyhátsál tower) for providing us information about the applied management practices. We thank Márta Birkás (Institute of Crop Production, Szent István University, Gödöllő, Hungary) for the invaluable help regarding the country-specific allocation data and the information about the fate of agricultural residues in Hungary. This research was supported by SEE-GRID-SCI (SEE-GRID eInfrastructure for regional eScience) project, funded by the European Commission through the contract nr RI-211338. The DNDC model calculations were supported by the NitroEurope Integrated Project of the European Commission's 6th R&D Framework Program and by the GVOP AKF 3.1.1. 2004–2005 0358/3.0 research projects. This research was partly supported by the Hungarian State Eötvös Fellowship.

References

Aber JD (1997) Why don't we believe the models. Bull Ecol Soc Am 78:232–233

Barcza Z, Kern A, Haszpra L, Kljun N (2009a) Spatial representativeness of tall tower eddy covariance measurements using remote sensing and footprint analysis. Agric Forest Meteorol 149:795–807

Barcza Z, Haszpra L, Somogyi Z, Hidy D, Lovas K, Churkina G, Horváth L (2009b) Estimation of the biospheric carbon dioxide balance of Hungary using the BIOME-BGC model. Időjárás Quart J Hung Meteorol Ser 113:203–219

Büttner G, Bíró M, Maucha M, Petrik O (2000) Land Cover mapping at scale 1:50.000 in Hungary: lessons learnt from the European CORINE programme. 20th EARSeL symposium, 14–16 June 2000. In: A decade of Trans-European remote sensing cooperation, pp 25–31

Chapin FS, Woodwell GM, Randerson JT, Rastetter EB, Lovett GM, Baldocchi DD, Clark DA, Harmon ME, Schimel DS, Valentini R, Wirth C, Aber JD, Cole JJ, Goulden ML, Harden JW, Heimann M, Howarth RW, Matson PA, McGuire AD, Melillo JM, Mooney HA, Neff JC, Houghton RA, Pace ML, Ryan MG, Running SW, Sala OE, Schlesinger WH, Schulze E-D (2005) Reconciling carbon-cycle concepts, terminology and methods. Ecosystems 9:1041–1050. doi:10.1007/s10021-005-0105-7

Chevallier F, Viovy N, Reichstein M, Ciais P (2006) On the assignment of prior errors in Bayesian inversion of CO_2 surface fluxes. Geophys Res Lett 33:L13802. doi:10.1029/2006GL026496

Churkina G, Tenhunen J, Thornton P, Falge EM, Elbers JA, Erhard M, Grunwald T, Kowalski AS, Sprinz D (2003) Analyzing the ecosystem carbon dynamics of four European coniferous forests using a biogeochemistry model. Ecosystems 6:168–184

Ciais P, Reichstein M, Viovy N, Granier A, Ogee J, Allard V, Aubinet M, Buchmann N, Bernhofer C, Carrara A, Chevallier F, De Noblet N, Friend AD, Friedlingstein P, Grunwald T, Heinesch B, Keronen P, Knohl A, Krinner G, Loustau D, Manca G, Matteucci G, Miglietta F, Ourcival JM, Papale D, Pilegaard K, Rambal S, Seufert G, Soussana JF, Sanz MJ, Schulze ED, Vesala T, Valentini R (2005) Europe-wide reduction in primary productivity caused by the heat and drought in 2003. Nature 437:529–533

Ciais P, Bousquet P, Freibauer A, Naegler T (2007) Horizontal displacement of carbon associated with agriculture and its impacts on atmospheric CO_2. Glob Biogeochem Cy 21:GB2014. doi:doi:10.1029/2006GB002741

Ciais P, Wattenbach M, Vuichard N, Smith P, Piao SL, Don A, Luyssaert S, Janssens I, Bondeau A, Dechow R, Leip A, Smith P, Beer C, van der Werf GR, Gervois S, Van Oost K, Tomelleri E, Freibauer A, Schulze ED, CARBOEUROPE Synthesis Team (2010) The European carbon balance. Part 2: croplands. Glob Change Biol 16:1409–1428. doi:doi: 10.1111/j.1365–2486.2009.02055.x

Gervois S, de Noblet-Ducoudré N, Viovy N, Ciais P, Brisson N, Seguin B, Perrier A (2004) Including croplands in a global biosphere model: methodology and evaluation at specific sites. Earth Interact 8:1–16

Gervois S, Ciais P, de Noblet-Ducoudré N, Brisson N, Vuichard N, Viovy N (2008) Carbon and water balance of European croplands throughout the 20th century. Glob Biogeochem Cy 22:GB2022, doi:10.1029/2007GB003018

Goudriaan J, Groot JJR, Uithol PWJ (2001) Productivity of agro-ecosystems. In: Roy J, Saugier B, Mooney HA (eds) Terrestrial global productivity. Academic Press, San Diego

Haszpra L, Barcza Z, Bakwin PS, Berger BW, Davis KJ, Weidinger T (2001) Measuring system for the long-term monitoring of biosphere/atmosphere exchange of carbon dioxide. J Geophys Res 106D:3057–3070

Haszpra L, Barcza Z, Davis KJ, Tarczay K (2005) Long-term tall tower carbon dioxide flux monitoring over an area of mixed vegetation. Agr Forest Meteorol 132:58–77

Heinsch FA, Zhao M, Running SW, Kimball JS, Neman RR, Davis KJ, Bolstad PV, Cook BD, Desai AR, Ricciuto DM, Law BE, Oechel WC, Kwon H, Luo H, Wofsy SC, Dunn AL, Munger JW, Baldocchi DD, Xu L, Hollinger DY, Richardson AD, Stoy PC, Siqueira MBS, Monson RK, Burns SP, Flanagan LB (2006) Evaluation of remote sensing based terrestrial productivity from MODIS using regional tower Eddy Flux network observations. IEEE Trans Geosci Remote Sens 44:1908–1924

Hollinger DY, Richardson AD (2005) Uncertainty in eddy covariance measurements and its application to physiological models. Tree Physiol 25:873–885

Hwang T, Kang S, Kim J, Kim Y, Lee D, Band L (2008) Evaluating drought effect on MODIS Gross Primary Production (GPP) with an eco-hydrological model in the mountainous forest, East Asia. Glob Change Biol 14:1–20

New M, Lister D, Hulme M, Makin I (2002) A high-resolution data set of surface climate over global land areas. Clim Res 21:1–25

Pásztor L, Szabó J, Bakacsi Z, László P, Dombos M (2006) Large-scale soil maps improved by digital soil mapping and GIS-based soil status assessment. Agrokémia és Talajtan 55:79–88

Peoples MB, Herridge DF, Ladha JK (1995) Biological nitrogen fixation: an efficient source for sustainable agricultural production? Plant Soil 174:3–28

Running SW, Hunt ERJ (1993) Generalization of a forest ecosystem process model for other biomes, Biome-BGC, and an application for global-scale models. In: Ehleringer JR, Field CB (eds) Scaling physiological processes: leaf to globe. Academic Press, San Diego (CA), pp 141–158

Sims DA, Rahman AF, Cordova VD, Baldocchi DD, Dennis, Flanagan LB, Goldstein AH, Hollinger DY, Misson L, Monson RK, Schmid HP, Wofsy SC, Xu L (2005) Midday values of gross CO_2 flux and light use efficiency during satellite overpasses can be used to directly estimate eight-day mean flux. Agr Forest Meteorol 131:1–12

Thornton PE (2000) User's Guide for Biome-BGC, Version 4.1.1. Numerical Terradynamic Simulation Group, School of Forestry, University of Montana, Missoula, MT 59812 USA

Thornton PE, Hasenauer H, White MA (2000) Simultaneous estimation of daily solar radiation and humidity from observed temperature and precipitation: an application over complex terrain in Austria. Agr Forest Meteorol 104:255–271

Tóth T, Pásztor L, Várallyay Gy, Tóth G (2007) Overview of soil information and soil protection policies in Hungary. In: Hengl T, Panagos P, Jones A, Tóth G (eds) Status and prospect of soil information in South-Eastern Europe: soil databases, projects and applications. Euro Comm, 86, p 77

Turner DP, Ritts WD, Cohen WB, Gower ST, Running SW, Zhao M, Costa MH, Kirschbaum AA, Ham JM, Saleska SR, Ahl DE (2006) Evaluation of MODIS NPP and GPP products across multiple biomes. Remote Sens Environ 102:282–292

Turner DP, Ollinger SV, Kimball JS (2004) Integrating remote sensing and ecosystem process models for landscape- to regional-scale analysis of the carbon cycle. BioScience 54:573–584

Vetter M, Churkina G, Jung M, Reichstein M, Zaehle S, Bondeau A, Chen Y, Ciais P, Feser F, Freibauer A, Geyer R, Jones C, Papale D, Tenhunen J, Tomelleri E, Trusilova K, Viovy N, Heimann M (2008) Analyzing the causes and spatial pattern of the European 2003 carbon flux anomaly using seven models. Biogeosciences 5:561–583

Verbeeck H, Samson R, Verdonck F, Lemeur R (2006) Parameter sensitivity and uncertainty of the forest carbon flux model FORUG: a Monte Carlo analysis. Tree Physiol 26:807–817

Vitousek PM, Mooney HA, Lubchenco J, Melillo JM (1997) Human domination of Earth's ecosystems. Science 277:494–499

Wang Q, Watanabe M, Ouyang Z (2005) Simulation of water and carbon fluxes using BIOME-BGC model over crops in China. Agr Forest Meteorol 131:209–224

Wang W, Davis KJ, Cook BD, Butler MP, Ricciuto DM (2006) Decomposing CO_2 fluxes measured over a mixed ecosystem at a tall tower and extending to a region: a case study. J Geophys Res 111:G02005. doi:10.1029/2005JG000093

White MA, Thornton PE, Running SW, Nemani RR (2000) Parameterization and sensitivity analysis of the Biome-BGC terrestrial ecosystem model: net primary production controls. Earth Interac 4:1–85

Yan H, Fu Y, Xiao X, Huang HQ, He H, Ediger L (2009) Modeling gross primary productivity for winter wheat-maize double cropping system using MODIS time series and CO_2 eddy flux tower data. Agr Ecosyst Environ 129:91–400

Zhang Y, Yu Q, Jiang J, Tang Y (2008) Calibration of Terra/MODIS gross primary production over an irrigated cropland on the North China Plain and an alpine meadow on the Tibetan Plateau. Glob Change Biol 14:757–767

Zhao M, Heinsch FA, Nemani RR, Running SW (2005) Improvements of the MODIS terrestrial gross and net primary production global data set. Remote Sens Environ 95:164–176. doi:10.1016/j.rse.2004.12.011

Zhao MS, Runnin GW, Nemani RR (2006) Sensitivity of Moderate Resolution Imaging Spectroradiometer (MODIS) terrestrial primary production to the accuracy of meteorological reanalyses. J Geophys Res 111:G01002. doi:10.1029/2004JG000004

Chapter 13
Model-Based Biospheric Greenhouse Gas Balance of Hungary*

Zoltán Barcza, Alberte Bondeau, Galina Churkina, Philippe Ciais, Szilárd Czóbel, Györgyi Gelybó, Balázs Grosz, László Haszpra, Dóra Hidy, László Horváth, Attila Machon, László Pásztor, Zoltán Somogyi, and Kristof Van Oost

*Citation: Barcza, Z., Bondeau, A., Churkina, G., Ciais, Ph., Czóbel, Sz., Gelybó, Gy., Grosz, B., Haszpra, L., Hidy, D., Horváth, L., Machon, A., Pásztor, L., Somogyi, Z., Van Oost, K., 2010: Modeling of biosphere–atmosphere exchange of greenhouse gases — Model-based biospheric greenhouse gas balance of Hungary. In: Atmospheric Greenhouse Gases: The Hungarian Perspective (Ed.: Haszpra, L.), pp. 295–330

Z. Barcza (✉) and Gy. Gelybó
Department of Meteorology, Eötvös Loránd University, H-1117
Budapest, Pázmány P. s. 1/A, Hungary
e-mail: bzoli@elte.hu

A. Bondeau
Potsdam Institute for Climate Impact Research, PIK, D-14412 Potsdam,
Telegrafenberg, P.O. Box 601203, Germany

G. Churkina
Leibniz-Centre for Agricultural Landscape Research, D-15374 Müncheberg,
Eberswalder Strasse 84, Germany

P. Ciais
Laboratoire des Sciences du Climat et de l'Environnement, CEA CNRS UVSQ,
F-91191 Gif sur Yvette, France

Sz. Czóbel
Institute of Botany and Ecophysiology, Szent István University, H-2103
Gödöllő, Páter K. u. 1., Hungary

B. Grosz
Department of General and Inorganic Chemistry, Eötvös Loránd University,
H-1117 Budapest, Pázmány P. s. 1/A, Hungary

L. Haszpra, D. Hidy, L. Horváth, and A. Machon
Hungarian Meteorological Service, H-1675 Budapest, P.O. Box 39, Hungary

A. Machon
Center for Environmental Science, Eötvös Loránd University, H-1117 Budapest,
Pázmány P. s. 1/A, Hungary
and
Department of Botany and Plant Physiology, Szent István University,
H-2103 Gödöllő, Páter K. u. 1., Hungary

L. Haszpra (ed.), *Atmospheric Greenhouse Gases: The Hungarian Perspective*,
DOI 10.1007/978-90-481-9950-1_13, © Springer Science+Business Media B.V. 2011

L. Pásztor
Research Institute for Soil Science and Agricultural Chemistry of the Hungarian Academy of Sciences, H-1022 Budapest, Herman O. út 15., Hungary

Z. Somogyi
Hungarian Forest Research Institute, H-1142 Budapest,
Stefánia út. 14., Hungary

K.Van Oost
Georges Lemaître Centre for Earth and Climate Research (TECLIM),
Earth and Life Institute (ELI), Université catholique de Louvain,
3, Place Louis Pasteur, B-1348 Louvain-la-Neuve, Belgium

Abstract We present the first comprehensive overview about the state-of-the-art estimates of the Hungarian biospheric greenhouse gas (GHG) balance. Biogeochemical models, statistical time series, and literature data are used together to describe grassland, forest, and arable land specific GHG fluxes. The estimates are aggregated to the country level to approximate the net biospheric greenhouse gas balance for the first time. The results suggest that the overall biospheric GHG balance of Hungary is negative (−13.8 Mt CO_2-equivalent year^{-1}), which means net GHG release to the atmosphere. According to this finding, the biosphere does not mitigate anthropogenic GHG emission in the country. More research is needed to constrain the estimates, and to provide reliable uncertainty estimates to the fluxes and stocks. We try to point out important research priorities that can help us to better understand the biogeochemical processes of the biosphere in Hungary, and to provide mitigation opportunities for biospheric GHG emission.

Keywords Net greenhouse gas balance • Carbon dioxide • Nitrous oxide • Methane • Net primary production • Net biome production • Biogeochemical models

13.1 Introduction

Estimation of a whole country's biospheric greenhouse gas (GHG) balance is a challenging task even if the country is relatively small and the land use is rather uniform. There is no comprehensive overview about the biospheric greenhouse gas balance of Hungary that includes estimates derived with techniques other than the IPCC method (see Part IV of this book). In this chapter, we provide the first estimates for as many biospheric GHG balance components for the whole country as possible using alternative methods. As uncertainty is the inherent property of all methods, the GHG balance should always be estimated based on as many data sources as possible (see e.g. Schulze et al. 2009).

The greenhouse gas balance of Hungary is not well constrained. There are only very few domestic studies dealing with the country-wide biospheric GHG emissions and removals for specific plant functional types (PFTs; Führer and Mátyás 2005; Somogyi 2008) or for the whole country (Mészáros and Molnár 1992; Barcza et al. 2009). Most of the results available for Hungary are provided by European synthesis studies (Janssens et al. 2005; Vuichard et al. 2007; Ciais et al. 2010). However, such large-scale studies cannot capture the smaller-scale variability of ecosystem processes caused by different soil types, meteorological conditions, and they also ignore country-specific management practices. There is a growing need to conduct country-specific analysis about the GHG emissions and removals using state-of-the-art scientific tools, and the existing knowledge and precise statistics about the special conditions.

The Hungarian GHG emission reports to the United Nations Framework Convention on Climate Change (UNFCCC) cover a wide range of emissions and removals caused by human activities (energy production, industry, waste management, agriculture, etc., see Part IV of the present book for details). The emissions and removals partly cover biospheric fluxes from the agriculture, land use, land use change, and forestry sectors. These fluxes have a large variability induced by changing weather and climate, and by changing management practice. The data reported to UNFCCC are not based on process-based models but rather on standardized conversion factors and statistical data, and partly on direct measurements (see Part IV of this book). Those data estimated by the IPCC methodology have special features in many senses as they estimate fluxes caused by anthropogenic activity. For example, in grassland category, carbon stock changes in mineral soils are reported that are estimated from changes in land use and management practices. In cropland category, only carbon stock changes in living (woody) biomass and in mineral soils are reported, while emissions from liming are also taken into account. Nonwoody croplands that are not subject to land use change are assumed to have zero CO_2 emission in the long term, which means steady state soil and biomass carbon pools (Barcza et al. 2009). Those estimates are not fully in accordance with the most recent European results (e.g. Gervois et al. 2008; Ciais et al. 2010). There are also uncertainties with the forest-related IPCC-based estimates (Somogyi 2008). Some recent, process-based models allow for estimates in carbon stock changes in the soil and litter pools, and relate the carbon sequestration and emission potentials of ecosystems as a function of weather and climate fluctuations (Churkina et al. 2003; Ciais et al. 2005; Bondeau et al. 2007; Gervois et al. 2008; Vetter et al. 2008). However, these models often use global soil and vegetation maps, land use and climate forcing datasets, and underpinning hypothesis that may be erroneous compared to with local data. As a consequence, at present it is problematic to compare the results of the IPCC-based carbon accounting methods used for UNFCCC and the process-based estimates (Barcza et al. 2009). Nevertheless, both the IPCC-based and the process-based estimates provide information about the GHG emissions and removals, and help us to critically evaluate their accuracy.

13.2 Data sources, Databases and Methods

Here, we describe the data sources, databases, measurements, and methods used for the country-specific simulations and estimations (Table 13.1). As a part of the presented data was not published yet (e.g. the results based on yield statistics, remote

Table 13.1 Models, databases, and other information sources used in the present study together with the greenhouse gas balance components estimated and their spatial and temporal resolution

Data source/model	Simulated/estimated GHG balance component	Spatial resolution	Temporal resolution
Biome-BGC	Net ecosystem exchange (NEE), gross primary production (GPP), total ecosystem respiration (R_{eco}), net primary production (NPP)	1/6° × 1/6°	2002–2009
DNDC	Soil N_2O emission, CH_4 flux	1/6° × 1/6°	2002–2004
Lund-Potsdam-Jena for managed Land (LPJmL)	Net biome production (NBP), NPP, NEE, harvest, fire CO_2 emission	0.5° × 0.5°	2002–2006
MOD17	GPP, NPP	1 km × 1 km	2000–2006
ORCHIDEE-STICS	NPP, NBP	1° × 1°	1990–1999
Spatially explicit erosion model	Carbon erosion, river carbon flux	1 km × 1 km	1995–2005 mean
Fuzzy logic model (Dechow and Freibauer, in prep.)	N_2O emission	Grasslands	1990–1999 mean
national crop yield statistics (HCSO data)	NPP, harvest	NUTS 3 level	2000–2009
FAOSTAT TradeSTAT	carbon export/import	Country	2000–2007
AGROTOPO	Soil carbon stocks	1:100,000 – 1:250,000 (100–500 m)	~1980
DKSIS	Soil carbon stocks	1:25,000 – 1:50,00 (25–100 m)	1938–1953
Herbaceous forest understorey carbon balance	NEE	N/A	2006
Führer and Mátyás 2005	NPP, forest carbon stocks, wood harvest	Hungarian forests	2002
Janssens et al. 2005	NBP, carbon export/import	Country	1990s
Ciais et al. 2010	N_2O emission	Croplands	1990–1999
Vuichard et al. 2007 (PASIM model)	N_2O emission	1° × 1°	1990–1999 (meteorology is from 1993)

sensing, and process-based models), it is very important to provide information about the data sources to enable repeatable calculations and verification of the results.

13.2.1 Land Cover Database

The determination of the land cover type was based on the CORINE Land Cover 2000 (CORINE2000) classification scheme (Büttner et al. 2002). CORINE2000 uses 50 land cover categories with 1:100,000 mapping scale. Biogeochemical models cannot distinguish among so many different land cover types because of the necessary simplifications and generalizations in the models. For the characterization of Hungary, the 50 land cover types were aggregated into four basic, essentially different categories. Using CORINE2000 nomenclature, we treated the "non-irrigated arable land" category (CLC code 211 in the CORINE2000 system) as cultivated arable land. "Pastures" (CLC code 231) and "natural grasslands" (CLC code 321) were handled together as grass. "Mixed forests" (CLC code 323) were treated as 50% broadleaf forest (CLC code 311) and 50% coniferous forest (CLC code 312).

For the definition of arable lands, we followed the convention of the Hungarian Central Statistical Office (HCSO; http://www.ksh.hu/pls/ksh/ksh_web.meta.objektum?p_lang=EN&p_menu_id=220&p_ot_id=200&p_obj_id=2144&p_session_id=64081067). Note that the HCSO-based arable land definition excludes kitchen gardens if they are used for crop production that is utilized by the persons living in the holding.

We found that the total area of the aggregated land cover types is not the same as it is reported by the Hungarian Central Statistical Office (HCSO, http://portal.ksh.hu/pls/ksh/docs/eng/agrar/html/tabl1_3_1.html). In order to provide accurate areas, we corrected the CORINE2000 based area estimates. As we did not have spatially explicit data, e.g. about the cultivated and the noncultivated areas for agriculture, we simply calculated the ratio of the CORINE2000 and the HCSO-based total areas, and we used the ratio to adjust the arable land, grassland, deciduous, and needleleaf forest areas inside the grid cells that were used for the GHG balance modeling. For the correction, we used the average areas of arable lands, grasslands, and forests for 2000 and 2009. The resulting land use database was applied in the simulations, and the same areas were taken into account for the country-specific GHG balance estimations and upscaling presented later (Table 13.2).

Table 13.2 Basic land cover types statistics of Hungary (CORINE2000 database and data from the Hungarian Central Statistical Office)

	Whole country	Arable land	Grassland	Evergreen needleleaf forest	Deciduous broadleaf forest	Other (settlements, wetlands, etc.)
Area [ha]	9,303,000	4,500,700	1,027,700	187,527	1,645,073	1,943,279
Percent of total area	100	48.38	11.05	2.02	17.68	20.89

13.2.2 National Yield Statistics

As agricultural net primary production (NPP; for definition of fluxes, see Chapin et al. 2005) can be estimated from crop yield data and information on crop allometry, we made full advantage of the crop yield statistics published by the Hungarian Central Statistical Office (2010) (HCSO, http://statinfo.ksh.hu/Statinfo/themeSelector.jsp?&lang=en). In the HCSO database, county level (Nomenclature of Territorial Units for Statistics (NUTS) 3 level region; for explanation on NUTS referencing system see http://epp.eurostat.ec.europa.eu/portal/page/portal/nuts_nomenclature/introduction) yield statistics are available annually together with the total harvested area and the total production of the main crops. In Hungary, the areas of NUTS 3 level regions vary between 8,445 km^2 (Bács–Kiskun county) and 2,265 km^2 (Komárom–Esztergom county).

In order to reduce the number of crop types and to simplify the calculations, we estimated the harvested area of the most important crop types. Based on the 2004 year data, it was found that the harvested areas of winter wheat, maize, sunflower, and barley were the largest among all crop types. The four crops constituted ~71% of the total arable land area. As the crop yield of barley was close to the yield of wheat, we treated barley as winter wheat, since no allometry data were available for barley for the NPP calculations. Based on this simplification, we simulated agricultural productivity as if it would have been a uniform mixture of 47% winter wheat, 38% maize, and 15% sunflower all over Hungary. This simplification causes uncertainty in the NPP calculations.

NPP can be estimated from yield data based on allometry, in particular the harvest index (HI defined as the ratio of yield to above-ground NPP) and other conversion ratios (Goudriaan et al. 2001; Ciais et al. 2005, 2007, 2010). We used country-specific HI and conversion data for the estimation of NPP similarly to the method described in Chapter 12. Dry matter content of biomass and carbon content of the different plant pools for wheat and maize were estimated from Goudriaan et al. (2001). For sunflower, we used conversion data presented by Ciais et al. (2007). As the crop yield is rather variable in Hungary, we used yield-dependent HI for winter wheat, maize, and sunflower (M. Birkás, personal communication, 2010). For the estimation of root production as percentage of total above-ground production, we also used yield-dependent allocation factors (M. Birkás, personal communication, 2010). In our case, yield dependence means that HI and root allocation are linear functions of crop yield, which corresponds to the observations and expert judgment in Hungary.

The NUTS 3 level data were interpolated to a regular 1/6° by 1/6° grid that was used for the country-specific simulations (see below). The interpolation was performed with a minimum curvature surface method.

Using the HI method, we estimated agricultural NPP for the whole country for the period of 2000–2009. Using the available information, we also estimated the amount of carbon removed from the fields due to harvest and residue management, and also carbon input to the soil (M. Birkás, personal communication, 2010).

For maize and sunflower, it was assumed that crop residues were ploughed into the soil (after chopping) according to the Hungarian practices. For wheat, we assumed that on 30% of the harvest area, 2/3 of the straw is removed and utilized elsewhere (typically as animal bedding). The rest 1/3 of the straw was assumed to be ploughed back to the soil or left at the soil surface. On the remaining 70% of the harvest area, it was assumed that no straw was removed from the fields. Note that the straw used for animal bedding might return to the fields after some time as manure; thus it could be a source of carbon to the soil. Based on the statistical data provided by HCSO, we can estimate the amount of carbon input to the agricultural lands due to application of manure. Using the conversion factors presented in Anthoni et al. (2004), the estimated carbon input for the total agricultural land is 16 gC m^{-2} $year^{-1}$ (2004–2008 mean). This carbon input is low as compared to the carbon input caused by residue management (see below), so in the present study we do not deal with manure input.

13.2.3 FAO Statistics

As discussed in Chapter 12, lateral carbon fluxes can substantially modify the carbon balance of a region. This is also true at national and continental scale (Ciais et al. 2007). In order to address this additional GHG component for Hungary, we made an attempt to quantify the amount of carbon that is exported or imported in the form of agricultural crop products and wood products.

For the quantification of exported and imported wood products, we used the data available at the Food and Agriculture Organization of the United Nations (FAO) (FAO 2010). At FAO web site, data were available for the period of 1997–2007. We calculated the carbon content of wood products according to Janssens et al. (2005).

For agricultural trade products, we used the TradeSTAT database of FAOSTAT (FAO 2010). Based on the weight of the trade products, we selected the top 95% products for Hungary for the period of 2000–2007. We calculated carbon content of the trade products using the conversion factors published by Ciais et al. (2007) if data were available. Conversion factors were not available for 16% of the exported and 37.5% of the imported products. Neglecting these products might cause uncertainties in the estimations. As a final step, we calculated the average carbon export/import for the 2000–2007 period.

13.2.4 Remote Sensing

As it was demonstrated in Chapter 12, remote sensing offers the possibility to estimate certain carbon cycle components of the terrestrial vegetation.

We used remote sensing data to characterize productivity of four (evergreen needleleaf forests, deciduous broadleaf forests, arable land, grassland) ecosystem types over Hungary. In the MOD17A3 light-use efficiency model product (Zhao et al. 2005), the annual sums of gross primary production (GPP) and NPP are available with 1 km spatial resolution for the period of 2000–2006. We used the 5.1 version of MOD17A3 product provided by University of Montana Numerical Terradynamic Simulation Group. To separate land cover categories in the country, we used the CORINE 2000 database introduced earlier (Büttner et al. 2002). The MOD17A3 NPP and GPP products of the MOD17 model were calibrated using both eddy covariance (EC) measurements at a flux tower in case of GPP and synthesized data for NPP described in Zhao et al. (2005).

Validation of the MOD17 cropland estimates for Hungary is presented in Chapter 12 of the present book. Validation of annual GPP estimations from the MOD17 model for forests was carried out using flux tower measurements performed in the Czech Republic over coniferous forest. (Since MOD17A3 data are only available for a limited time period mentioned above, data from eddy covariance measurement sites that has been initiated in recent years cannot be used in the validation.) The data were acquired from the CarboEurope-IP database (http://gaia.agraria.unitus.it/database) as standard level 4 data files from the Bíly Kríz forest site (49°30′9.4″N, 18°32′18.3″E, 908 m asl) between 2004 and 2006 (Marek et al. 2006). We compared MODIS estimations to flux-tower-based GPP using two different methods available in level 4 data files. The MODIS product usually underestimated the annual GPP from the flux measurements. GPP estimations derived by artificial neural network technique (GPP_{ANN}) gave lower estimations than the GPP based on the marginal distribution sampling technique (GPP_{MDS}). The three years average differences were 103 and 129 gC m^{-2} $year^{-1}$, respectively. The validation indicated that MOD17 performs well (the error is close to the uncertainty of GPP derived from the flux measurements), at least for coniferous forests in the vicinity of Hungary.

13.2.5 Soil Carbon Inventories

Two different databases have been used to quantify the soil carbon stocks in Hungary. Those are the AGROTOPO database and the so-called Digital Kreybig Soil Information System.

AGROTOPO (1994) is a nationwide spatial soil information system whose spatial resolution corresponds to about 1:250,000 – 1:100,000 scale. It is the digital version of the 1:100,000 scaled agrotopographic map series, which delineates soil mapping units (SMUs) on the 85 map sheets covering the country. The polygons of the agrotopographic maps are coded with ten-digit polygon identifiers which provide information on the main soil properties. AGROTOPO pedologically characterizes the total area of Hungary (including forests, arable lands, and grasslands) with almost 4,000 soil patches, which are considered as homogeneous soil units at the spatial resolution of the system.

The estimation of organic matter (OM) and soil organic carbon (SOC) content can be based on two parameters of AGROTOPO database. "Organic matter resource" gives the average SOC concentration in tC ha^{-1} for the topsoil in six distinct categories (<50, 50–100, 100–200, 200 300, 300–400, >400 tC ha^{-1}). The other soil feature is the "Depth of humus layer" giving its thickness in five distinct classes (<20, 20–40, 40–70, 70–100, >100 cm). Volumetric OM and SOC quantities may be inferred by proper combination of the two parameters describing "Organic matter resource" and "Depth of humus layer".

In Hungary, the national soil mapping project initiated and led by Kreybig (1937) was a national survey based on field and laboratory soil analyses, and at the same time serving practical purposes. Overall chemical and physical soil properties of the soil root zone featuring soil mapping units were identified and displayed on maps; further soil properties were determined and measured in soil profiles. The Kreybig legacy represents a valuable treasure of soil information, which is developed in the Digital Kreybig Soil Information System (DKSIS; Pásztor et al. 2010).

DKSIS simultaneously contains two types of geometrical datasets. Soil mapping units are represented by polygons and simply characterized by three attributes: (i) combined texture and water management categories, (ii) overall soil chemical properties, (iii) areas with shallow depth are also distinguished (yes/no).

The soil conditions were identified on agricultural areas, and a simplified land use categorization is also provided delineating temporarily waterlogged areas, forests, lakes, marshes, rivers, and settlements. Detailed soil properties were determined and measured in soil profiles. The surveys applied various pits and boreholes, some of which were deepened to 10 m or to the groundwater level. There is representative profile description in the database for about 22,000 plots. This profile information is transferred for further locations, which sums up in approximately 250,000 plots.

Volumetric OM and SOC quantities for the DKSIS database may be inferred from humus content [%] and the depth of humus layer [cm], which are measured in soil profiles.

13.2.6 Extrapolation from Point Measurements

The role of the carbon cycling of herbaceous forest undergrowth vegetation in the Hungarian carbon balance was estimated on the basis of *in situ* investigation. The measurement was carried out in the understorey of a natural oak forest stand of the Botanical Garden of Szent István University (Gödöllő, Hungary), and covered the whole vegetation period of common forest spring geophytes. The seasonal dynamics of phenology (including leaf area), soil respiration, dark respiration, net ecosystem exchange (NEE), biomass and its carbon content in each spring geophyte-dominated stand were studied simultaneously on a weekly basis in 2006. Static chamber measurements of NEE were made from three permanent plots, dominated by widely distributed European vernal species like *Corydalis solida* L., *Ranunculus ficaria* L., and *Anemone ranunculoides* L. The stands of the investigated species are important parts of the Central European spring pattern in several deciduous forest communities.

To measure NEE at stand level, a homemade, portable, nondestructive open canopy chamber system was used. During the usual hourly measurements, a Ciras 2 infrared gas analyzer (PP Systems, Hitchin, UK) was connected to the chamber (60-cm diameter, made from plexiglass) taking air samples from the connecting (inner and outer) tubes (Czóbel et al. 2008). Gas exchange rates were calculated from the rate of CO_2 concentration change. Twilight and nighttime respirations were simulated by covering the chamber (partly or fully) with a black plastic sheet. The calculated plot level sink activity was also cross-calibrated by biomass carbon content data. The upscaling process of stand level CO_2 sequestration potential of herbaceous forest understorey vegetation to national (deciduous) forest level was performed based on the average CO_2 fluxes of the geophyte plots and the HCSO-based forest area estimates.

13.2.7 Literature Review

There are a few publications in the Hungarian and international literatures where GHG balance components are presented for Hungary. Though some of them turned out to be unreliable (e.g. the cropland carbon source intensity is overestimated in the previous studies; see Ciais et al. 2010), one might find it useful to collect all estimates and describe their reliability according to the state-of-the-art knowledge. Future studies might rely on the data presented here.

We utilized data from the Janssens et al. (2005) study where net biome production (NBP) was estimated for Hungarian grasslands, arable lands, forests, and wetlands. In addition, carbon export due to wood and food trade was also estimated in the same study. For methodological details, we simply refer to Janssens et al. (2005).

We also used the Ciais et al. (2010) study where cropland and grassland related N_2O emission data was provided for Hungary using so-called fuzzy logic empirical model developed by Dechow and Freibauer (2010, in prep.; for details see Ciais et al. 2010).

We also used grassland-related N_2O emission data from the Vuichard et al. (2007) study. This N_2O emission was calculated at a spatial resolution of $1° \times 1°$ for year 1992 using the PASIM grassland model (Vuichard et al. 2007).

13.2.8 Spatially Explicit Modeling of Erosion

We used the tillage and water erosion models described in Van Oost et al. (2009) to estimate lateral carbon fluxes from erosion processes (Ec) on agricultural land.

High-resolution topography (SRTM data; CIAT 2004), land use (CORINE2000; Büttner et al. 2002), and soil (Soil Geographical Database of Europe; http://eusoils.jrc.ec.europa.eu/esbn/SGDBE.html) databases were used as model input. The model was run at a spatial resolution of 100 m and the soil erosion estimates (Es) were aggregated to 1 km^2. Lateral carbon fluxes as a result of cropland erosion was than calculated as Ec = Es*OC/100, where OC is the topsoil carbon content (%) derived from the OCTOP map (Jones et al. 2005).

Note that wind erosion is neglected by the model, though, especially in the Hungarian Great Plain, this kind of erosion is a well-recognized process and should be taken into account in future studies.

13.2.9 Process-Based Ecosystem Models

Point measurements provide invaluable information about the GHG balance of small regions where the spatial representativeness depends on the measurement technique (see Part II of this book). The spatial scale ranges from ~10 to 20 cm (chamber method) to a few kilometers (tall tower based eddy covariance method). Atmosphere-based inversion techniques have the ability to estimate the surface/atmosphere exchange of GHGs on large spatial scales (in country or continental level, see e.g. Schulze et al. 2009). There is a spatial gap between the results of the small-scale studies (bottom-up approach) and the large-scale, inversion-based fluxes (top-down approach). In order to fill the gaps, we need mathematical tools that can capture the variability of GHG-related biospheric processes due to varying environmental conditions.

Process-based biogeochemical models (ecosystem models) are useful tools to estimate the GHG balance of larger regions. Those models describe the spatial and temporal variability of the processes that drive plant photosynthesis, microbial respiration, autotrophic respiration, nitrification/denitrification, etc., and most recently human management as well (White et al. 2000; Bondeau et al. 2007; Gervois et al. 2008; Vetter et al. 2008).

In order to better constrain the Hungarian biospheric GHG balance, we used results from European and global scale ecosystem models.

We used results from the LPJmL (Lund–Potsdam–Jena for managed Land) model for the time period of 2002–2006 (A. Bondeau, personal communication , 2010). The LPJmL model is a process-based dynamic vegetation model describing the biogeochemical fluxes for the actual vegetation. The dynamic global vegetation model LPJ simulates the dynamic of the structure of natural vegetation, the carbon stocks in the vegetation and the soil, and the coupled water and CO_2 fluxes. Disturbances by natural fires are accounted for. The representation of agriculture has been implemented within the LPJ model (LPJmL; Bondeau et al. 2007) through the use of 11 crop functional types and two types of managed grasses. The modeling of crop growth requires the accounting for specific phenology and allocations rules, including the allocation to the storage organs, which are harvested. The LPJmL results presented here have been obtained using the CRU climate data as input for the time period of 1901–2006 at 0.5° spatial resolution (Mitchell and Jones 2005). The land use data set used is presented in Fader et al. (2010). For each grid cell, the model simulates the carbon and water fluxes of the natural vegetation (its composition is simulated by the model itself), and the carbon and water fluxes of each crop functional type (rain fed or irrigated) present in the grid cells according to the land use data set. Beside the respiration fluxes, carbon is removed from the natural vegetation through fires, and – from the agricultural stands – through harvest. There is neither burning of agriculture residues nor timber harvest from managed forests within the model version used. In LPJmL, the yield for each crop is scaled over each country by an arbitrary factor to be equal to the FAO value. Therefore, the LPJmL results are not independent from the FAO statistics and can only be used to predict the spatial distribution of NPP within the country.

We also used results from the ORCHIDEE-STICS model (Gervois et al. 2008). ORCHIDEE-STICS is a process-based ecosystem model that was developed from the ORCHIDEE model. The crop model STICS was coupled with ORCHIDEE in order to provide more reliable carbon balance estimations about managed agricultural land. STICS is parameterized as default with winter wheat and maize varieties grown in France, which may not account for the phenology and yield of these crops in Hungary. Further, sunflower is not modeled despite of its importance in the crop production of Hungary. We used data from the ORCHIDEE-STICS model in 0.5° spatial resolution for the period of 1990–1999. As the ORCHIDEE-STICS domain does not cover Hungary completely, data for the missing parts were estimated using the method proposed by Ciais et al. (2010).

We used two additional, process-based models in the country-specific, GHG balance related research (Biome-BGC and DNDC). Those models were adapted to the Hungarian conditions in the frame of national research projects.

DNDC was developed to estimate the GHG fluxes of different type of crops. (The adaptation of the DNDC model for grasslands and arable lands is described in Chapters 10 and 12 of this book.) Basically, the model was developed for agricultural crops, while another model, Forest-DNDC, could be used for forests. Unfortunately, at the moment, there is no database available for the application of the Forest-DNDC model in Hungary. DNDC offers the opportunity of re-parameterization of plant information. Together with the Hungarian Forest Research Institute, we set up a database for the Hungarian forest-related simulations. We re-parameterized DNDC for three forest types: coniferous forest, board-leaf forest, and mixed-forest. These groups are rather ambiguous and not really detailed but we could give some specific average properties to the model about the forests, at least.

The adaptation of the Biome-BGC model and the results of the country-wide simulations are described in more detail below in separate sections.

13.3 Estimation of the Biospheric Carbon Dioxide Balance with Biome-BGC

13.3.1 Extension of Biome-BGC to Country Scale

Most ecological models used in Europe or worldwide are of coarse spatial resolution, and more importantly, they only use general parameterization for specific plant functional types uniform in the entire region (Janssens et al. 2005; Gervois et al. 2008; Vetter et al. 2008; Ciais et al. 2010). To achieve a higher accuracy in the carbon dioxide budget estimation, the special features of smaller geographical regions should also be taken into account.

We present here the results of an attempt to estimate the biospheric carbon dioxide balance of Hungary. For this purpose, the Biome-BGC process oriented ecological system model (version 4.1.1) was adapted to the Hungarian conditions. The present study is based on our previous work (Barcza et al. 2009) though the results differ from the former estimates due to the recently implemented changes in the modeling design and the more accurate calibration procedure described in Chapter 9 in the present book.

In order to use Biome-BGC for the simulation of the carbon balance of large areas, the internal model parameters (ecophysiological parameters) are held constant while the spatially varying soil and geographical parameters, and the temporally and spatially varying meteorological data are supplied to the model to reflect regional differences in the functionality of the different ecological systems. Application of constant model parameters is only feasible if the spatial variability of the ecophysiological parameters is expected to be low. In case of Hungary, due to the relatively uniform climate, this assumption seems reasonable.

For the determination of the biospheric carbon dioxide balance, the area of Hungary was covered by a grid of 1/6° × 1/6° spatial resolution (466 pixels in total). The fraction of the different plant functional types, the characteristic soil type, and weather condition were determined for each grid cell. The determination of the land cover types was based on the CORINE2000 database (Büttner et al. 2002). Note that a significant amount of land area (vineyards, kitchen gardens, fruit trees, berry plantations, marshes, shrublands, urban areas, etc.), which constitutes ~20% of the country area (Table 13.1), is ignored by the model. At present, due to the lack of available simulation methodology, we can assume that those areas are carbon neutral, which means that the carbon stocks are in equilibrium with the atmosphere.

For the spin-up phase of the model run (estimation of equilibrium carbon and nitrogen pools; see Chapter 9 for explanation), the CRU TS 1.2 database was used. In the normal phase of the modeling (2002–2009), the grid-interpolated measurements of the Hungarian meteorological network were used. The interpolated data fields were produced by the Hungarian Meteorological Service applying the MISH method (Szentimrey et al. 2005).

The calibration procedure of Biome-BGC for Hungarian grasslands is presented in Chapter 9. We calculated ecophysiological parameters for a typical Hungarian grassland using the results of the calibration for three sites (Bugac, Hegyhátsál, Mátra). For grassland-related simulations, we used a modified version of the Biome-BGC model in order to account for the problems related to the stomatal conductance and other environmental controls (for methodological developments, see Chapter 9).

As we also implemented changes in the modeling of arable lands, the results presented here substantially differ from those published in the Barcza et al. (2009) study. As it was mentioned in Barcza et al. (2009), biomass removal was not handled by the model; thus the total ecosystem respiration was overestimated. As a consequence, the arable land NEE data presented in Barcza et al. (2009) might be considered as an upper limit for the carbon balance of arable land. In fact, as we proved it in Chapter 12 of the present book, the arable land related model results are biased toward higher respiration values as compared to the measurements. To address this issue, we implemented a simple bias correction to account for biomass removal due to harvest and residue management. This bias correction (on the order of 50–100 gC m^{-2} $year^{-1}$) substantially changes the overall picture of the carbon dioxide balance of arable lands if it is upscaled to the country area. This is because the former simulations showed that NEE was in the same order of magnitude as the most recently calculated bias (Barcza et al. 2009) and because of the high share of agricultural land in Hungary. We believe that the new analysis provides better estimates for the Hungarian agricultural carbon dioxide balance.

There are no eddy covariance based measurements available either for the deciduous or the coniferous forests in Hungary; therefore, the model calibration cannot be performed as it was implemented for grasslands and arable lands. After a literature review, we decided to use the ecophysiological data of Pietsch et al. (2005) recommended for Biome-BGC. Their parameter sets were determined on the basis of measurements in oak (*Quercus robur/petraea*) and Scotch pine (*Pinus sylvestris*) forests. The measurements were carried out in low elevation regions of Austria and Czech Republic, close to Hungary, where the climatic conditions are similar to those in Hungary.

Although oak is the dominant species among the Hungarian deciduous trees, significant areas are covered by Black locust (*Robinia pseudoacacia*) and other species of higher or lower productivity. Thus, the generalized application of the oak parameter set may be considered as an oversimplification to characterize all Hungarian deciduous forests, but the parameterization cannot be refined without measurements on other species yet. A better parameterization could be achieved using Hungarian forest inventory data (see e.g. Somogyi 2007); however, the calibration methodology for Biome-BGC is not developed yet.

We also improved our modeling design in relation with forests. It means that the present results are different from those published in the original Barcza et al. (2009) study. As it was pointed out by Cienciala and Tatarinov (2006), during the spin-up phase of the simulations, the steady-state carbon pools should be in accordance with those estimates available from the literature. If some of the equilibrium pools (e.g. soil or litter carbon pool) are over/underestimated, it may cause bias in the simulations. To address this issue, we made modifications in the ecophysiological parameters during the spin-up phase according to Cienciala and Tatarinov (2006). The modifications were made in the "annual whole-plant mortality fraction" and "annual fire mortality fraction" parameters. We also changed the method for the simulation of phenology. In the present work, we used Biome-BGC's default phenological estimations instead of the user-supplied growing season start and end dates originally used in the Barcza et al. (2009) study. Additionally, biological nitrogen fixation for the new simulations was set according to Cienciala and Tatarinov (2006).

13.3.2 Results

Using the input data and parameter sets presented above, we estimated NEE, GPP, total ecosystem respiration (R_{eco}), and NPP at country level using daily time steps for 2002–2009. Carbon dioxide fluxes are expressed in carbon mass equivalent (1 gC = 3.67 gCO_2 or 1 gCO_2 = 0.27 gC) per unit area and per unit time. Nevertheless, we do not address the complete carbon exchange of the ecosystem as we do not deal with methane, biogenic volatile organic compounds (VOC), and leaching of dissolved organic and inorganic carbon (DOC/DIC) (note that DOC + DIC for Europe is ~7 gC m^{-2} $year^{-1}$ according to Schulze et al. 2009).

The daily data were aggregated for each year, for each grid cell, and for the entire area of Hungary taking into account the spatial extent of the different plant functional types in each individual grid cell.

The mean NEE for 1 m^2 of land area for the different land cover types and for the whole country are presented in Table 13.3. The data presented suggest that forests, grasslands, and also arable lands are usually sinks of carbon dioxide.

The magnitude of NEE is the highest for forests (mean annual NEE for 2002–2009 is −180.5 gC m^{-2} $year^{-1}$), which is followed by arable lands (−118.4 gC m^{-2} $year^{-1}$). Grasslands seem to act as almost carbon neutral (−8 gC m^{-2} $year^{-1}$). The balance shows that biosphere is usually a net sink of CO_2 in Hungary from the point of view of the atmosphere (country mean NEE is −93.7 gC m^{-2} $year^{-1}$).

Figure 13.1 shows the total fluxes for the entire country in MtC $year^{-1}$. Though arable lands act as relatively weak sinks of CO_2 per unit land area, they form the strongest sink of atmospheric CO_2 due to their high share in Hungary. Though the CO_2 sink strength is high, this flux of CO_2 entering the vegetation is not sequestered by the biosphere due to human intervention (harvest, residue management) but returns to the atmosphere soon after animal and human consumption and digestion (Ciais et al. 2010). As lateral carbon flux associated with agriculture can be quantified using yield statistics and assumptions about residue management, NBP can be approximated. NBP estimates are presented later in this chapter.

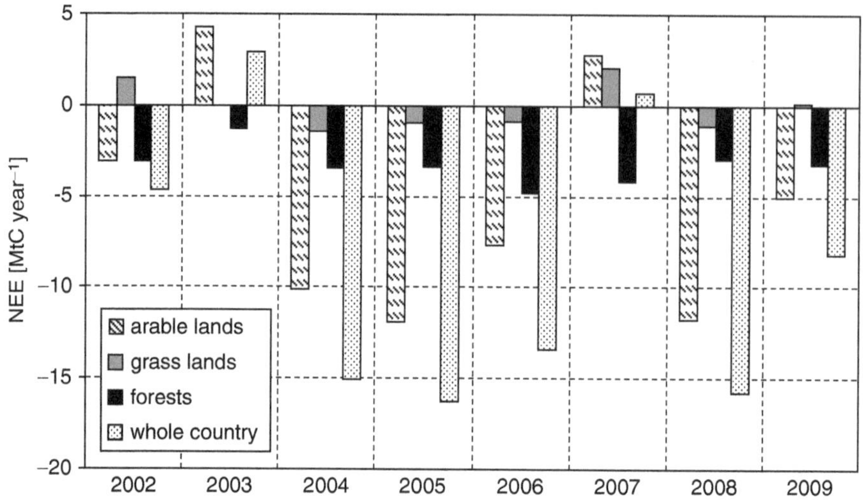

Fig. 13.1 Biome-BGC based net ecosystem exchange (NEE) results from the country-specific simulation for 2002–2009

The climate fluctuations significantly influence the activity of the biosphere and, thus, also its carbon balance. The climate variation in the study period (2002–2009) offered a unique possibility to study the effect of these fluctuations. 2003 was an extremely hot and dry year (not only in Hungary but in large areas of Europe; Ciais et al. 2005), which ended a long, increasingly warm, and dry period followed by cooler and wetter years. In 2007, a very strong and long heat wave occurred in July, which affected plant growth, caused severe leaf-loss in the country, and caused yield loss in case of summer crops (Kern et al. 2008). The results presented in Table 13.3 show that 2003 and 2007 were extreme years. In 2003, arable lands turned to net sources of CO_2 and the resulting country-wide NEE was positive. In 2007, both grasslands and arable lands were net sources of CO_2 to the atmosphere. The resulting country-wide average NEE was close to zero (7 gC m^{-2} $year^{-1}$). It is interesting to note that forests were affected only in 2003, most probably due to the long-lasting warm period and drought. Grasslands were significant sources of CO_2 in 2002, 2007, and also in 2009. The interannual variability of the fluxes suggest that process-based models provide invaluable information about the carbon cycle components. It means that the models can be potentially used to quantify the uncertainty of the mean GHG balance of large geographical areas, and the response of the biosphere to the changing climate and extreme events.

The extension of the present study with the quantitative estimate of lateral carbon transport may provide a useful tool for constraining the total biospheric carbon budget of Hungary. We present an approach for a broader GHG accounting later in this chapter.

Table 13.3 Results (gC m^{-2} year^{-1}) of the Biome-BGC simulation for the period of 2002–2009. Mean fluxes are given for 1 m^2 of biome type (arable lands, grasslands, and forests) per year, and for 1 m^2 of country area per year for the whole country. Negative NEE values indicate CO_2 uptake from the atmosphere. NPP, GPP, and R_{eco} are positive by definition. Note that NEE≠R_{eco}–GPP due to the applied bias correction

	Flux	2002	2003	2004	2005	2006	2007	2008	2009
Arable lands	NEE	–68	95	–225	–265	–171	62	–262	–112
Grasslands		145	–3	–139	–95	–82	200	–103	12
Forests		–170	–71	–189	–184	–264	–227	–160	–178
Whole country		–50	32	–162	–175	–144	7	–170	–88
Arable lands	GPP	827	702	746	833	842	806	866	806
Grasslands		1,342	1,380	1,704	1,838	1,747	1,335	1,760	1,519
Forests		1,532	1,357	1,540	1,666	1,713	1,558	1,637	1,556
Whole country		850	759	852	934	937	844	936	864
Arable lands	R_{eco}	733	664	713	760	794	791	823	776
Grasslands		1,487	1,377	1,565	1,743	1,665	1,536	1,656	1,532
Forests		1,362	1,286	1,350	1,482	1,449	1,331	1,477	1,378
Whole country		787	727	784	852	853	814	872	816
Arable lands	NPP	515	413	469	523	516	484	536	482
Grasslands		815	831	1,086	1,141	1,047	752	1,116	878
Forests		834	709	886	942	941	782	900	810
Whole country		503	431	522	565	550	471	560	490

13.4 Synthesis of the Available Estimates for the Biospheric Greenhouse Gas Balance of Hungary

As this is the first attempt to synthesize the available results for the quantification of the Hungarian biospheric GHG balance, we generally present fluxes and stocks without addressing their uncertainty. However, as there are usually more than one estimate for the fluxes and stocks, the presented values can be used to give an impression about the related uncertainties and confidence intervals. The uncertainty of the fluxes at the European level are described in Luyssaert et al. (2009), Schulze et al. (2009), and Ciais et al. (2010), and we can adopt those estimates to provide approximate uncertainty levels for Hungary, as well.

For the description of the carbon cycle components, we used the terminologies presented in Chapin et al. (2005). GPP is defined as a flux with positive sign, while NEE is negative if carbon dioxide is taken up from the atmosphere. NEE is calculated as R_{eco}–GPP. NBP (defined as -NEE minus horizontal carbon removal including erosion, while other fluxes are generally ignored) is positive if the ecosystem gains carbon. Similarly, NBP of the whole country is positive if there is carbon accumulation on country level. NPP can only be positive by definition.

In order to better understand the carbon cycle and GHG balance components, the reader should be familiar with the conceptual flow chart of GHG fluxes presented in Schulze et al. (2009) (their Fig. 2).

In case of different PFTs (grassland, forests, arable lands), we provide mean fluxes per square meter for the area of the specific PFT (Table 13.1; note that peatlands are not considered in the present study due to the lack of information). In order to describe fluxes and stocks for the whole country, we provide mean data per square meter of total land area (93,030 km^2).

For the quantification of the net greenhouse gas balance (NGB), we convert N_2O and CH_4 emission/uptake to CO_2 equivalent using the global warming potential (GWP) conversion factors. For consistency with the current National Inventory Reports submitted to UNFCCC, we use the GWP values defined by IPCC Second Assessment Report (IPCC 1996). It means that GWP of nitrous oxide is 310, while GWP of methane is 21 on 100-year time horizon. NGB is positive if the biosphere is a net sink of GHGs.

13.4.1 Grasslands

Grasslands constitute about 11% of the total area of Hungary. As there are three monitoring sites in Hungary that measure NEE over grasslands (Bugac, Mátra, and Hegyhátsál), and there is also a considerable amount of data about the non-CO_2 GHG exchange between grassland soils and the atmosphere (see Chapter 6), we can state that Hungarian grasslands are in the main focus of the national research. Through the adaptation of the Biome-BGC, MOD17, and DNDC models, we also gained spatially explicit estimations about certain GHG fluxes. Results from the international literature can help to constrain the estimates. Table 13.4 shows the available estimates for grassland GHG fluxes.

13.4.1.1 GPP and NPP

GPP can be estimated by the Biome-BGC ecological model and by the MOD17 remote sensing based model. Biome-BGC results in higher GPP than MOD17. Mean GPP from the two models is 1,314 gC m^{-2} year^{-1} (corresponding to 13.5 MtC year^{-1} on country level; in the following country totals will be given for some of the fluxes in parenthesis, next to the values relating to 1 m^2), which is close to the mean European estimate (1,343 ± 269 gC m^{-2} year^{-1}; Schulze et al. 2009).

NPP estimate from Biome-BGC is larger than the MOD17-based estimate, which is not surprising taking into account the GPP estimates. Mean NPP from the two models is 768 gC m^{-2} year^{-1} (7.9 MtC year^{-1}). The European estimate for NPP is 750 ± 150 gC m^{-2} year^{-1} according to Schulze et al. (2009).

Table 13.4 GHG balance components for total grassland area as estimated by the different models and methods. The fluxes are mean values for the grasslands and are given in gC m^{-2} $year^{-1}$ in all cases except for non-CO_2 GHGs where the unit is grams of GHG m^{-2} $year^{-1}$. Negative NEE indicates CO_2 uptake from the atmosphere. Negative NBP means carbon source. Stocks are given in kg m^{-2}

	Biome-BGC	MOD17	DNDC	Fuzzy logic model	HCSO yield data	Vuichard et al. (2007)	Janssens et al. (2005)	AGRO-TOPO	DKSIS
NEE	–8.0								
GPP	1,578.2	1,049.7							
R_{eco}	1,570.3								
NPP	958.3	577.7							
Harvest					28.3				
NBP	–20.3[a]						57.0		
Carbon stock in soil								11.76	13.71
N_2O emission			0.1142	0.0963		0.148			
CH_4 exchange			–0.057						

[a]Calculated as NBP = –NEE – harvest

13.4.1.2 NEE and R_{eco}

For NEE, we only have one estimate, which is based on the adapted Biome-BGC model: -8 gC m^{-2} $year^{-1}$. It means that grasslands are approximately CO_2 neutral. Considering the total grassland area in the country, the country-level NEE is –0.082 MtC $year^{-1}$.

Annual total ecosystem respiration is estimated to be 1,570 gC m^{-2} $year^{-1}$ (Biome-BGC data; corresponding to 16.1 MtC $year^{-1}$ for the whole territory of the country), and there is no other estimate for R_{eco} at the moment. As GPP might be lower than the Biome-BGC data (see MOD17 result above), it means that R_{eco} might also be smaller.

13.4.1.3 NBP and Carbon Stocks

For the quantification of NBP, a major problem is the lack of accurate management-related data. Meadows are frequently cut and the removed grass is utilized as fodder or animal bedding. Large grassland areas are grazed by animals in Hungary. This horizontal carbon transport and carbon input – similarly to arable lands – modifies the carbon balance of grassland soils, and may eventually shift grassland NBP from net CO_2 sink to net source. At present, we only have information about the grass yield from HCSO (http://statinfo.ksh.hu/Statinfo/themeSelector.jsp?&lang=en) for the period of 2000–2008. According to the data, estimated mean yield is 28.3 gC m^{-2} $year^{-1}$ (0.29 MtC $year^{-1}$). Due to the lack of information, we cannot quantify carbon input via manure. The European results suggest that harvest (removed carbon) is on the order of 217 ± 269 gC m^{-2} $year^{-1}$, while carbon input via manure is 40 gC m^{-2} $year^{-1}$ (Schulze et al. 2009).

We can provide a first estimate to NBP by combining the Biome-BGC-based NEE and the harvest data. The estimated NBP is –20.3 gC m^{-2} $year^{-1}$ (–0.21 MtC $year^{-1}$), which suggests that carbon is being lost from Hungarian grassland soils. As carbon input to soil via manure is most likely not zero, the carbon loss is smaller than the above estimate. If we accept the European estimate, NBP is close to zero. At present, we may assume that grasslands are carbon neutrals, and we can use this assumption for the estimation of the carbon balance of the whole country. In contrast, according to the estimate of Janssens et al. (2005) for Hungary, NBP is 57 gC m^{-2} $year^{-1}$, which means that the soil carbon stock in grasslands is increasing. This estimate is exactly the same as the European mean NBP for grasslands (the uncertainty is ±34 gC m^{-2} $year^{-1}$; Schulze et al. 2009). Further studies are needed to constrain the NBP results.

Grasslands store a considerable amount of carbon in soil. AGROTOPO and DKSIS provide 11.76 and 13.71 kgC m^{-2} for SOC, respectively, for Hungarian grasslands. Total SOC of Hungarian grasslands based on the average of the two methods is 130.9 MtC.

13.4.1.4 Non-CO_2 GHGs and NGB

N_2O emission from grassland soils is estimated by three independent methods. The highest estimate is given by the PASIM model (Vuichard et al. 2007), while the lowest source is provided by the fuzzy logic model. The average N_2O emission is 0.1195 $gN_2O\ m^{-2}\ year^{-1}$ (1.23 $ktN_2O\ year^{-1}$), which is close to the Hungarian, DNDC-based data (0.1142 $gN_2O\ m^{-2}\ year^{-1}$). The difference between the highest and lowest estimate is 0.0517 $gN_2O\ m^{-2}\ year^{-1}$ (0.53 $ktN_2O\ year^{-1}$), which means that the uncertainty is rather high.

According to the DNDC model adapted for Hungary, methane is being taken up from the atmosphere by grasslands. The mean CH_4 flux is −0.057 $gCH_4\ m^{-2}\ year^{-1}$, which is equivalent with −0.59 $ktCH_4\ year^{-1}$ for the territory of the country.

Using NBP and the two non-CO_2 GHG fluxes, we may provide a first estimate for the net greenhouse gas balance (NGB) of Hungarian grasslands. N_2O emission and CH_4 uptake are converted to CO_2-equivalent and then to carbon mass equivalent. Assuming zero NBP together with the mean N_2O emission and methane removal, total NGB is −9.78 gC-eq $m^{-2}\ year^{-1}$ (−0.1 MtC-eq $year^{-1}$), which means a small net source to the atmosphere. Of course, this estimate is rather uncertain because of the lack of information about carbon input and other uncertainties (lack of information about management, fertilizer use, model structure, etc.); thus, currently we can only state that grasslands are close to be GHG neutral biomes. The European best estimate for total NGB is currently 14 ± 18 gC $m^{-2}\ year^{-1}$ (Schulze et al. 2009), which means that European grasslands are small net sinks or close to neutral even after taking into account the major non-CO_2 GHGs.

13.4.2 Forests

As forests can store a significant amount of carbon in standing biomass and also in soils, they have significant carbon sequestration potential. Generally, this potential cannot be fully utilized due to wood harvest. Forest understorey herbaceous vegetation, generally ignored by forest surveys, can also act as a net carbon sink, and most probably contributes to soil carbon sequestration in forests.

Traditionally, climate change mitigation is focused on the forestry sector; thus, the first eddy covariance measurement sites were also established in forests inside the FLUXNET network. In spite of the international efforts, there is no eddy covariance site over forest in Hungary. In the lack of such measurements, we based our forest-related estimates on data from forestry and process-based models.

Table 13.5 shows the estimates collected for Hungarian forests. Deciduous broadleaf and evergreen needleleaf forest related data were aggregated, and the results are discussed for forests in general.

Table 13.5 GHG balance components for total forest area (deciduous + evergreen needleleaf) as estimated by the different models and methods. The fluxes are mean values for the forests and are given in gC m^{-2} $year^{-1}$ in all cases except for non-CO_2 GHGs where the unit is grams of GHG m^{-2} $year^{-1}$. Negative NEE indicates CO_2 uptake from the atmosphere. Negative NBP means carbon source. Stocks are given in kg m^{-2}

	Biome-BGC	MOD17	DNDC	Forest understorey[a]	Janssens et al. (2005)	Führer and Mátyás (2005)	AGRO-TOPO	DKSIS
NEE	−180.5			−8.73				
GPP	1,569.8	1220.3						
R_{eco}	1,389.3							
NPP	850.4	551.1				376.6		
Harvest						104.8		
NBP[b]	83.5				190.4			
Carbon stock in biomass						7.41		
Carbon stock in soil						13.17	7.91	10.63
N_2O emission			0.0764					
CH_4 exchange			−0.098					

[a] Forest understory estimate is only applicable to deciduous broadleaf forests

[b] Calculated with the combination of NEE from Biome-BGC, forest understorey and wood harvest data

13.4.2.1 GPP and NPP

GPP is estimated by Biome-BGC and MOD17 models. The magnitude of GPP based on Biome-BGC is higher than that produced by MOD17. Mean GPP is 1,395 gC m^{-2} $year^{-1}$ (25.56 MtC $year^{-1}$). The European mean GPP for forests is 1,107 ± 55 gC m^{-2} $year^{-1}$ (Schulze et al. 2009), which is lower than our estimate.

Forest NPP is estimated by three methods. NPP is the largest according to Biome-BGC, while it is the lowest according to the inventory-based results published in Führer and Mátyás (2005). Mean NPP is 592.7 gC m^{-2} $year^{-1}$ (10.86 MtC $year^{-1}$). According to the European synthesis, NPP is 518 ± 67 gC m^{-2} $year^{-1}$ (Schulze et al. 2009), which is close to our MOD17-based estimate.

13.4.2.2 NEE and R_{eco}

We only have one estimate for forest NEE based on the Biome-BGC model. According to the model, NEE is −180.5 gC m^{-2} $year^{-1}$ (−3.31 MtC $year^{-1}$). NEE was also estimated for forest herbaceous understorey plants in deciduous broadleaf forests. Yearly average NEE was −8.73 gC m^{-2} $year^{-1}$ for forest floor vegetation (corresponding to −0.14 MtC $year^{-1}$).

R_{eco} was also only estimated by the Biome-BGC model. R_{eco} equals 1,389.3 gC m^{-2} $year^{-1}$ (25.46 MtC $year^{-1}$). As the magnitude of our best estimate for GPP is lower than the Biome-BGC result, this value most likely overestimates R_{eco}.

13.4.2.3 NBP and Carbon Stocks

Wood harvest is estimated by Führer and Mátyás (2005) for Hungarian forests. It is estimated to be 104.8 gC m^{-2} $year^{-1}$ (1.92 MtC $year^{-1}$). The European mean wood harvest is 63 ± 11 gC m^{-2} $year^{-1}$ (Schulze et al. 2009).

We have two estimates for forest NBP. The Biome-BGC-based NBP is calculated with the combination of NEE and the wood harvest data published by Führer and Mátyás (2005). We also incorporated the herbaceous forest understorey NEE in the NBP estimate taking into account that it only refers to deciduous forests. The resulting NBP is 83.5 gC m^{-2} $year^{-1}$ (1.53 MtC $year^{-1}$). Janssens et al. (2005) also estimated forest NBP, which is 190.4 gC m^{-2} $year^{-1}$ (3.5 MtC $year^{-1}$). It is also possible to estimate NBP from the European mean carbon sequestration efficiency (NBP/NPP ratio) presented in Luyssaert et al. (2009). If we consider the inventory-based estimate as the most reliable data source for NPP, we get an NBP estimate of 56.4 gC m^{-2} $year^{-1}$ (1.03 MtC $year^{-1}$), which is close to the European mean (74 ± 12 gC m^{-2} $year^{-1}$; Schulze et al. 2009). The mean NBP calculated from the three estimates (Biome-BGC, Janssens et al. (2005) and the NBP/NPP-based estimate) is 110.1 ± 71 gC m^{-2} $year^{-1}$ (2.02 ± 1.3 MtC $year^{-1}$; uncertainty is estimated as standard deviation of the three data), and we consider this value as our best estimate so far.

Carbon stock in biomass is estimated by Führer and Mátyás (2005). The data suggest that 7.41 kgC is stored in 1 m^2 of forest area as dendromass on average. This is equivalent with 135.8 MtC on country level.

The relatively undisturbed forest soils store 13.17 kgC m^{-2} according to Führer and Mátyás (2005). Soil carbon inventory data provide two additional estimates. According to AGROTOPO, the forest soil carbon stock is 7.91 kgC m^{-2}, while it is 10.63 kgC m^{-2} based on the DKSIS database. Total organic carbon content of Hungarian forest soils is 193.7 MtC based on the average from the three methods.

13.4.2.4 Non-CO_2 GHGs and NGB

Considering non-CO_2 GHGs, we only have simulation data from the DNDC model. At present, our best estimate for forest soil N_2O emission is 0.0764 gN_2O m^{-2} $year^{-1}$, which corresponds to 1.4 ktN_2O $year^{-1}$.

According to the DNDC results, methane is being taken up by forest soils at a mean rate of 0.098 gCH_4 m^{-2} $year^{-1}$, which sums up to 1.8 $ktCH_4$ $year^{-1}$ on country level.

Net greenhouse gas balance is calculated from our best estimate for NBP and the CO_2-equivalent N_2O and CH_4 exchange data. As simulated N_2O emission and CH_4 uptake is rather small in forests, NGB (104.2 gC m^{-2} $year^{-1}$; that is 1.91 MtC $year^{-1}$ for the whole country) is close to NBP. It means that Hungarian forests are sinks of GHG at present. The European estimate for forest NGB is 74 ± 22 gC m^{-2} $year^{-1}$ (Schulze et al. 2009).

13.4.3 Arable Lands

The GHG cycle of agricultural land is basically driven by human decisions and interventions. There is no other biome type in Hungary where the soil and vegetation are disturbed by humans so severely and frequently as in case of arable lands. As a consequence, the GHG cycle of arable lands cannot be quantified without information about land management. In other words, we cannot handle arable lands as natural vegetation in any way, which means that at least lateral carbon movement has to be taken into account to quantify the fluxes and stocks.

For the estimation of the country's arable land GHG balance, we used four different models (Biome-BGC, MOD17, DNDC, and ORCHIDEE-STICS). Biome-BGC and DNDC were parameterized with available country-specific information. We also used yield data to quantify NPP and laterally transported carbon.

The available results are presented in Table 13.6. Note that due to the bias correction of NEE data produced by Biome-BGC, the combination of GPP and R_{eco} does not give the NEE presented.

Table 13.6 GHG balance components for total arable land area as estimated by the different models and methods. The fluxes are mean values for the arable lands and are given in gC m^{-2} $year^{-1}$ in all cases except for non-CO_2 GHGs where the unit is grams of GHG m^{-2} $year^{-1}$. Negative NEE indicates CO_2 uptake from the atmosphere. Negative NBP means carbon source. Stocks are given in kg m^{-2}

	Biome-BGC	MOD17	DNDC	ORCHIDEE-STICS	Erosion model	HCSO yield data	Janssens et al. (2005)	Ciais et al. (2010)	AGRO-TOPO	DKSIS
NEE	−118.4									
GPP	803.6	940.1								
R_{eco}	756.6									
NPP	492.4	522.1		940.1		507.1				
Harvest (including residue removal)						193.2				
NBP	−78.5[a]			16.7			−126.2			
Erosion induced carbon sink					−3.03					
Erosion induced carbon loss via rivers					2					
Soil carbon stock									13.26	13.68
N_2O emission			0.9679					0.3886		
CH_4 exchange			−0.1979							

[a] Estimated as NBP = −NEE − harvest

13.4.3.1 GPP and NPP

We have two estimates for GPP of arable lands. The MOD17-based estimate gives higher GPP than Biome-BGC. The average GPP is 871.9 gC m^{-2} $year^{-1}$ (39.24 MtC $year^{-1}$). The Schulze et al. (2009) study provides higher estimate for the European mean (1,120 ± 224 gC m^{-2} $year^{-1}$; Schulze et al. 2009), but our estimate is close to the lower end of the uncertainty range.

NPP based on yield statistics is 507.1 gC m^{-2} $year^{-1}$, which is very close to the estimates by Biome-BGC (492.4 gC m^{-2} $year^{-1}$) and that of MOD17 model (522.1 gC m^{-2} $year^{-1}$). The estimate of ORCHIDEE-STICS is higher than those of the three others (940.1 gC m^{-2} $year^{-1}$), most likely because of the limitations of the model (the model is parameterized as default with winter wheat and maize varieties grown in France, which may not be representative for the phenology and yield of these crops in Hungary). Mean NPP calculated from the results of the three models is 507.2 ± 15 gC m^{-2} $year^{-1}$ (22.83 ± 0.67 MtC $year^{-1}$). We can conclude that NPP is well constrained for the Hungarian arable lands. The European mean (550 ± 50 gC m^{-2} $year^{-1}$; Schulze et al. 2009) is very close to our estimate, which supports that the European results are plausible.

13.4.3.2 NEE and R_{eco}

Mean NEE is −118.4 gC m^{-2} $year^{-1}$ as simulated by the Biome-BGC model, (−5.33 MtC $year^{-1}$). It means that arable lands are net sinks of CO_2 most probably due to the spatial separation between plant uptake and respiration caused by yield consumption and digestion. It is interesting to note that multi-annual mean NEE from the Hegyhátsál tall tower measurements is around −105 to −136 gC m^{-2} $year^{-1}$ (1997–2008, without 2000; depending on the gap-filling algorithm used). It means that the Hegyhátsál measurements might be representative to the arable lands of the whole country. Previous estimate suggested that the NEE of arable lands is +96 gC m^{-2} $year^{-1}$, but it was pointed out that this estimate is most likely an overestimation of NEE (Barcza et al. 2009). The present estimate is calculated from the Biome-BGC model results via a simple bias correction (see Chapter 12), and it is more realistic than the previous estimate (see the discussion below).

Total ecosystem respiration estimated by the Biome-BGC model is 756.6 gC m^{-2} $year^{-1}$ (34.05 MtC $year^{-1}$).

13.4.3.3 NBP and Carbon Stocks

Lateral carbon transport due to harvest and residue removal is 193.2 gC m^{-2} $year^{-1}$ (8.7 MtC $year^{-1}$), which is lower than the European estimate (257 ± 23 gC m^{-2} $year^{-1}$; Schulze et al. 2009). Carbon input to soils from residue management (not shown in Table 13.6.) was also estimated as the difference between NPP and removed carbon, which is 313.8 gC m^{-2} $year^{-1}$ (14.12 MtC $year^{-1}$).

NBP of arable lands describe the carbon stock changes in soils taking into account all vertical (atmosphere/biosphere) and lateral carbon fluxes. We have three independent estimates for NBP of arable lands. One estimate is provided by the combination of the Biome-BGC model output and the carbon removal estimated from yield statistics. The second estimate is provided by the ORCHIDEE-STICS model, and the third one is taken from the literature (Janssens et al. 2005). NBP provided by Janssens et al. is most likely an overestimation of NPB due to methodological problems (Ciais et al. 2010). Therefore, we only use the two model results. NBP from Biome-BGC is −78.5 gC m^{-2} $year^{-1}$, which means that arable land soils are losing C. For the entire arable land area, NBP sums up to −3.53 MtC $year^{-1}$. The ORCHIDEE-STICS-based NBP is 16.7 gC m^{-2} $year^{-1}$ (0.75 MtC $year^{-1}$), which predicts slight carbon accumulation in the soils. The mean NBP from the two estimates is −30.9 gC m^{-2} $year^{-1}$ (−1.39 MtC $year^{-1}$; this is our best estimate for the time being). The European mean arable land NBP is −10 ± 9 gC m^{-2} $year^{-1}$ (Schulze et al. 2009; Ciais et al. 2010), which means a small net release of carbon from European soils.

Though NBP of arable lands is still uncertain, the results suggest that arable land soils are still carbon sources in Hungary. Note that using the previous estimate for NEE (Barcza et al. 2009) and taking into account the removed carbon, NBP would be around −300 gC m^{-2} $year^{-1}$, which means that 3 kgC would be lost from arable lands soils in a 10-year time horizon. As most of the arable lands in Hungary have been cultivated for a long time, the carbon loss from soils was most likely slowed after the initial disturbance (conversion of land). It means that the 300 gC loss per year is unrealistic and there is no signal for such a large overall carbon loss based on soil inventory data (Szabó et al. 2008). Our present estimates are closer to the equilibrium assumption, which means that the results are in accordance with the expectations.

Using our best estimates for NBP and NPP, the carbon sequestration efficiency of Hungarian arable lands (NBP/NPP) is −0.06, which is smaller than the European mean (Ciais et al. 2010).

Soil erosion causes mobilization of soil carbon and can lead to changes in the agricultural NBP. According to our results, tillage erosion contributed to 49% of the total amount of carbon being eroded on arable lands, while the remaining part was eroded by water erosion processes. Using an average carbon enrichment ratio of 1.3 for water erosion (Jacinthe et al. 2004) and a delivery of 30% to the fluvial network (Van Oost et al. 2009), we estimate the carbon export to the rivers to be 2 gC m^{-2} $year^{-1}$ (0.09 MtC $year^{-1}$) and the carbon re-burial in depositional zones to be 9.63 gC m^{-2} $year^{-1}$ (0.433 MtC $year^{-1}$). The net effect on the carbon budget is a loss of 2 gC m^{-2} $year^{-1}$ as a result of fluvial export and a net uptake of 3.03 gC m^{-2} $year^{-1}$ (0.134 MtC $year^{-1}$) on arable lands as a result of erosion and burial (Van Oost et al. 2009).

These estimates exclude wind erosion, which means that the mobilized carbon is most probably underestimated by the model. Nevertheless, the results indicate that according to the most recent findings, erosion does not cause carbon loss from Hungarian soils but rather the result is a small net sink in the eroding soils due to the so-called replacement process and the re-burial of carbon (Van Oost et al. 2007). River flux is a net loss from the Hungarian soils, but as the sink is greater than the river loss, the net effect is a small (~1 gC m^{-2} $year^{-1}$) increase in the soil carbon stocks in Hungary. This value slightly modifies the previous NBP estimate.

Arable land soils store a vast amount of carbon. AGROTOPO and DKSIS provide 13.26 and 13.68 kgC m^{-2} for mean SOC for Hungarian arable lands, respectively. Note that the average of the two values are 27.5% higher than that for forests, which is surprising as we expect lower carbon stocks in arable lands than in forests due to disturbance caused by land conversion and regular ploughing. SOC in arable lands is 606.24 MtC based on the average of the two estimates.

13.4.3.4 Non-CO_2 GHGs and NGB

N_2O emission from arable land soils is estimated by DNDC model and by Ciais et al. (2010). The Hungarian estimate is ~2.5 times higher than the other one, most probably due to the assumptions about the fertilizer data used (see Chapter 12). Note that the amount of fertilizer application in DNDC is based on expert judgment (B. Grosz, personal communication) and is not justified by data published by the Hungarian Central Statistical Office. Fertilizer supply data provided by the HCSO is based on commercial data of the Research Institute of Agricultural Economics (https://www.aki.gov.hu/index.php). As the amount of fertilizer used in DNDC is most likely overestimation, N_2O emission is also likely overestimated by the Hungarian adaptation of DNDC. Nevertheless, we have decided to use the N_2O emission data presented in Chapter 12.4 keeping in mind the above problems. If we rely on those data, N_2O emission is 0.9679 gN_2O m^{-2} $year^{-1}$, which corresponds to 43.6 ktN_2O $year^{-1}$ for the country.

DNDC suggests that CH_4 is being taken up by arable lands in Hungary. Mean CH_4 uptake is −0.1979 gCH_4 m^{-2} $year^{-1}$, which adds up to 8.91 ktCH_4 $year^{-1}$ on country level.

NGB is calculated from our best NBP estimate (corrected by the erosion sink) and the non-CO_2 GHG exchange data using the GWP conversion factors. NGB is −110.6 gC-eq m^{-2} $year^{-1}$ (−4.98 MtC-eq $year^{-1}$). It means that arable lands are significant sources of GHGs, and the magnitude of the source is very close to the magnitude of sink capacity of the forests (per unit are). Note that the carbon equivalent of soil N_2O emission alone is 81.8 gC m^{-2} $year^{-1}$. Schulze et al. (2009) suggest that the European mean cropland NGB is −40±40 gC-eq m^{-2} $year^{-1}$, which also means a source to the atmosphere. If our estimates are reliable, it means that the specific emission of the Hungarian arable lands is more than twice as large as the European mean.

13.4.4 Whole Country

The most challenging task is to quantify the GHG balance of a whole country. Hungary is a small country, where the majority of the landscape is managed through agriculture and forestry. Due to the human dominance, the lateral movement of carbon (and to a lesser extent nitrogen via fertilizer application and harvest) severely modifies the spatial distribution of sources and sinks. We have to take into account all lateral carbon movement in order to establish a full GHG accounting system.

Full GHG accounting includes the recognition of related processes if biospheric and anthropogenic fluxes are handled together to provide better estimates for a country. For example, the Hungarian GHG emission report sent annually to UNFCCC includes GHG emission from waste, but it does not include all GHG emission originating from human excrement where the carbon content is partly originating from arable lands. Biospheric GHG accounting ends with the quantification of the yield, but no attempt is made to follow the fate of the yield inside the food processing and industry chain. Therefore, it is hard to match the results of the biospheric GHG balance estimates and the anthropogenic emissions as they are related and overlap. The results presented here and later in the book (Part IV) address only part of the above-mentioned processes.

Table 13.7 summarizes the available data for the biospheric GHG fluxes and carbon stocks for Hungary.

13.4.4.1 GPP and NPP

Our best estimate for GPP is 841.6 gC m^{-2} $year^{-1}$, which is the average from Biome-BGC and MOD17 (78.3 MtC $year^{-1}$). The result shows that the overall GPP is determined by arable lands, although forests and grasslands have higher GPP values per unit area (Tables 13.4 and 13.5). Country-wide NPP is 503.1 gC m^{-2} $year^{-1}$ (46.8 MtC $year^{-1}$), which is the average of the Biome-BGC, MOD17 and LPJmL results.

13.4.4.2 NEE and R_{eco}

Mean NEE for the whole area of the country is −115.4 gC m^{-2} $year^{1}$ (Biome-BGC and LPJmL mean; corresponding to −10.74 MtC $year^{1}$ on country level). According to Biome-BGC, R_{eco} is 813 gC m^{-2} $year^{-1}$ (75.63 MtC $year^{-1}$), but this might be an upper limit due to the lower GPP caused by averaging the two estimates.

13.4.4.3 NBP and Carbon Stocks

Lateral carbon fluxes are estimated from the crop yield, grass yield, and forestry data. The mean lateral flux (carbon movement inside the country, which is a redistribution of carbon in fact) is rather large: it is 117.3 gC m^{-2} $year^{-1}$ (10.91 MtC $year^{-1}$). Lateral carbon transport due to export/import is estimated to be around 13 gC m^{-2} $year^{-1}$ and 10.09 gC m^{-2} $year^{-1}$ according to FAOSTAT and Janssens et al. (2005), respectively. The mean flux is 11.56 gC m^{-2} $year^{-1}$ (1.08 MtC $year^{-1}$). Food export exceeds food import by a factor of 4, while wood import is larger than

Table 13.7 GHG balance components for the whole country as estimated by the different models and methods. The fluxes are mean values for the country area and are given in gC m^{-2} $year^{-1}$ in all cases except for non-CO_2 GHGs where the unit is grams of GHG m^{-2} $year^{-1}$. Negative NEE indicates CO_2 uptake from the atmosphere. Negative NBP means carbon source. Stocks are given in kg m^{-2}

	Biome-BGC	MOD17	DNDC	Lund-Potsdam-Jena for managed Land (LPJmL)	FAOSTAT Trade-STAT	HCSO yield data + Führer and Mátyás, (2005)	Janssens et al. (2005)	AGRO-TOPO	DKSIS
NEE	–93.7			–137.1					
GPP	872.2	811.0							
R_{eco}	813.0								
NPP	511.5	424.9		572.9					
Lateral carbon removal due to agriculture and forestry						117.3			
NBP	6.7[b]			0.14			–17.3		
Carbon stock in soils								11.45	12.81
Fire emission				7.2					
Trade					13.03[a]		10.09[a]		
N_2O emission			0.5244						
CH_4 exchange			–0.107						

[a] As carbon export is larger than carbon import, this is a net sink for the country (carbon content will return to the atmosphere elsewhere) [b] Calculated from the Biome-BGC based NBP estimates for grasslands, forests, and arable lands

export. The sum is a net carbon export from Hungary. As the exported carbon was originally taken up by vegetation in Hungary, but consumed elsewhere, it causes a net carbon sink for the country from the atmospheric point of view. The lateral carbon transport caused by international trade has implications on the atmospheric mixing ratio of CO_2 and these sources/sinks have to be taken into account for proper, atmosphere-based studies (Ciais et al. 2007).

NBP for the whole country is estimated from three sources (note that NBP is calculated here from the biome-specific NBP estimates using area weighting; other fluxes, including international trade, will be incorporated later to refine NBP and to estimate NGB). Based on Janssens et al. (2005), NBP is −17.3 gC m^{-2} $year^{-1}$, which means net carbon loss on average. The LPJmL-based NBP is 0.14 gC m^{-2} $year^{1}$, but it is important to keep in mind that this estimate includes fire emission (7.2 gC m^{-2} $year^{-1}$), which is only estimated by LPJmL. Biome-BGC and management-related lateral carbon transport data can also be combined and aggregated to estimate NBP. The resulting estimate is 6.7 gC m^{-2} $year^{--1}$, which is very close to the LPJmL data if fire emission is not included. Mean NBP from the three independent estimates is −1.1 ± 14 gC m^{-2} $year^{-1}$ (−0.1 ± 1.3 MtC $year^{-1}$) (fire emission is not considered for LPJmL). The results suggest that country-wide NBP is close to zero.

Soil carbon stocks in total land area of Hungary are 11.45 kgC m^{-2} and 12.81 kgC m^{-2} according to AGROTOPO and DKSIS, respectively. The average of the two data is 12.13 kgC m^{-2}, which sums up to 1,128.5 MtC for the whole territory of the country.

13.4.4.4 Non-CO_2 GHGs and NGB

The overall biospheric N_2O emission of Hungary is estimated by the DNDC model. DNDC handles grasslands, arable lands, and forests, and additionally it provides emission estimates for berries, fruits, grapes, and fallow. Mean N_2O emission for the entire area of the country is 0.5244 gN_2O m^{-2} $year^{-1}$ (48.78 ktN_2O $year^{-1}$). This emission is equivalent to 44.3 gC m^{-2} $year^{-1}$. Note that this value probably overestimates N_2O emission (see the discussion in Section 13.4.3.4) due to the governing role of arable lands in the budget.

As we discussed earlier, the DNDC results suggest that CH_4 is generally being taken up by soil in Hungary. Mean CH_4 flux for Hungary is −0.107 gCH_4 m^{-2} $year^{-1}$, which adds up to −9.95 $ktCH_4$ $year^{-1}$ for the whole territory of the country. Using the GWP of methane, it is equivalent to −0.6 gC m^{-2} $year^{-1}$ (uptake), which is rather small relative to the other terms.

Our best estimate for the country total net greenhouse gas balance is calculated from the mean biospheric NBP taking into account fire emission, the trade-related carbon export, and the major non-CO_2 GHG fluxes. The resulting NGB is −40.5 gC-eq m^{-2} $year^{-1}$ (−3.77 MtC-eq $year^{-1}$), which means net GHG source to the atmosphere. If the estimates are reliable, it means that at present international trade seems to mitigate biospheric GHG emission of the country by ~30%.

13.5 Conclusions and Outlook

The presented results suggest that the overall biospheric GHG balance of Hungary expressed in CO_2-equivalent is negative, which means that biosphere is a net source of GHGs in Hungary. It means that the, biosphere does not mitigate anthropogenic GHG emission in the country.

Grasslands have the potential to sequester carbon dioxide from the atmosphere. According to the results of the most recent synthesis, the CO_2 removal capacity of grasslands is comparable with those of forests (Schulze et al. 2009). Management practices can significantly alter the carbon sequestration capacity of grasslands (see e.g. Chapter 6). In order to utilize grasslands for GHG mitigation, research projects are needed that can provide full GHG budget of grasslands under different soil and climate conditions in Hungary. The measurement results introduced in Chapter 6 and the model results presented in Chapter 10 suggest that the carbon sequestration capacity of grasslands is rather vulnerable: grazing and cutting can shift the grasslands from net carbon dioxide sink to net carbon dioxide source. This vulnerability needs to be addressed and quantified in the future in more detail. The three existing eddy covariance sites are good candidates for such concentrated effort, where the GHG mitigation potential of Hungarian grasslands is studied. Future research projects have to find a methodology to assess grassland management and lateral carbon fluxes associated with grasslands in order to estimate grassland NBP at country level.

Forests can play a significant role to improve the GHG balance of Hungary, i.e. to foster CO_2 removal from the atmosphere. At present, Hungarian forests act as net sinks of CO_2 due to the sustainable level of tree harvest. As it was pointed out in Chapter 11 with the CASMOFOR model, there are good opportunities in carbon sequestration through reforestation and afforestation on abandoned agricultural lands or grasslands. Due to land abandonment and reforestation efforts in the past 20 years, about 200,000 ha forests have been established in Hungary (estimated as the forest area in 2009 minus that in 1990; HCSO, http://portal.ksh.hu/pls/ksh/docs/eng/agrar/html/tabl1_3_1.html). The above-presented approach ignores land use change because of the lack of available methodology and land use change information. As land use conversion clearly affects the GHG balance of Hungary, these activities should be taken into account in future studies.

There is a need to establish eddy covariance measurements over forested areas in Hungary in order to constrain the estimates of the carbon balance components. Future research activities should focus on the establishment of such integrated monitoring stations, probably with the help of existing intensive forest surveys.

As croplands dominate the land area of Hungary, the GHG mitigation options of agriculture should be considered in the future. At present, arable lands act as net sources of CO_2 to the atmosphere, which means that soil carbon is still decreasing. Although methane is taken up by soils, nitrous oxide shifts the overall GHG budget to even higher emission values (by ~44 gC m^{-2} $year^{-1}$ CO_2-equivalent emission) due to the application of fertilizers and management practices. Due to their large area, policymakers should consider the introduction of alternative, environment-friendly practices in Hungarian agriculture.

The field studies presented in Chapter 8 indicated that management practices (conventional tillage, minimum tillage, direct drilling, mulching, residue management, etc.) can significantly alter the carbon sequestration potential of arable lands. Therefore, at the first glance, it seems that introduction of alternative management practices could help us to mitigate GHG emission. However, as it is pointed out by Smith et al. (2001), non-CO_2 GHGs have to be also considered when planning mitigation efforts. Recent studies pointed out that N_2O emission can increase in no-tillage systems or alternative management systems as compared to conventional tillage (Freibauer et al. 2004). At present, there are no measurement projects that address the effect of agricultural management on the full GHG balance of arable land soils. New monitoring projects are needed that can fully utilize the available expertise of the researchers.

As we mentioned earlier, we did not present uncertainties for all the estimated fluxes and stocks. Future studies are supposed to describe the uncertainty of the GHG balance results. The results presented should be further constrained, and the final aim is to get together the results of the different methodologies and techniques to provide the most reliable GHG balance for Hungary.

Acknowledgments Biome-BGC version 4.1.1 was provided by the Numerical Terradynamic Simulation Group (NTSG) at the University of Montana. NTSG assumes no responsibility for the proper use of Biome-BGC by others. We thank Márta Birkás (Institute of Crop Production, Szent István University, Hungary) for the invaluable help regarding the country-specific allocation data and the information about the fate of agricultural residues in Hungary. We acknowledge the support from the Hungarian National Scientific Research Fund (OTKA T23811, T42941, OTKA-NKTH CK77550), the Hungarian Ministry of Economy and Transport (GVOP-3.2.1.-2004-04-0107/3.0, GVOP-AKF-3.1.1.-2004-05-0358/3.0), the INTERREG IIIB CADSES program (5D038), and the European Commission's 5th and 6th R&D Framework Programmes (EVK2-CT-1999-00013, EVK2-CT-2002-00163, GOCE-CT-2003-505572, GOCE-037005, IMECC, Project No. 026188). The authors highly appreciate the help and advice of M. Zhao (Numerical Terradynamic Simulation Group, University of Montana, Missoula, USA), N. Vuichard (Laboratoire des Sciences du Climat et de l'Environnement, IPSL-LSCE, CEA-CNRS-UVSQ, Gif sur Yvette, France), I. Janssens (University of Antwerpen, Antwerpen, Belgium), T. Major (Hungarian Central Statistical Office, Budapest, Hungary), T. Szentimrey, R. Hodossyné Rétfalvi, B. Birszki (Hungarian Meteorological Service, Budapest, Hungary), J. Szabó (Research Institute for Soil Sciences and Agricultural Chemistry, Budapest, Hungary), K. Trusilova, E. Tomelleri, M. Jung (Max Planck Institute for Biogeochemistry, Jena, Germany), Z. Nagy, K. Pintér, Z. Tuba, Z. Németh (Szent István University, Gödöllő, Hungary), A. Kern (Hungarian Academy of Sciences, Budapest, Hungary), B. Balázs (Eötvös Loránd University, Budapest, Hungary), S. Schaphoff (Potsdam Institute for Climate Impact Research, Potsdam, Germany).

References

AGROTOPO (1994) Spatial Soil Information System. RISSAC HAS, Budapest. http://www.mta-taki.hu/en/departments/gis-lab/databases

Anthoni PM, Freibauer A, Kolle O, Schulze E-D (2004) Winter wheat carbon exchange in Thuringia, Germany. Agric For Meteorol 121:55–67

Barcza Z, Haszpra L, Somogyi Z, Hidy D, Lovas K, Churkina G, Horváth L (2009) Estimation of the biospheric carbon dioxide balance of Hungary using the BIOME-BGC model. Időjárás – Q J Hung Meteorol Serv 113:203–219

Bondeau A, Smith PC, Zaehle S, Schaphoff S, Lucht W, Cramer W, Gerten D, Lotze-Campen H, Müller C, Reichstein M, Smith B (2007) Modelling the role of agriculture for the 20th century global terrestrial carbon balance. Global Change Biol 13:679–706. doi:10.1111/j.1365-2486.2006.01305.x

Büttner G, Feranec J, Jaffrain G (2002) Corine Land Cover Update 2000. Technical Report 89 (European Environment Agency, 2002). http://www.eea.europa.eu/publications/technical_report_2002_89Accessed Dec 2007

Chapin FS, Woodwell GM, Randerson JT, Rastetter EB, Lovett GM, Baldocchi DD, Clark DA, Harmon ME, Schimel DS, Valentini R, Wirth C, Aber JD, Cole JJ, Goulden ML, Harden JW, Heimann M, Howarth RW, Matson PA, McGuire AD, Melillo JM, Mooney HA, Neff JC, Houghton RA, Pace ML, Ryan MG, Running SW, Sala OE, Schlesinger WH, Schulze E-D (2005) Reconciling carbon-cycle concepts, terminology and methods. Ecosystems 9:1041–1050. doi:10.1007/s10021-005-0105-7

Churkina G, Tenhunen J, Thornton P, Falge EM, Elbers JA, Erhard M, Grunwald T, Kowalski AS, Sprinz D (2003) Analyzing the ecosystem carbon dynamics of four European coniferous forests using a biogeochemistry model. Ecosystems 6:168–184

Ciais P, Reichstein M, Viovy N, Granier A, Ogee J, Allard V, Aubinet M, Buchmann N, Bernhofer C, Carrara A, Chevallier F, De Noblet N, Friend AD, Friedlingstein P, Grunwald T, Heinesch B, Keronen P, Knohl A, Krinner G, Loustau D, Manca G, Matteucci G, Miglietta F, Ourcival JM, Papale D, Pilegaard K, Rambal S, Seufert G, Soussana JF, Sanz MJ, Schulze ED, Vesala T, Valentini R (2005) Europe-wide reduction in primary productivity caused by the heat and drought in 2003. Nature 437:529–533

Ciais P, Bousquet P, Freibauer A, Naegler T (2007) Horizontal displacement of carbon associated with agriculture and its impacts on atmospheric CO_2. Global Biogeochem Cycles 21:GB2014. doi:10.1029/2006GB002741

Ciais P, Wattenbach M, Vuichard N, Smith P, Piao SL, Don A, Luyssaert S, Janssens I, Bondeau A, Dechow R, Leip A, Smith Pc, Beer C, van der Werf GR, Gervois S, Van Oost K, Tomelleri E, Freibauer A, Schulze ED, CARBOEUROPE Synthesis Team (2010) The European carbon balance. Part 2: croplands. Global Change Biol 16:1409–1428. doi:10.1111/j.1365-2486.2009.02055.x

CIAT (2004) Hole-filled Seamless SRTM Data V1. International Centre for Tropical Agriculture, Cali Palmira

Cienciala E, Tatarinov FA (2006) Application of BIOME-BGC model to managed forests. 2. Comparison with long-term observations of stand production for major tree species. For Ecol Manage 237:252–266

Czóbel S, Szirmai O, Nagy J, Balogh J, Ürmös Z, Péli ER, Tuba Z (2008) Effects of irrigation on the community composition, and carbon uptake in Pannonian loess grassland monoliths. Community Ecol 9:91–96

Dechow R, Freibauer A (2010) Modelling nitrous oxide emissions of agricultural soils with fuzzy logic based inference schemes. Ecol Model (unpublished data)

Fader M, Rost S, Müller C, Bondeau A, Gerten D (2010) Virtual water content of temperate cereals and maize: present and potential future patterns. J Hydrol 384(3–4):218–231

FAO (2010) Food and Agriculture Organization of the United Nations. TradeSTAT database. http://faostat.fao.org/. Accessed Jan 2010

Freibauer A, Rounsevell M, Smith P, Verhagen A (2004) Carbon sequestration in the agricultural soils of Europe. Geoderma 122:1–23

Führer E, Mátyás Cs (2005) A klímaváltozás hatása a hazai erdők szénmegkötő képességére és stabilitására (The impact of climate change on the carbon sequestration capacity and stability of Hungarian forests). Magy Tud 50:837–841

Gervois S, Ciais P, de Noblet-Ducoudré N, Brisson N, Vuichard N, Viovy N (2008) Carbon and water balance of European croplands throughout the 20th century. Glob Biogeochem Cycles 22:GB2022. doi:10.1029/2007GB003018

Goudriaan J, Groot JJR, Uithol PWJ (2001) Productivity of agro-ecosystems. In: Roy J, Saugier B, Mooney HA (eds) Terrestrial global productivity. Academic, San Diego, CA

Hungarian Central Statistical Office (2010). http://statinfo.ksh.hu/Statinfo/themeSelector.jsp?&lang=en. Accessed Feb 2010

IPCC (1996) In: Houghton JT, Meira Filho LG, Callander BA, Harris N, Kattenberg A, Maskell K (eds) Climate change 1995: the science of climate change, Contribution of working group I to the second assessment report of the intergovernmental panel on climate change. Cambridge University Press, Cambridge, United Kingdom and New York, NY

Jacinthe PA, Lal R, Owens B, Hothem DL (2004) Transport of labile carbon in runoff as affected by land use and rainfall characteristics. Soil Tillage Res 77:111–123

Janssens IA, Freibauer A, Schlamadinger B, Ceulemans R, Ciais P, Dolman AJ, Heimann H, Nabuurs G-J, Smith P, Valentini R, Schulze E-D (2005) The carbon budget of terrestrial ecosystems at country-scale – a European case study. Biogeosciences 2:15–26

Jones RJA, Hiederer R, Rusco E, Loveland PJ, Montanarella L (2005) Estimating organic carbon in the soils of Europe for policy support. Eur J Soil Sci 56:655–671

Kern A, Barcza Z, Bartholy J, Pongrácz R, Timár G, Ferencz Cs (2008) Analysis of MODIS NDVI time series for Hungary in 2007: detecting the phenological impacts of the summer heatwave. Geophysical Research Abstracts, 10: EGU2008-A-05083, 2008. SRef-ID: 1607-7962/gra/EGU2008-A-05083

Kreybig L (1937) The survey, analytical and mapping method of the Hungarian Royal Institute of Geology (in Hungarian and German). M Kir Földtani Intézet Évkönyve 31:147–244

Luyssaert S, Ciais P, Piao SL, Schulze E-D, Jung M, Zaehle S, Schelhaas MJ, Reichstein M, Churkina G, Papale D, Abril G, Beer C, Grace J, Loustau D, Matteucci G, Magnani F, Nabuurs GJ, Verbeeck H, Sulkava M, van der Werf GR, Janssens IA and members of the CARBOEUROPE-IP synthesis team (2009) The European carbon balance: part 3: Forests. Global Change Biol 16:1429-1450. doi: 10.1111/j.1365-2486.2009.02056.x

Marek M, Zverinová Z, Janous D (2006). Level 4 dataset CEIP_EC_L4_CZBK1_2006_v02 in CarboeuropeIP Ecosystem Component Database. http://gaia.agraria.unitus.it/database Accessed Oct 2007

Mészáros E, Molnár Á (1992) Energy production, economy and greenhouse gas emissions in Hungary. Időjárás 96:14–21

Mitchell TD, Jones PD (2005) An improved method of constructing a database of monthly climate observations and associated high-resolution grids. Int J Climatol 25:693–712. doi:10.1002/joc.1181

Pásztor L, Szabó J, Bakacsi Z (2010) Digital processing and upgrading of legacy data collected during the 1:25 000 scale Kreybig soil survey. Acta Geodaetica Geophys Hung 45:127–136

Pietsch SA, Hasenauer H, Thornton PE (2005) BGC-model parameters for tree species growing in central European forests. For Ecol Manage 211:264–295

Schulze E-D, Luyssaert S, Ciais P, Freibauer A, Janssens IA, Soussana JF, Smith P, Grace J, Levin I, Thiruchittampalam B, Heimann M, Dolman AJ, Valentini R, Bousquet P, Peylin P, Peters W, Rödenbeck C, Etiope G, Vuichard N, Wattenbach M, Nabuurs GJ, Poussi Z, Nieschulze J, Gash JH, the CarboEurope Team (2009) Importance of methane and nitrous oxide for Europe's terrestrial greenhouse-gas balance. Nat Geosci 2:842–850. doi:10.1038/ngeo686

Smith P, Goulding KWT, Smith KA, Powlson DS, Smith JU, Falloon PD, Coleman K (2001) Enhancing the carbon sink in European agricultural soils: including trace gas fluxes in estimates of carbon mitigation potential. Nutr Cycl Agroecosyst 60:237–252

Somogyi Z (2007) A Kyotoi Jegyzőkönyv és az erdők (The Kyoto protocol and the forests). Erdészeti Lapok 142:152–154

Somogyi Z (2008) A hazai erdők üvegház hatású gáz leltára az IPCC módszertana szerint (Greenhouse gas inventory of forests in Hungary using the IPCC methodology). Erdészeti Kutatások 92:145–162

Szabó J, Dombos M, Pásztor L, Bakacsi Zs, László P (2008) Practical problems of soil state assessment; experiments in the Bodrogköz sample area. Tájökológiai Lapok 6:27–41

Szentimrey T, Bihari Z, Szalai S (2005) Meteorological interpolation based on surface homogenized data basis (MISH). Geophysical Research Abstracts, 7: 07310. SRef-ID: 1607-7962/gra/EGU05-A-07310

Van Oost K, Quine TA, Govers G, De Gryze S, Six J, Harden JW, Ritchie JC, McCarty GW, Heckrath G, Kosmas C, Giraldez JV, da Silva JRM, Merckx R (2007) The impact of agricultural soil erosion on the global carbon cycle. Science 318:626–629

Van Oost K, Cerdan O, Quine TA (2009) Accelerated sediment fluxes by water and tillage erosion on European agricultural land. Earth Surf Process Land 12:1625–1634. doi:10.1002/esp.1852

Vetter M, Churkina G, Jung M, Reichstein M, Zaehle S, Bondeau A, Chen Y, Ciais P, Feser F, Freibauer A, Geyer R, Jones C, Papale D, Tenhunen J, Tomelleri E, Trusilova K, Viovy N, Heimann M (2008) Analyzing the causes and spatial pattern of the European 2003 carbon flux anomaly using seven models. Biogeosciences 5:561–583

Vuichard N, Ciais P, Viovy N, Calanca P, Soussana JF (2007) Estimating the greenhouse gas fluxes of European grasslands with a process-based model: 2. Simulations at the continental level. Global Biogeochem Cycles 21:GB1005. doi:10.1029/2005GB002612

White MA, Thornton PE, Running SW, Nemani RR (2000) Parameterization and sensitivity analysis of the Biome-BGC terrestrial ecosystem model: net primary production controls. Earth Interact 4:1–85

Zhao M, Heinsch FA, Nemani RR, Running SW (2005) Improvements of the MODIS terrestrial gross and net primary production global data set. Remote Sens Environ 95:164–176. doi:10.1016/j.rse.2004.12.011

Part IV
Greenhouse Gas Emissions and Removals in Hungary Based on IPCC Methodology

Chapter 14
Methodological Introduction and Overall Trends in Anthropogenic Emissions in Hungary*

Gábor Kis-Kovács, Katalin Lovas, Edit Nagy, and Klára Tarczay

Abstract The latest estimations of emissions and removals of greenhouse gases (GHG) are presented based on the Intergovernmental Panel on Climate Change (IPCC) methodology and in line with the requirements of the Kyoto Protocol. After a short methodological introduction, the whole time-series of emissions between 1985 and 2008 is analyzed and key drivers of the underlying trends are provided. To better understand the Hungarian emission trends, the time interval of the inventory is split into three periods with different emission-relevant economic processes in the background. The transition from a centrally planned economy to a free market one in Hungary in 1989–1990 caused significant reduction in emissions; then, after a period of about 14 years of relatively stagnant emission level (1992–2005), GHG emissions fell again quite significantly, by 8.4%, between 2005 and 2008.

Keywords Greenhouse gas inventory • Anthropogenic emissions and removals • IPCC methodology • Trends in emissions

14.1 Introduction

In this chapter, the latest estimations of emissions and removals of greenhouse gases (GHG) will be presented based on the Intergovernmental Panel on Climate Change's (IPCC) methodology (IPCC 1997, 2000, 2006) and in line with the

---

*Citation: Kis-Kovács, G., Lovas, K., Nagy, E., Tarczay, K., 2010: Greenhouse gas emissions and removals in Hungary based on IPCC methodology — Methodological introduction and overall trends in anthropogenic emissions in Hungary. In: Atmospheric Greenhouse Gases: The Hungarian Perspective (Ed.: Haszpra, L.), pp. 333–344.

G. Kis-Kovács (✉), K. Lovas, E. Nagy and K. Tarczay
Hungarian Meteorological Service, H-1525 Budapest, P.O. Box 38, Hungary
e-mail: kiskovacs.g@met.hu

L. Haszpra (ed.), *Atmospheric Greenhouse Gases: The Hungarian Perspective*,
DOI 10.1007/978-90-481-9950-1_14, © Springer Science+Business Media B.V. 2011

requirements of the United Nations Framework Convention on Climate Change (UNFCCC). This Convention, that took effect in 1994, had the aim of returning to the 1990 levels of the anthropogenic emissions of carbon dioxide and other GHG not controlled by the Montreal Protocol. Parties of the Convention committed to develop, periodically update, and publish national inventories of anthropogenic emissions by sources and removals by sinks of all GHG, using comparable methodologies. Later in 1997, the Kyoto Protocol (http://unfccc.int/kyoto_protocol/items/2830.php) set binding targets for 37 industrialized countries and the European Community for reducing GHG emissions by 5.2% below 1990 levels over the 5-year period 2008–2012. (Hungary's reduction commitment was 6% below the average of 1985–1987.) To ensure the effectiveness of this international agreement, the Parties' compliance with their commitments has to be monitored, for which purpose reliable emission data are needed. Therefore, the Kyoto Protocol requires Annex I Parties (including Hungary) to submit annual greenhouse gas inventories at regular intervals, using agreed methodologies that shall be those accepted by the IPCC. The submitted information is reviewed by expert review teams regularly.

The information in this chapter is based on the officially submitted national inventory reports of Hungary. In contrast with other sections of this book, GHG emissions and removals presented here are not all inclusive but confine to more or less anthropogenic processes as required by the above-mentioned international agreements. This means that greenhouse gas emissions and removals included in the national inventories are a result of human activities. Another important difference is that carbon dioxide from the combustion or decay of short-lived biogenic material is reported zero. Furthermore, increase in biomass stocks in annual crops in a single year is assumed equal to biomass losses from harvest and mortality in that same year – thus there is no net accumulation of biomass carbon stocks. As regards relevant territory, national inventories include greenhouse gas emissions and removals taking place within national territory. The Kyoto Protocol also defines the list of GHG (carbon dioxide (CO_2), methane (CH_4), nitrous oxide (N_2O), hydrofluorocarbons (HFCs), perfluorocarbons (PFCs), sulfur hexafluoride (SF_6)) and the emitting sectors (*Energy, Industrial Processes, Solvent and Other Product Use, Agriculture, Waste*) to be incorporated in the inventory. Every sector consists of source categories and subcategories (e.g., industrial processes sector – mineral products source category – cement production subcategory).

All in all, the inventory database contains time series of yearly values of emissions and removals broken down by gases and source categories. At the time of writing, the latest Hungarian inventory published spans an interval of 24 years (1985–2008) (NIR 2010). A national total is calculated by summing up emissions and removals for each gas and each source category. For adding up amounts of different GHG, the concept of global warming potential (GWP) is used. GWP is an index comparing the integrated radiative forcing over a specified period (in case of Kyoto Protocol 100 years) from a unit mass pulse emission relative to CO_2. GWP is reevaluated every now and then; nevertheless, current inventories are based on the values published in the IPCC Second Assessment Report (IPCC 1996), and are the following: CO_2 (1), CH_4 (21), N_2O (310), HFCs (140-11700, depending on compounds), PFC (6500–7400, depending on compounds), SF_6 (23900). National totals are then expressed in carbon dioxide equivalents.

14.2 Methodological Background

The IPCC Guidelines provide methodologies for estimating emissions and removals of GHG (IPCC 1997, 2000, 2006). However, the basic idea is not greenhouse gas specific; the same approach is used for other pollutants, and other emission inventories, as well (e.g., see the EMEP/EEA air pollutant emission inventory guidebook (EEA 2009)). The basic equation is as simple as this:

$$Emission = AD \times EF, \tag{14.1}$$

where *AD* stands for activity data, which represents some human activity (e.g., fuel use, industrial production, animal population, dwellings supplied with public sewerage, area of vineyard abandonment), whereas *EF* is the emission factor that quantifies the emission (or removal) per unit of activity. For example, in energy industry, which is the most important source category, emission factors for combusting natural gas or lignite are 56.1 t CO_2 TJ^{-1} and 107.5 t CO_2 TJ^{-1}, respectively; the importance of the mix of fuels used to produce energy becomes apparent at a glance.

Emission factors are usually dependent on several other factors, used technologies, etc., which leads us to the concept of tiers. A tier represents a level of methodological complexity. In the Guidelines, usually three tiers are provided. Tier 1 is the basic method, where activity data are usually aggregated national statistics and the emission factors are default values representing typical process conditions. Higher tier methodologies are more demanding in terms of complexity and data requirements as they require country-specific information on technologies used, facility-level data whenever possible, or use of complex models. For key categories, i.e., categories that have a significant influence on a country's total inventory of GHG in terms of the absolute level of emissions and removals, the trend in emissions and removals, or uncertainty in emissions and removals, it is required to apply higher-tier methods. Accordingly, the compilers of the Hungarian inventory aim at taking into account the technologies available in Hungary to the extent possible. For example, the emission trading system of the European Union makes possible to have access to facility-level activity and verified emission data.

Although this basic equation (14.1) can widely be used, in some source categories other approaches are used. For example, mass balance method is used for estimating the change in carbon content of living biomass in forests or, in case of solid waste disposal sites, a calculation method is applied which assumes that the degradable organic component in waste decays slowly throughout a few decades.

To ensure that the national inventory fulfils its main purpose, namely monitoring the country's compliance with its commitments, it has to meet certain quality standards. In other words, it has to be accurate, complete, consistent, comparable, and transparent (ACCCT). The first two requirements need no special explanation: an inventory is accurate, if it has no systematic bias towards under- or overestimations, whereas a complete inventory covers all relevant sources and sinks, and gases within the borders of the country. The next two criterions are closely linked to the requirements of the UNFCCC. Consistency ensures that the trends in the times-series of the inventory reflect real differences in emissions, and they are not caused by any methodological changes. National greenhouse gas inventories of all countries shall be comparable; therefore, the submitted information shall be compiled in accordance with the UNFCCC reporting guidelines and the IPCC guidelines and good practice guidance. Beside the inventory, a quite comprehensive document, the national inventory report has to be submitted to ensure transparency; i.e., to demonstrate that the inventory meets the good practice requirements for national greenhouse gas emissions inventories.

14.3 Level of GHG Emissions in Hungary in 2008

In 2008, total emissions of GHG in Hungary were 73.1 million ton carbon dioxide equivalent (NIR 2010). This is by far the lowest value in the whole time series (1985–2008). Taking also into account the mostly carbon-absorbing processes in the LULUCF (land use, land use change, and forestry) sector, the net emissions of Hungary were 68.6 million ton CO_2-eq in 2008. As recalculations can occur at any time, it is important to note that this value represents the state-of-the art estimation from April 2010. Nevertheless, with about 7 ton, the Hungarian per capita emissions are below the European average.

The most important anthropogenic greenhouse gas is carbon dioxide accounting for 76.9% of total GHG emissions. The main source of CO_2 emissions is burning of fossil fuels for energy purposes, including transport. CO_2 emissions have decreased by 33.8% since the middle of the 1980s. Methane represents 11.6% in the GHG inventory. Methane is generated mainly in waste disposal sites and animal farms, but the fugitive emissions of natural gas are also an important source. CH_4 emissions are 28.5% lower than in base year. Nitrous oxide contributes 10.3% to the total GHG emissions. Its main sources are agricultural soils, and manure management. N_2O emissions are 56.9% lower compared to the base year (1985–1987). The total emissions of fluorinated gases (F-gases) amount to 1.3%. F-gas emissions are showing a fluctuating, slightly growing tendency, especially due to their applications in the cooling industry.

By far, the biggest emitting sector was the *Energy* sector contributing 75.9% to the total GHG emission in 2008. *Agriculture* was the second largest sector with 12.0%, while emissions from *Industrial Processes* (with *Solvent and Other Product Use*) accounted for 7.0%, and the *Waste* sector contributed 5.1%. Compared to the base year, emissions were significantly reduced in the *Energy* (–33.1%), *Agriculture* (–49.8%), and *Industrial Processes* (–56.5%) sectors. In contrast, emissions in the *Waste* sector have increased since 1985 (+25.3%). *Solvent and Other Product Use* and *Land Use, Land-Use Change and Forestry* (LULUCF) sectors show fluctuating behavior.

14.4 Trends in Greenhouse Gas Emissions

By ratifying the Kyoto Protocol, Hungary committed to reducing its GHG emissions by 6%. In 2008, our emissions were 36.1% lower than in the base year (average of 1985–1987). For the most part, this significant reduction was mainly a consequence of the political changes that led to the transition from a centrally planned economy to a free market one in Hungary in 1989–1990. Then between 2005 and 2008, after a period of about 14 years of relatively stagnant emission level (1992–2005), GHG emissions fell again quite significantly, by 8.4% (Fig. 14.1).

Consequently, to better understand the Hungarian emission trends, the time interval of the inventory should be split into three periods with different emission-relevant economic processes in the background. The first period (1985–1995) would be the years of the regime change in Hungary, whereas in the second period (1995–2005) the rules of the market economy became decisive. The second period can also be characterized by the decoupling of GDP growth from the GHG emission trend, which is undoubtedly an important development. By 1999, the GDP reached the pre-1990 level; however, emission remained significantly below the levels of the preceding years. Thus, the emissions per GDP were decreasing (Fig. 14.2).

In the third period, after 2005, Hungary experienced an emission reduction of about 8%, basically due to mild winters, higher energy prices, and modernization in the chemical industry.

Starting with the first period (1985–1995), the process of transition into market economy brought in its train radical and painful decline in the output of the national economy. The production decreased in almost every economic sector including also the GHG relevant sectors (energy, industry, and agriculture). Consequently, GHG emissions decreased substantially in these years by around 35 million ton CO_2 equivalent. Between the mid-1980s and the mid-1990s, emissions fell back in the

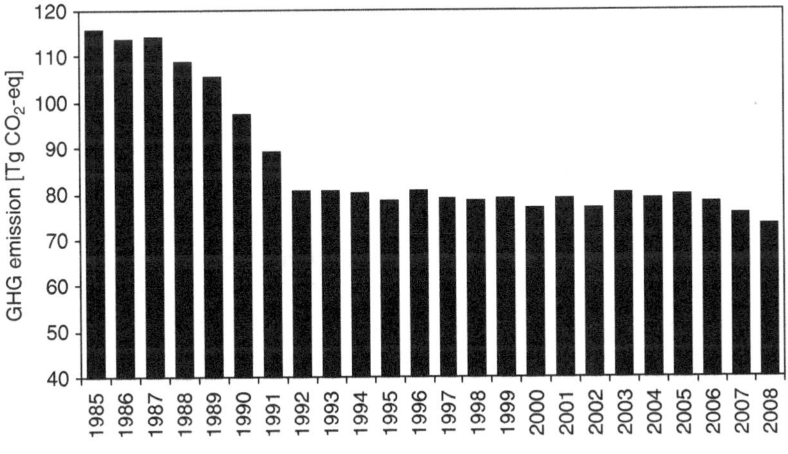

Fig. 14.1 Trends in total GHG emissions (Tg CO_2-eq)

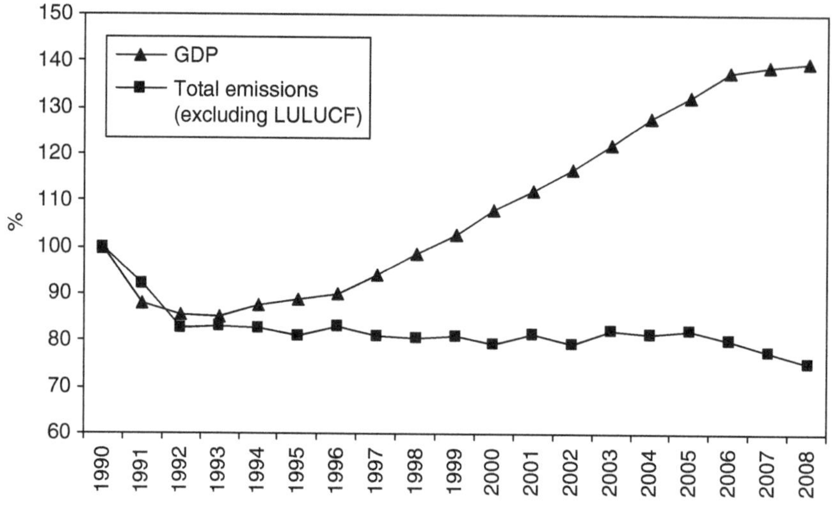

Fig. 14.2 Comparison of trends in GDP and GHG emissions (1990 = 100%)

Energy sector by around 25%, and even more, by around 50% in the *Industrial Processes* and *Agriculture* sectors.

The most significant drop in energy use occurred in the industry, especially in energy-intensive industrial sectors (manufacture of basic metals and machinery, mining, etc.). The industrial output in 1992 was two third of that in 1989. Several factories were closed down, capacity utilization was reduced; consequently, the production decreased drastically in each industrial sector. Some examples:

- Cement production dropped by 40%
- Iron and steel production: two of the three plants of the country were provisionally closed down
- Aluminum: two of the three plants of Hungary were closed down in 1991 (aluminum production stopped in 2006 eventually)
- Ferroalloys: ceased to exist (1991)
- Ammonia: four of the five plants of the country were closed down (1987, 1991, 1992, and 2002)
- Nitric acid: three of the four plants of the country were closed down (1988, 1991, and 1995)

The agricultural sector suffered a similar decline. As the result of the political and economic processes, the number of agricultural farms was reduced by more than 30%, the number of employees by more than 50%, the volume index of the gross agricultural production by more than 30%, the livestock by about 50%, and the use of fertilizers by more than 60%. As a consequence, the share of the agricultural sector in total GHG emissions decreased from 16.7% to 12.6%.

The small increase of emissions in the *Waste* sector is exceptional among all the sectors, and it is attributable to the slightly increasing quantities of waste generated and collected but more importantly to the applied calculation method, which

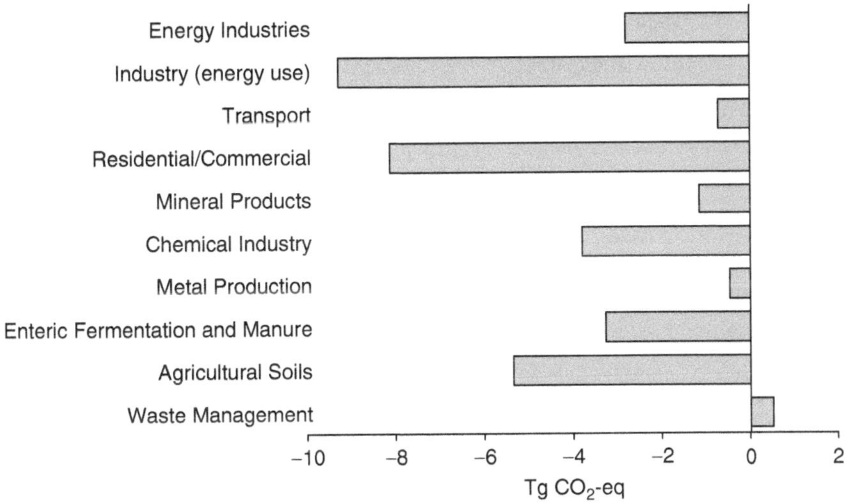

Fig. 14.3 Changes in emissions due to regime change (Tg CO_2-eq)

assumes that the degradable organic component in waste decays slowly throughout a few decades (Fig. 14.3.).

After the mid-1990s, emissions seemed to have been stabilized around 79 million ton CO_2 equivalent. However, behind the quite stable emission level, opposite processes could be observed, which can be illustrated by the relatively bigger changes in the *Energy* sector. The fuel use of industry decreased further and had only a 15% share in CO_2 emissions. In contrast, emissions from transport increased significantly by more than five million ton CO_2 equivalent, which represented a growth of 75% between 1995 and 2005 (Fig. 14.4).

In the third period, say after 2005, emissions fell by 6.7 million ton or 8.4%. The decreasing energy use by other sectors and manufacturing industries, and the diminishing process-related emissions in the chemical industry were the main drivers of these changes. Most importantly, total fuel consumption in the residential sector decreased by almost 20% (including a 13% decrease in natural gas consumption – mainly due to the extreme mild winter in 2007, but probably the growing energy prices and the support for modernization and insulation of buildings might have played a role as well). Decreased production volumes and modernization in the chemical industry led to an emission reduction of more than 80%. In contrast, emissions from energy industries and transport grew further (Fig. 14.5).

Emissions (excluding LULUCF) decreased by 3.4% (−2.6 million ton) between 2007 and 2008. In comparison with 2007, emissions in 2008 were lower in all major sectors. The highest relative reduction (−20.6%) occurred in the *Industrial Processes* sector mainly due to lower production volumes and modernization in chemical industry (−71.4%). Out of the 2.6 million ton reduction, chemical industry was responsible for about 1.2 million ton. Further decrease of 0.9 million ton was mainly due to favorable changes in the fuel-mix used by the energy industries (less fossil fuel consumption).

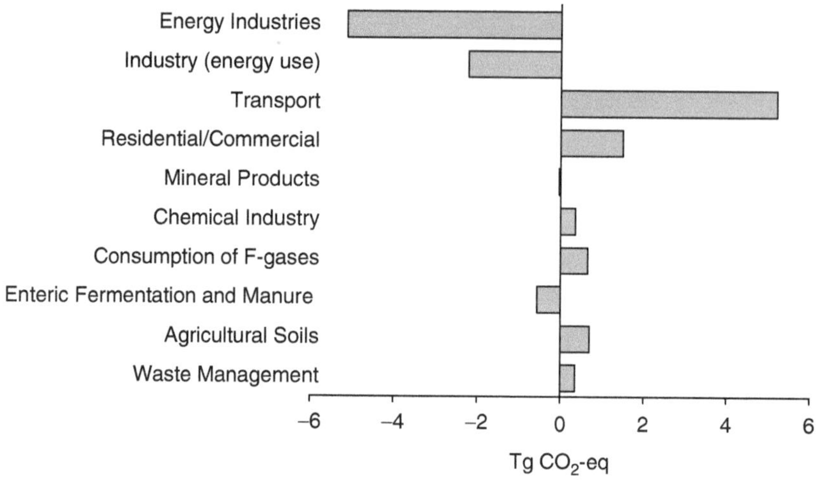

Fig. 14.4 Changes in emissions between 1995 and 2005 (Tg CO_2-eq)

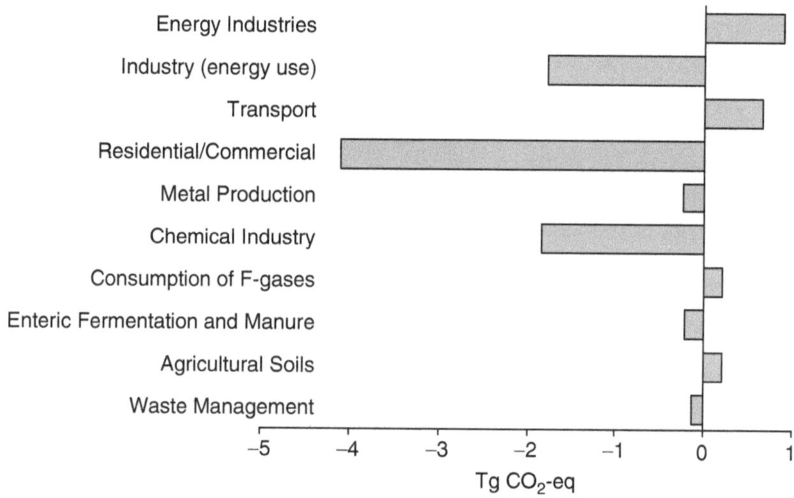

Fig. 14.5 Changes in emissions between 2005 and 2008 (Tg CO_2-eq)

14.5 Description and Interpretation of Emission Trends by Source Categories

Emissions by the *Energy* sector decreased in the first place as a result of reduced energy consumption and use of fuels with more favorable composition. The significant reduction in emissions between 1987 and 1992 was mainly due to the transition to market economy, which caused sudden decrease in energy demand. Besides, ongoing changes in the fuel-structure, i.e., solid fuel as the most important source in the 1980s has been

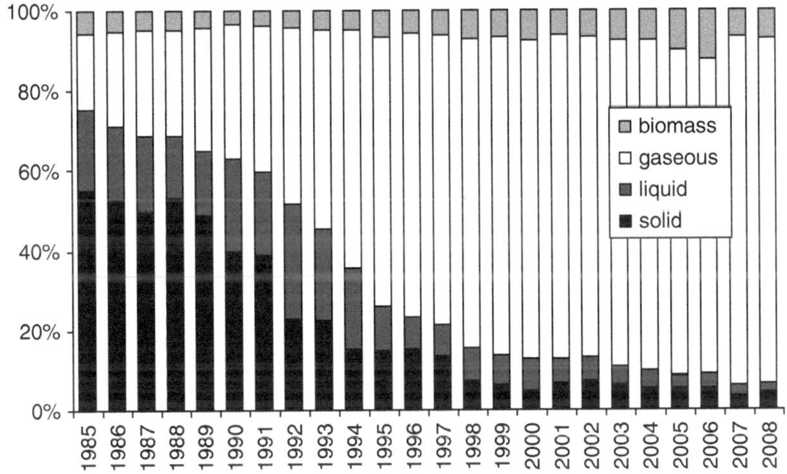

Fig. 14.6 Fuel switch in the residential sector (TJ)

replaced by natural gas, led to further decrease in the total emission. Maybe the most pronounced fuel switch could be observed in the residential/commercial sector. Solid fuels almost disappeared, while the share of natural gas reached 85% (Fig. 14.6).

Between 2005 and 2008, growing emissions from energy industries and transport could be observed, which were more than offset by the drastic reduction of emissions by the residential sector and manufacturing industries. Overall emissions from the *Energy sector* decreased by 2.1% between 2007 and 2008. Although slightly more electricity was generated in domestic power stations, the lower share of fossil fuels and the higher share of nuclear and renewable energy for electricity and heat production resulted in 4.3% decrease in emissions from energy industries. The growth in GHG emissions from transport was quite moderate (0.4%) after a more than 80% increase between 1995 and 2007. The almost sixfold rise in biofuel use could nearly meet the increased energy demand of transport in 2008. The tertiary sector used some 8% less energy, and the residentials' energy consumption remained below expectations. Taking into account the growing energy prices (e.g., the price of pipelined gas increased by 70% between 2006 and 2008), energy-saving measures must have contributed to the trend of emissions.

In 2008, *Agriculture* was the second largest source of greenhouse gas emissions in Hungary. Emissions from agriculture include CH_4 and N_2O gases: 81.2% of total N_2O emissions were generated here. Emissions from agriculture decreased by 49.8% over the period 1985–2008. The bulk of this decrease occurred in the years between 1985 and 1995, when agricultural production fell by more than 30%, and livestock numbers underwent a drastic decrease. The contribution of agriculture to total emissions decreased over the period 1985–2008 from 15.3% to its present share of 12.0%.

Between 1996 and 2008, agricultural emissions showed a slightly decreasing trend with fluctuations of 5% around 9 Mt. Behind this trend, there were compensatory processes. While the number of livestock decreased further leading to lower emission, the use of fertilizers increased by about 60%, which caused growing nitrous oxide emissions from agricultural soils.

Agricultural emissions fell by 1.4% between 2007 and 2008. This reduction was mainly driven by the 9.3% decrease in swine population due to high forage prices in 2008. Besides, rising fertilizer prices led to 8% lower fertilizer use and thus lower N_2O emissions from agricultural soils, which could not be offset by increased emissions from crop residues.

The *Industrial Processes* sector was the third largest contributing 6.5% to the total GHG emissions in 2008. (*Solvent and Other Product Use* added further 0.6% to the total emissions.) Process-related industrial emissions decreased by 56.5% between the base year and 2008, and by 32.1% between 2005 and 2008.

The key driver of the 20.6% reduction between 2007 and 2008 was the chemical industry. Ammonia and nitric acid production decreased by 28% and 18%, respectively, which was also reflected in the lower energy use of chemical industry. On top of this, the new nitric acid plant, thanks to a Joint Implementation (JI) project (see Article 6 of the Kyoto Protocol), almost abolished the factory's nitrous oxide emission in the magnitude of one million ton CO_2 equivalent. Besides, brick production decreased, and the cement industry was able to lower its emissions by using more additives to lower the fraction of clinker in the cement. Although emissions of F-gases represent only 1.3% of the total GHG emissions, their trend requires special attention. As these gases are harmless for the ozone layer, the use of HFCs in the refrigeration and air-conditioning industry has got widespread; thus their emission increased tenfold.

The *Waste* sector represented 5.1% of the total national GHG emissions in 2008. In contrast with other sectors, the emissions of *Waste* sector showed significant increase from the base year (+25.3%). However, the growth of emissions seemed to be stopping in recent years; moreover, a reduction of 3.4% could be observed between 2005 and 2008. In all the years, the largest category was solid waste disposal on land, representing 78.7% in 2008, followed by wastewater handling (19.5%) and waste incineration (1.8%). Emissions from wastewater handling have a pronounced decreasing trend due to a growing number of dwellings connected to the public sewerage network, whereas emissions from waste disposal sites have increased until the middle of this decade.

The LULUCF sector was a net sink of carbon because of the huge amount of carbon uptake of forests due to the continuous afforestation efforts and the sustainable forest management. In the inventory period, the forest area increased by 350,000 ha, and the amount of the current annual increment exceeded the annual harvest in all years. The complex dynamics of the land use and land-use changes lead to highly fluctuating estimates of sectoral removals. Our estimates indicate 3.4 million ton average annual removal with fluctuations in the range of ±97% in the inventory period (Fig. 14.7).

In 2008, the LULUCF sector accounted for 4.5 million ton carbon dioxide removals. The removals by forests amounted to almost five million ton, while the living biomass of orchards and vineyards were a net source of carbon, because of the continuous decrease of vineyard areas in Hungary. In 2008, the emission of the living biomass of vineyards and orchards accounted for net 0.17 million ton CO_2.

The following tables summarize the changes in emissions and removals of GHG both by gas and by source category (Tables 14.1 and 14.2).

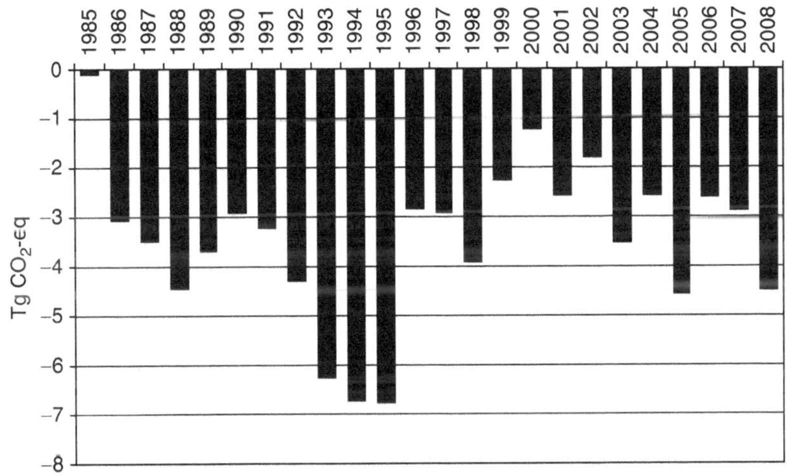

Fig. 14.7 Net removals by the LULUCF sector (Tg CO_2-eq)

Table 14.1 Trends in emissions of greenhouse gases (GHG) without LULUCF (Gg CO_2-eq)

	CO_2	CH_4	N_2O	HFC	PFC	SF_6	Total
BY[a]	84,938	11,862	17,394	0	268	81	114,545
1990	72,496	11,318	13,240	0	271	40	97,365
1995	61,435	9,292	7,686	1	167	70	78,650
2000	58,542	9,435	8,540	211	211	140	77,080
2005	60,940	8,891	9,067	538	209	201	79,846
2006	59,650	8,814	8,860	592	2	244	78,162
2007	57,885	8,660	8,374	621	2	172	75,714
2008	56,223	8,480	7,498	703	2	232	73,139

[a] BY= base year, average of 1985–1987

Table 14.2 Trends of sectoral GHG emissions in Hungary (Gg CO_2-eq)

	A	B	C	D	E	F	Total
BY[a]	82,869	10,908	285	17,511	−2,248	2,972	112,296
1990	70,496	8,851	226	14,501	−2,923	3,291	94,442
1995	60,794	5,455	205	8,685	−6,760	3,511	71,891
2000	57,806	6,277	214	9,075	−1,226	3,708	75,853
2005	59,825	6,996	366	8,804	−4,570	3,855	75,276
2006	58,732	6,415	335	8,846	−2,610	3,834	75,552
2007	56,679	5,979	366	8,906	−2,876	3,784	72,838
2008	55,476	4,748	406	8,783	−4,515	3,725	68,623

[a] BY= base year, average of 1985–1987
A = Energy; B = Industrial Processes; C = Solvent and Other Product Use; D = Agriculture; E = LULUCF; F = Waste

References

EEA (European Environment Agency) (2009) EMEP/EEA air pollutant emission inventory guidebook – 2009, Technical report No 6/2009. Available online at: http://www.eea.europa.eu/publications/emep-eea-emission-inventory-guidebook-2009

IPCC (1996) Climate Change 1995, The Science of Climate Change (eds: Houghton JT, Meira Filho LG, Callander BA, Harris N, Kattenberg A, Maskell K) Intergovernmental Panel on Climate Change. Available online at: http://www.ipcc.ch/ipccreports/sar/wg_I/ipcc_sar_wg_I_full_report.pdf

IPCC (1997) Revised 1996 IPCC Guidelines for National Greenhouse Gas Inventories, Intergovernmental Panel on Climate Change, Organisation for Economic Cooperation and Development, and International Energy Agency. (IPCC/OECD/IEA), UK Meteorological Office, Bracknell. Available online at: http://www.ipcc-nggip.iges.or.jp/public/gl/invs1.htm

IPCC (2000) Good Practice Guidance and Uncertainty Management in National Greenhouse Gas Inventories, Intergovernmental Panel on Climate Change National Greenhouse Gas Inventories Programme, Institute for Global Environmental Strategies, Japan. Available online at: http://www.ipcc-nggip.iges.or.jp/public/gp/english/

IPCC (2006) 2006 IPCC Guidelines for National Greenhouse Gas Inventories. In: Eggleston HS, Buendia L, Miwa K, Ngara T, Tanabe K (eds) Intergovernmental Panel on Climate Change National Greenhouse Gas Inventories Programme. Institute for Global Environmental Strategies, Japan

NIR (2010) National Inventory Report for 1985–2008 – Hungary (ed: Kis-Kovács G), Hungarian Meteorological Service, Budapest, April 2010. Available online at: http://unfccc.int/files/national_reports/annex_i_ghg_inventories/national_inventories_submissions/application/zip/hun-2010-nir-26may.zip

Chapter 15
Terrestrial Biosphere*

Zoltán Somogyi, György Borka, Katalin Lovas, and József Zsembeli

Abstract Terrestrial biosphere is rather intensively managed in Hungary resulting in more emissions than removals. The major sources of CO_2 emissions are agricultural soils. CH_4 and N_2O emissions are released through management of agricultural soils, enteric fermentation, and manure management. Most CO_2 removals occur in forests. All emissions and removals are estimated by adapting the various methodological guidances of IPCC. This chapter provides information on country-specific assumptions, data related to human-induced activities, and factors applied in the estimation. The estimates obtained demonstrate declining emissions over the period 1985–1995, fluctuating ones between 1996 and 2008, and varying degree of interannual variability.

Keywords Land use • Land use change • Sinks • Carbon stock change • Non-CO_2 emissions

*Citation: Somogyi, Z., Borka, Gy., Lovas, K., Zsembeli, J., 2010: Greenhouse gas emissions and removals in Hungary based on IPCC methodology – Terrestrial biosphere. In: Atmospheric Greenhouse Gases: The Hungarian Perspective (Ed.: Haszpra, L.), pp. 345–365.

Z. Somogyi (✉)
Hungarian Forest Research Institute, H-1142 Budapest, Stefánia út. 14., Hungary
e-mail: som9013@helka.iif.hu

G. Borka
Research Institute for Animal Breeding and Nutrition, H-2053 Herceghalom, Gesztenyes utca 1., Hungary

K. Lovas
Hungarian Meteorological Service, H-1525 Budapest, P.O. Box 38, Hungary

J. Zsembeli
Karcag Research Institute, Research Institutes and Study Farm, Centre For Agricultural and Applied Economic Sciences, University of Debrecen, H-5300 Karcag, Kisújszállási út 166., Hungary

L. Haszpra (ed.), *Atmospheric Greenhouse Gases: The Hungarian Perspective*,
DOI 10.1007/978-90-481-9950-1_15, © Springer Science+Business Media B.V. 2011

15.1 Introduction

This chapter covers greenhouse gas (GHG) emissions and removals related to the terrestrial biosphere. In Hungary, most land is managed; therefore, we use the land use sector (land management and animal husbandry) as a proxy for the "terrestrial biosphere." The estimation of the amount of GHG emissions and removals for the terrestrial biosphere as reported here is thus follows the logic of the "managed land" proxy and is an adaptation of the methodology suggested by the Good Practice Guidance (GPG) of IPCC (2000), GPG for Land Use, Land Use Change and Forestry (LULUCF) of IPCC (2003), and the IPCC 2006 Guidelines (IPCC 2006).

The methodology, which can be applied to obtain estimates for the entire country, separates estimates by GHG and carbon pools (in case of CO_2). Also, as the methodology of the estimation of emissions and removals is land use category specific, the broad IPCC land use categories, i.e., *Forest Land* (FL), *Cropland* (CL), *Grassland* (GL), *Wetlands* (WL), *Settlements* (SL), and *Other Land* (OL) are retained in this study (for definitions, see below). We also apply the category of land remaining in the same land use category over an inventory period, and land converted to another land use category. However, only aggregate area data on the land remaining in the same land use categories are available for WL and SL. These categories, whose share is low in the GHG fluxes, only amount to approximately 10% of the area of the country and are considered unmanaged.

Emissions and removals that are indirectly associated with managing land, e.g., using fossil fuels in various machines and buildings and burning of organic residuals, are covered by the estimates for other economic sectors as defined by IPCC.

15.2 Definition of the Land Use Categories and Pools

As the IPCC methodology is based on the area of the various land categories, the definition of these categories have been adjusted to the available country-specific data so that the area estimates are as accurate as possible. FL is defined in Hungary as land larger than 0.5 ha with trees potentially higher than five meters of height and a canopy closure of potentially more than 30%. The qualifier "potentially" means that all those areas are included where an afforestation or regeneration is under way, i.e., the trees are small or not even existent in the area at the time of the inventory, or the canopy has been opened up, but trees are deemed to be able to reach the mentioned thresholds *in situ*. Land predominantly under agricultural or urban land use is not included; however, forest roads and other areas under forest management, which are not covered by trees, are included to account for all area. Emissions and removals are estimated using carbon related inventory data such as tree volume, which are, in turn, estimated on land covered by trees. We also note that all forests in Hungary can be regarded as managed, i.e., human interventions are typically intensive and frequent.

CL area contains all arable lands, kitchen gardens, orchards, and vineyard areas as reported in the "Land area of Hungary by land use categories" statistics of the Hungarian Central Statistical Office (HCSO, 2009). Arable lands are any land area under regular agricultural management, irrespective of the soil cultivation and whether the area is temporarily covered by inland waters or is fallow. Orchards include land area covered by nonforest fruit trees (e.g., apples, pears, and cherries) and bush species. Vineyard can include more grape varieties, and unproductive areas, as well.

GL area refers to the grassland (meadow and pasture) area as reported in the "Land area of Hungary by land use categories" statistics of HCSO (2009). Meadow is a land covered by grass (artificial planting included) from which the yield is utilized by cutting, irrespective of whether it is occasionally used for grazing or not, whereas pasture is a land used for grazing. Only improved GL are classified here; nonimproved GL and abandoned meadows and pastures are taken into account in the OL category.

The WL area matches the wetlands and water body categories of the CORINE land cover database and contains the inland marshes (low-lying land usually flooded in winter, and more or less saturated by water all year round), peat bogs (peat land consisting mainly decomposed moss and plant biomass, whether exploited or not), water courses (natural or artificial water courses including those serving as water drainage), and water bodies (natural or artificial lakes, ponds, etc.).

For SL, the "Artificial surfaces" definition of the CORINE land-cover inventory is applied, which includes the urban areas, industrial, commercial and transport units, mines, dump and construction sites, and artificial nonagricultural vegetated areas.

Finally, OL is the residual land area of the country after all other land has been accounted for in the above categories. It includes, e.g., the uncultivated (abandoned) agricultural lands and other unmanaged lands in Hungary, which do not fall into any of the above five categories.

Concerning carbon pools, above-ground forest biomass is defined as the total biomass above stump of trees taller than two meters of height, including all branches, leaves, and bark. Below-ground biomass comprises stump and living roots thicker than 2 mm. For CL and GL, biomass is defined as in the IPCC Guidelines. The soil carbon pool in all relevant land use categories contain all organic carbon in the uppermost 30 cm layer below the litter layer, including dead roots and living roots finer than 2 mm. Litter and dead wood together include all dead leaves and all fallen branches and other parts of trees that are found above the surface in various states of decomposition. Litter also includes the fumic and humic layers. Litter and deadwood have not been separated, as there is no data yet for these pools.

15.3 Identification of Land Use Categories

The identification of the IPCC land use categories required the compilation of data from different statistical surveys in Hungary. The main sources of activity data were the National Forestry Database (NFD 2010), the land use statistics of the HCSO (2009), and the CORINE land-cover inventories.

The NFD provides the data for the estimates of FL. The NFD comprises of data for the whole forested area of the country, irrespective of ownership. The survey is continuous; approximately 10% of the whole forested area is revised annually, and the whole forested area is surveyed in 10-year-long cycles. The inventory is stand-based, the average size of a forest compartment being about 5 ha, and the spatial resolution of mapping of forests is 0.1 ha.

The main data source for CL and GL is the database of HCSO based on agricultural surveys. According to the IPCC method, the estimation of emissions from mineral soils for a year requires area data for the previous 20 years. Therefore, a consistent area database is needed. It was possible to create as the HCSO data set goes back to as early as 1853, although there have been changes in the methodology since the beginning of the data collection (Kecskés 1997).

Because of the differences in the definitions of the various land use categories between the IPCC guidelines and those applied in the country, the data in the various datasets was adjusted to achieve consistency for FL, CL, and GL.

The areas of SL and WL were determined by the CORINE land cover inventories of 1990, 2000, 2006, and an auxiliary land-cover database for 1985. (This data set is similar to the CORINE data sets, and it was produced via processing satellite images specifically for GHG inventory purposes by the Institute of Geodesy, Cartography and Remote Sensing.) The WL category contains all wetlands in Hungary. Finally, the area of OL was obtained by totaling all previously listed land uses and subtracting them from the total area of the country. The distribution of the various land use categories in 2008 is demonstrated in Fig. 15.1.

With respect to trends, a significant decrease of the area of croplands was characteristic for the last four decades, and roughly 800,000 ha were converted to another land use category. However, cropland is still the main land use category in Hungary. In contrast, about 250,000 ha of new forests were established in the last

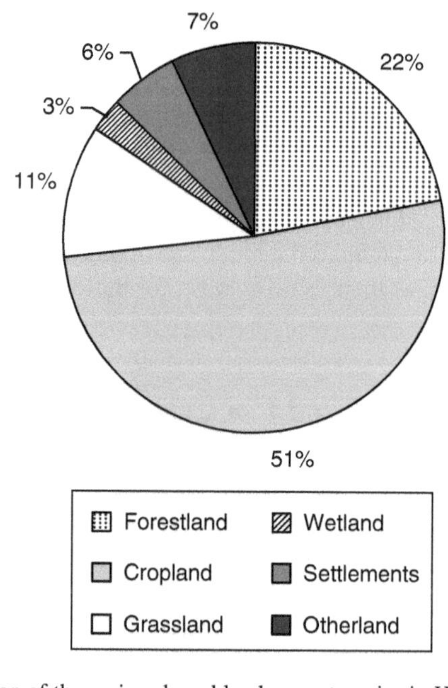

Fig. 15.1 The distribution of the various broad land use categories in Hungary in 2008

quarter of a century, yet, forests only occupy some one fifth of the terrestrial land area of the country.

Emissions and removals do not proportionately result from the above area changes; rather, they depend on the size of the biomass and other carbon pools affected and on other factors. Emissions and removals of the terrestrial biosphere are reported below in the order of their estimated scale, by gas, carbon pool (where relevant), and land use.

15.4 CO_2 Emissions and Removals of the Biomass Carbon Pool

Of all GHGs, CO_2 is by far the most important one in terms of rate of fluxes. In a similar fashion, biomass is where the largest fluxes occur. Because of its inherent biomass stock, emissions and removals are the largest on forest land. CO_2 emissions and removals of some other nonforest biomass pools are also estimated.

The main processes that result in emissions and removals in a sustainably managed forest include the increase of biomass carbon stocks due to tree photosynthesis and the reduction of biomass carbon stocks due to a variety of natural factors (e.g., decomposition and fires) and human-induced ones (forestry operations). Due to the nature of the Hungarian forestry statistics, forestry data describing some of these processes, such as estimated increment and the amount of harvest, are available in the country annually. However, these processes were not separately considered using the above data; rather, the stock-change method as suggested by the IPCC (2006) was applied. This method requires tree volume data in two subsequent points in time, from which the change of the volume stock between the two time points can be calculated. We can apply this method as the estimates of total volume of all forests are also available annually. The volume stock data is most probably estimated with a higher accuracy and is less prone to the effects of various annual disturbances in forests; thus, it provides a more robust estimate of biomass carbon stock changes for any year and for the trend of the estimates.

When selecting the stock change method, it was considered that the uncertainties of neither of the above statistics are not known; however, bias is considerably reduced when consecutive growing stock values are deducted to obtain stock changes. We note, however, that since growing stocks and their changes incorporate the effects of all processes, no particular inferences can be made with respect to any of these processes if the stock change method is used.

Equation 3.2.3 of the GPG for LULUCF (IPCC 2003) has been modified to adapt it to the Hungarian conditions. The following equations were used to estimate carbon stock changes:

$$\Delta C_{BF} = (C_{t2} - C_{t1}) / (t_2 - t_1)$$

and

$$C_t = [V_t * D] * (1 + R) * CF$$

whereΔC_{BF} = carbon stock changes of forest biomass (ton C); C_t = carbon stocks at time t (ton C); V_t = volume stocks at time t (m^3); D = basic wood density (ton m^{-3}); R = root-to-shoot ratio (dimensionless); CF = carbon fraction of biomass (ton C ton biomass^{-1}); t_2 and t_1 are consecutive years, so $t_2 - t_1 = 1$ year.

The application of these equations is possible because, as it was mentioned above, the forest inventory is continuous to enable the preparation of forest management plans. The inventory is compiled by surveying each stand, the average size of which being about 5 ha, once in every 10 years. Each stand is identified on management plans, and the inventory data are stored in a computerized database. During the survey, which has been conducted applying rather standard forest inventory methodology, the main stand measures (such as tree height, breast height diameter, basal area, and density) are estimated for each species by various measurement methods that are selected based on the age, site, volume stock, and quality of the stand. Usually, more accurate methods are used for stands of higher volume stocks and higher economic value. In years between surveys, yield functions are used to update volume stocks. For the carbon stock change estimation, volume stock data aggregated for all main species and species categories were used for each inventory year.

Tree volume in the forest inventory is calculated from breast height diameter and height of sample trees using volume functions by Sopp and Kolozs (2003).

Concerning basic wood density, country-specific data by main species and species groups are available (Table 15.1), as they have recently been revised and completed by Somogyi (2008).

Note that no biomass expansion factor is applied, because all wood volume (m^3) values in Hungary are estimated and expressed as total above-ground volume of trees including stem, all branches, twigs, and bark (i.e., the volume of all above-ground parts of the trees above stump). Therefore, to convert the total above-ground volume to total above-ground biomass, only conversion is done. However, this is equivalent with applying the same conversion factor for each part of the trees. Since twigs and branches may have density that is different from that of wood, this method may introduce an unknown, but little bias for individual trees and stands of above- or below-average share of twigs and branches (such as coppice stands of low density on poor site) but not for the entire forests. With respect to the below-ground biomass, a general root-to-shoot ratio (R) is applied. In lack of country-specific data, a conservative value of 0.25 is used, which is based on expert judgment, and is within the default range suggested by the IPCC (2006).

Finally, the IPCC default value, i.e., 0.48 (for broadleaves) and 0.51 ton C ton biomass^{-1} (for conifers) is used for the carbon fraction of dry wood.

Figure 15.2 shows the trend of the net removals estimated using the above methodology. Forests are net sinks in each year, this sink amounting to a few percent of the total emissions in the country. The interannual variability of the estimated values is also rather high, at least as compared to the estimates by most countries that apply the IPCC methodology. This is partly due to the high frequency of data collection in Hungary, but also due to the annual variability of tree growth and the amount of harvest by species.

Table 15.1 Wood density values for the main species and species groups in Hungary (Somogyi 2008)

Species or species group	Basic wood density (tons of dry matter m^{-3} of fresh wood)
Quercus robur L.	0.57
Quercus petraea L.	0.61
Quercus cerris L.	0.639
Other Quercus sp.	0.55
Fagus silvatica L.	0.583
Carpinus betulus L.	0.67
Robinia pseudoacacia	0.59
Acer sp.	0.52
Ulmus sp.	0.58
Fraxinus sp.	0.56
Other hardwood broadlives	0.5
Populus hybrids	0.34
Indigenous Populus species	0.36
Salix alba L.	0.36
Alnus sp.	0.43
Other softwood broadleaves	0.48
Pinus silvestris L.	0.42
Pinus nigra L.	0.47
Larix deciduas L.	0.49
Picea abies L.	0.39
Other conifers	0.37

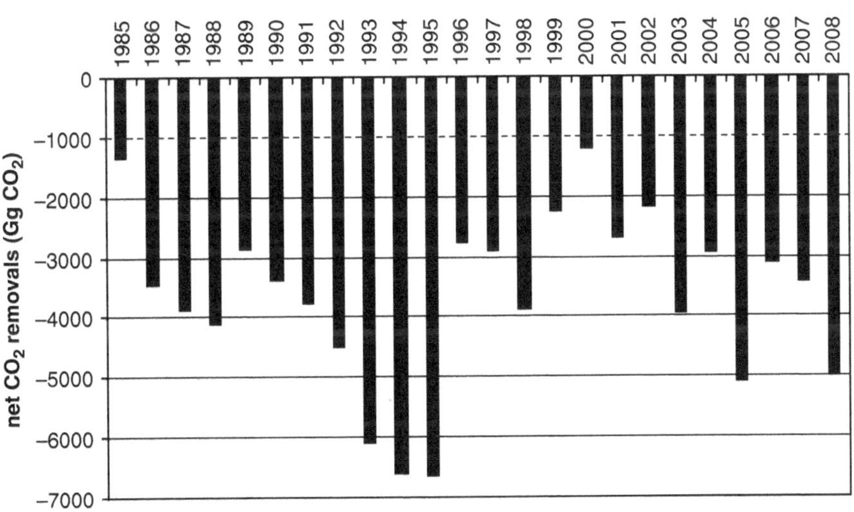

Fig. 15.2 Total net removals by the forests (1985–2008)

In the cropland remaining in cropland category, only orchards and vineyards have perennial woody vegetation that was considered biomass. The carbon stock change in CL woody biomass, ΔC_{BC}, was estimated from the biomass gains and losses by adapting the Tier 1 method of the GPG for LULUCF (IPCC 2003):

$$\Delta C_{BF} = \Delta C_G - \Delta C_L$$

where ΔC_{BF} = carbon stock changes of perennial woody biomass (ton C); ΔC_G = annual increase in carbon stocks of perennial woody biomass due to biomass growth (ton C year^{-1}); ΔC_L = annual decrease in carbon stocks of perennial woody biomass due to biomass loss (ton C year^{-1})and

$$\Delta C_G = A_G * G$$

$$\Delta C_L = A_L * L$$

where A_G = area of perennial woody cropland (ha); G = IPCC default value for perennial crops carbon accumulation rate = 2.1 tC ha^{-1} year^{-1}; A_L = area of cropland on which perennial woody crops are removed; L = IPCC default value for carbon loss of perennial crops = 63 tC ha^{-1} year^{-1}.

Figure 15.3 demonstrates the trend and the rate of the estimated emissions for the CL woody biomass. In the long term, these net emissions are rather small; however, if land use changes are intensive, the associated fluxes in individual years can be rather high as compared to those in the FL.

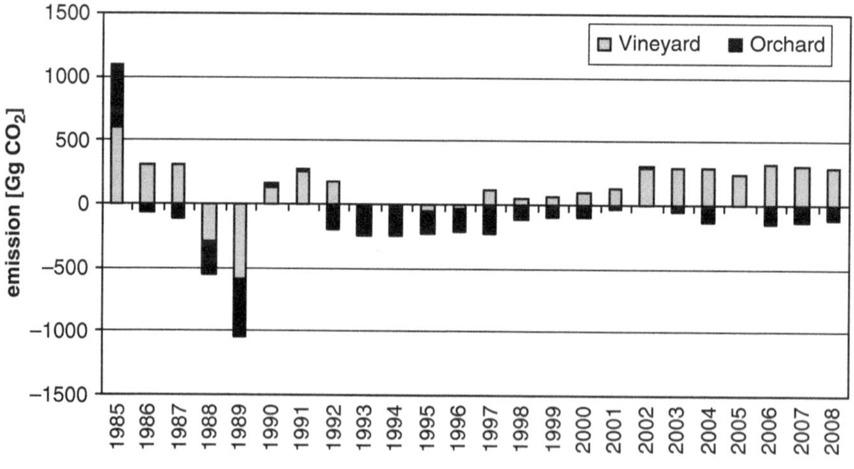

Fig. 15.3 CO_2 emissions and removals associated with vineyards and orchards (1985–2008)

15.5 Carbon Stock Changes in Mineral Soils in FL

In Hungary, the area of land of organic soils is small; thus, only emissions and removals of mineral soils are estimated. These emissions and removals are calculated from estimated carbon stock changes.

Due to lack of proper data, carbon stock change in mineral soils for the entire terrestrial biosphere cannot be estimated. For FL, three case studies have produced country-specific values. Somogyi et al. (2005) analyzed the possible effects of afforesting former croplands with Black locust in a sandy and dry region in the middle of the country. In this study, soil samples were taken from 12 places in the region on a random basis from a spatial point of view. Ten of the places were Black locust stands of different ages, and the rest were control areas, i.e., areas that were not afforested, rather used as croplands or fallow, and that were next to the afforested stands, seemingly having similar site conditions to the adjacent stands. By setting the age of these control areas to zero, a complete false time series could be established from the age of zero to 20 years. Within each sampled stand and in each control areas, five randomly selected places were established where soil samples were taken. Because of the high variation of soil characteristics, a soil pit was made at each place where samples were taken from four layers (i.e., 0–5 cm, 5–10 cm, 10–20 cm, and 20–30 cm) by each cardinal direction. Thus, the total number of soil samples was 12 * 5 * 4 * 4 = 960. The results of standard laboratory analysis of the soil samples demonstrated that the total carbon stock changes in all layers studied amounted to 0.09 tC ha^{-1} annually. This is the resultant of increase in two layers (0–5 and 5–10 cm), and decrease in the other two layers. It must also be added that it is only the increase in the 0–5 cm layer (0.236 tC ha^{-1} annually) that can be regarded as significant at the 95% probability level ($p = 0.0167$), whereas the changes in the three other layers are not significant ($p > 0,5$). In these layers, the carbon stock changes can be zero or even negative in the first decade from a statistical point of view.

In the study performed by Horvath (2006), the carbon stock (forest floor and mineral soil 0–60 cm) development after afforestation was investigated in chronosequences at six different sites. It was found that afforestation of croplands, on average, results in a total carbon sequestration of 43 tC ha^{-1} over a period of 150 years. Afforestation of grasslands does not lead to carbon sequestration in the long term; rather, an initial net carbon emission occurs in the first three decades after the afforestation, which is then offset by a net sequestration over the subsequent 100 years.

Finally, Somogyi (2010) estimated the potential effects of typical regeneration methods on the soil carbon pool in typical Hungarian forests. In this study, six paired *Quercus robur* L. stands were sampled where one stand of the pair, having a rather old stand before regeneration, represented the "reference" carbon stock, whereas the other stand of the pairs, right next to the reference stand of the pairs, was regenerated one or two years before taking the soil samples. The regeneration included soil preparation, i.e., a relatively large disturbance, and even stumping in some cases. Ten samples were taken in each stand using the methodology outlined above. The basic statistics of the results are reported in Table 15.2.

Table 15.2 The main statistics of the soil samples taken by Somogyi (2010). In each stand of each pair, the number of samples is 10

		Organic carbon content in the 0–30 cm topsoil layer (tC ha^{-1})				
		Regenerated stand		Reference stand		
Pair #	Method of regeneration	Mean	Std	Mean	Std	Differences of the carbon stocks between the regenerated and reference stand
1	Unstumped, first year	32.6	6.8	36.5	6.7	−3.9
2	Unstumped, first year	67.0	38.0	66.3	50.2	0.7
3	Unstumped, first year	40.4	13.0	77.1	48.9	−36.7
4	Stumped, second year	35.3	10.0	48.4	7.2	−13.0
5	Stumped, first year	27.6	12.6	35.3	10.0	−7.7
6	Stumped, first year	43.8	12.1	44.5	8.7	−0.7

The main conclusions of the study can be summarized as follows:

- Five values out of six are negative, and there is only one positive value.
- The differences between the soil carbon content of regenerated and reference stands are considerable in four, and negligible in two cases.
- All considerable values are negative, indicating a significant carbon loss from the soil, with values between 4 and 37 tC ha^{-1}.
- No major differences can be found with respect to the technology of the regeneration.
- The standard deviations are rather high in all cases.

15.6 Carbon Stock Changes in Mineral Soils in Land Other than FL

The carbon stock of mineral soils has been changing especially for the last decade due to increasing changes in the principles and subsequent practice of soil cultivation. The need for environmentally friendly and energy saving soil tillage systems is increasing as the consequences of improper soil cultivation practice that characterized the last decades manifested in rather unfavorable soil properties (Birkás 2002; Birkás et al. 2007). In accordance with efforts to reduce soil degradation due to the improper soil use, the conventional soil cultivation methods are prospectively replaced by conservation tillage, including different versions of reduced till, mulch-till, and crop residue management (Forgács et al. 2005), which aim at the decrease of the depth of the regularly cultivated soil layer and the formation of a topsoil rich in organic matter, hence, they considerably affect soil carbon stocks in croplands. Although there are no extensive measurements in Hungary yet, there are already some results concerning the effect of reduced tillage systems on the CO_2 emission from the soil that provide valuable information with respect to soil utilization (Gyuricza et al. 2005; Tóth and Koós 2006; Zsembeli et al. 2005, 2006; Zsembeli and Kovács 2007).

The estimation is based on the Tier 1 methodology of GPG for LULUCF (IPCC 2003) for Approach 2 of representing land areas. Equation 3.3.4 (B) was applied for CL, GL, and OL. First, annual carbon stock changes on a unit land area are estimated separately for each land use change type using the following formulas:

$$\Delta C_{Mineral} = (SOC_0 - SOC_{0-T})^* / T$$

where $\Delta C_{Mineral}$ = annual carbon stock change in mineral soils for a certain land use change type (tC ha^{-1} year^{-1}); SOC_0 = average soil organic carbon stock in any calendar year (tC ha^{-1}); SOC_{0-T} = average soil organic carbon stock T years prior to the above calendar year (tC ha^{-1}); T = length of time period for equilibrium stock change to take place (the IPCC GPG for LULUCF IPCC, 2003 default value, i.e., 20 years is used).

SOC (on unit area) in turn is calculated from a reference carbon stock and factors that apply to different situations reflecting the type and intensity of the land use:

$$SOC = SOC_{ref} * F_{LU} * F_{MG} * F_{I}$$

where SCO_{ref} = reference carbon stock (tC ha^{-1}, see Table 15.4 below); and F_{LU}, F_{MG}, F_{I} = factors reflecting different soil carbon stock levels by land use (LU), management (MG), and sewage input (I) types (dimensionless; see Tables 15.5–15.7 below).

In accordance with the GPG for LULUCF (IPCC 2003), the area of each of the above land use change types were stratified by soil type, climate, management, and input types. Moreover, CL and GL were also stratified into areas that are actively managed, and which, although they remain in the same land use category, have been abandoned and which are thus not actively managed. Thus, SOC and $\Delta C_{Mineral}$ were estimated for each of these strata. The various strata are denoted by the following indices:

Index *LU* is for land use types (managed CL, abandoned CL, managed GL, abandoned GL, OL)
Index *c* is to represent climate zones
Index *s* is for the soil type
Index *m* represents the different management systems on CL and GL
Index *i* is for input type.

Total soil carbon stock changes for a land use change type are estimated by multiplying the carbon stock changes per unit land with the total area of the various land use change types by climatic zones, soil type, management types and input type, and by applying various factors specific to these types. Finally, total soil carbon stock changes in the mineral soils of all nonforest land are calculated by summing up:

$$\Delta C_{total} = \sum_{LU,c,s,m,i} (A_{LU,c,s,m,i} * \Delta C_{LU,c,s,m,i})$$

where ΔC_{total} = total carbon stock changes in the mineral soils of the nonforest land (tC); $A=\sum_{LU,c,s,m,i} A_{LU,c,s,m,i}$ = total area of nonforest land (ha); $A_{LU,c,s,m,i}$= area of each *LU* change category by *c*, *s*, *m*, and i (hectare).

The categorization of CL is partly based on expert judgment due to the lack of sufficient statistics on the recent Hungarian land use practice, mainly with respect to management type and level of input. Nevertheless, the input factors can be estimated fairly well based on the actual composition of annual crops, while the change in the management practice can be tracked using the activity data itself.

For the stratification HCSO's area, data were harmonized with CORINE land cover (CLC) data (1990, 2000) and the climate data provided by the Hungarian Meteorological Service (HMS). As land use data are not available for the periods before 1990 and after 2000, the stratification was estimated by the interpolation and extrapolation of the available data.

The soil types were determined on the base of the digital soil map of Hungary (called AGROTOPO database). This database was harmonized with the land use types of CLC to determine the rate of land use types on different soil types (this harmonization was done by the GIS Lab of the Research Institute for Soil Science and Agricultural Chemistry of the Hungarian Academy of Sciences, or RISSAC). We note here that the Hungarian national soil classification system classifies soils by genetic types, and these types are not compatible with the types identified by the applied by the IPCC; therefore, the Hungarian soil types were reclassified.

The climate classification and the determination of the spatial distribution of climate zones were made by HMS. Two categories were determined, i.e., the Cold Temperate Dry (CTED) zone where the mean annual temperature (MAT) is below 10°C, and Warm Temperate Dry (WTED) zone where MAT is above 10°C. In both categories, the annual precipitation is less than the evapotranspiration.

After determining the climate zones, they were harmonized with the soil classes: the four soil types were grouped into the two climate categories (performed by RISSAC) according to their spatial distribution in Hungary. As a result, the proportions indicated in Table 15.3 were gained.

Table 15.3 The distribution of the areas of the various climate zones and soil types for Cropland

Climate zone	Soil	Management	Input	Area in 2008 (1000 ha)
Cold dry	HAC	Full till	Low	744.6
Cold dry	HAC	Full till	Medium	579.7
Cold dry	HAC	Full till	High with no manure	74.5
Cold dry	HAC	Reduced till	Medium	90.4
Warm dry	HAC	Full till	Low	1101.0
Warm dry	HAC	Full till	Medium	855.2
Warm dry	HAC	Full till	High with no manure	110.1
Warm dry	HAC	Reduced till	Medium	135.6
Cold dry	LAC	Full till	Low	31.6
Cold dry	LAC	Full till	Medium	23.0
Cold dry	LAC	Full till	High with no manure	2.9
Warm dry	LAC	Full till	Low	25.2
Warm dry	LAC	Full till	Medium	18.3
Warm dry	LAC	Full till	High with no manure	2.3
Cold dry	Sandy	Full till	Low	61.7
Cold dry	Sandy	Full till	Medium	44.9
Cold dry	Sandy	Full till	High with no manure	5.6
Warm dry	Sandy	Full till	Low	74.2
Warm dry	Sandy	Full till	Medium	54.0
Warm dry	Sandy	Full till	High with no manure	6.7
Cold dry	Aquic	Full till	Low	160.7
Cold dry	Aquic	Full till	Medium	116.8
Cold dry	Aquic	Full till	High with no manure	14.6
Warm dry	Aquic	Full till	Low	245.5
Warm dry	Aquic	Full till	Medium	178.6
Warm dry	Aquic	Full till	High with no manure	22.3
Total				4780.0

HAC/LAC is for high/low activity clay. For details, see IPCC (2006)

Concerning management, full tillage was the only applied cultivation system in Hungary until the end of the 1990s. This is still the dominant practice, although the area cultivated with reduced tillage methods has been increasing year by year. To account for changes in soil carbon stocks of CL, we estimated the areas of the two main cultivation types at the beginning and at the end of the inventory time period. As there is no sufficient data available to estimate the precise actual area of reduced tillage, the calculation is based on expert judgment. The principle of the calculation is that the total area of cereals (winter wheat, barley, and maize) can be considered stable (approximately 2.6 million hectare) with insignificant fluctuation. The newly introduced soil protective cultivation methods are used mainly in the case of cereal production. By considering the cumulative number of machines and tools suitable for reduced tillage that has been sold since 1998 (source: KITE Ltd., the biggest company in agricultural service and commerce in Hungary), the extent of the area where these machines and tools can be applied was estimated.

With respect to input, the characteristics of crop rotations were taken into consideration to select the appropriate input factors that represent the agricultural practice in Hungary. According to the GPG for LULUCF (IPCC 2003), the input factors represent the effect of changing carbon input to the soil as a function of crop residue yield, bare-fallow frequency, cropping intensity, or applying amendments. Therefore, the four soil types representing the Hungarian croplands were divided further into three input categories (Table 15.3).

Low residue return is due to removal of most residues, which is characteristic of the growing technology of cereals (wheat, rye, barley) and a certain fraction of maize. As the total area of cereals, except for maize, is approximately 1.4 million hectares, the proportion of the low-input category is significant. We also considered that crop residues are typically removed from a certain amount of the area of the crops listed under medium input. Medium input cropping systems represent annual cropping with crops where crop residues are returned to the field. This way of growing is typical mainly for maize, sunflower, and sugar beet production. Finally, high-input (without manure) rotations are not widely used in Hungary and are practically limited to the use of green manures and cover crops. Therefore, no area was allocated to the high-input (with manure) category.

For the estimation, the SOC_{ref}, F_{LU}, F_{MG}, and F_I values were taken from the relevant tables of the GPG for LULUCF (IPCC 2003) and the values applied are presented in Tables 15.4–15.7. Figure 15.4 demonstrates that the resulting CO_2 emissions, which reflect both the effect of the abandonment of agricultural areas and the changes in management practices.

Table 15.4 SOC_{ref} values by region and soil type

Region	High activity clay soils	Low activity clay soils	Sandy soils	Aquic soils
Cold temperate, dry	50	33	34	87
Warm temperate, dry	38	24	19	88

Table 15.5 Land use factor (F_{LU}) by land use type

Land use type	Factor
Forest land	1.00
Cropland	0.82
Grassland	1.00

Table 15.6 Management factor (F_{MG}) by land use type and management regime

Land use type	Management regime	Factor
Cropland	Full till	1.00
	Reduced till	1.03
Grassland	Nondegraded	1.00
	Improved	1.14

Table 15.7 Input factor (F_I) for cropland by the amount of input of organic matter

Land use type	Input of organic matter	Factor
Cropland	Low	0.92
	Medium	1.00
	High – with no manure	1.07

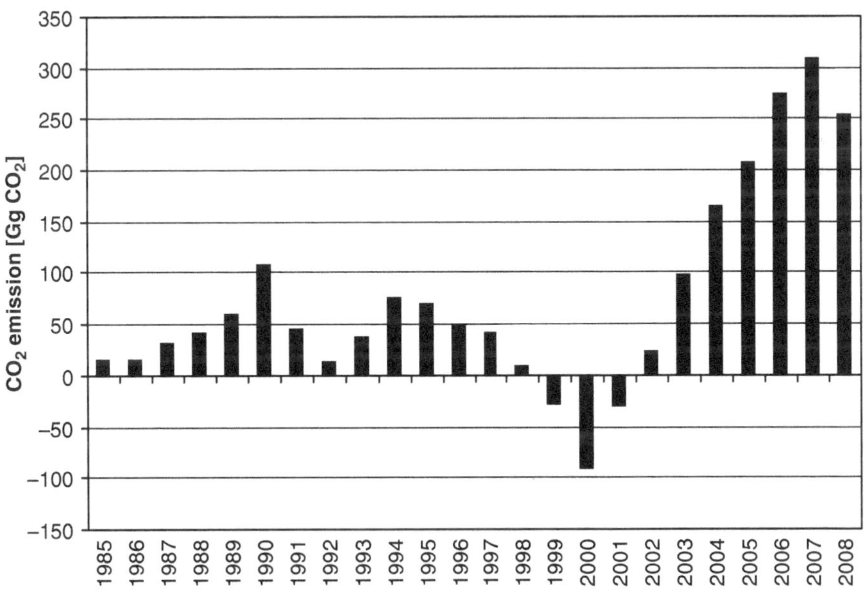

Fig. 15.4 CO_2 emissions/removals from cropland, grassland, and abandoned agricultural area soils (1985–2008)

15.7 CO_2 Emission from Liming

For the estimation of CO_2 emissions from liming, the Tier 1 methodology of the IPCC (2003) Guidelines was applied. Liming has shown a decreasing tendency in Hungary for the last decade. There is no reliable data on the amount of lime applied in Hungary. Therefore, the estimated amount is based on expert judgment by the extent of reclaimed areas for which data are available (AERI, 2010). Also, the estimates cannot be redistributed for each land use; rather, the estimation is made for the entire managed land area.

The estimation involved the following assumptions:

- Two thirds of the acidic soils are reclaimed with limestone containing amendments, whereas 27% are reclaimed with dolomite.
- Half of the salty soils were estimated to be reclaimed with limestone or other carbonate containing material.
- The average quantity of $CaCO_3$ applied is 8 t ha^{-1} and that of $CaMg(CO_3)_2$ is 7 t ha^{-1}.

The default emission factor of 0.12 was used for carbonate containing lime, and 0.122 for dolomite.

15.8 CH_4 and N_2O Emissions from Agricultural Production

Land use is responsible for various further emissions that are accounted for here. These emissions are mainly associated with animal husbandry; however, following the terminology of the IPCC (2003; 2006) Guidelines, this source of emissions is termed here as "agricultural production." As mentioned before, CO_2 emissions from using fossil fuels in various machines, buildings, etc. and burning of organic residuals are covered in Chapter 14.

Non-CO_2 GHG emissions from agricultural production are rather high as they represent, on average, 12.7% of the emissions of the country between 1985 and 2008, and amounted to 8,783 Gg CO_2-equivalent in 2008. The emissions, however, have been decreasing in both absolute terms and their share in total emissions.

The estimation is based on the basis of the methodology recommended by the GPG and Uncertainty Management in National Greenhouse Gas Inventories (IPCC 2000). This methodology takes the number of animals and production yield by livestock categories, harvested production, harvested area, and fertilizer use in plant production (mainly from official agricultural production data of HCSO). These activity data are converted to emissions by the appropriate emission factors. In addition to country-specific factors based on literature, and production statistics data and expert judgments, default values of the IPCC (2000) were also used (for details see Borka and Lovas, 2010).

More than 99% of above the non-CO_2 emissions come from three sources: enteric fermentation by livestock (CH_4, 19.3%), manure management (CH_4 and N_2O, 24.3%), and agricultural soils (N_2O, 68.9%). Negligible amount of methane is generated in rice cultivation. Field burning of agricultural residues has been essentially eliminated in Hungary since 1986, as it is only permitted for plant health emergency reasons.

The trend of all non-CO_2 emissions (in CO_2eq) between 1985 and 2008 is demonstrated in Fig. 15.5. Emissions from agricultural production decreased by 49.8% over the period of 1985–2008. The bulk of this decrease occurred in the years between 1985 and 1995, when agricultural production fell by more than 30%, and livestock numbers underwent a drastic decrease. The contribution of agricultural production to total emissions decreased over the period 1985–2008 from 15.3% to its present share of 12.0%.

Between 1996 and 2008, emissions from agricultural production stagnated around 9 Mt CO_2 equivalent per year with fluctuations of up to 5%. The number of livestock decreased that resulted in lower emission; on the other hand, the use of fertilizers increased by about 60%, which increased the nitrous-oxide emissions from agricultural soils. Note, however, that the fertilizer use of the Hungarian agriculture sector is still only slightly higher than the half of the amount in the period between 1980 and 1985.

In case of the agricultural production, it is also possible to estimate uncertainties of the emission estimates. The procedure is based on IPCC (2000). Table 15.8

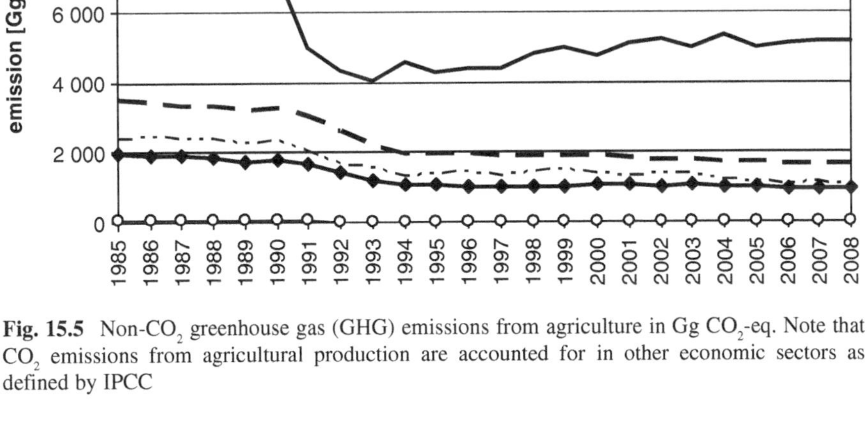

Fig. 15.5 Non-CO_2 greenhouse gas (GHG) emissions from agriculture in Gg CO_2-eq. Note that CO_2 emissions from agricultural production are accounted for in other economic sectors as defined by IPCC

Table 15.8 Uncertainty of activity data, emission factors, and estimated emissions for non-CO_2 emissions from the agricultural sector.

Source of emission	Greenhouse gas (GHG)	Combined uncertainty of activity data (%)	Uncertainty of emission factor (%)	Combined uncertainty of emissions (%)
Enteric fermentation	CH_4	1.0–16.3	50	28.6
Manure management	CH_4	10.1–25.8	50	42.2
	N_2O	51.0–56.3	150	70.1
Rice cultivation	CH_4	201.6	80	216.9
Direct soil emissions	N_2O	34.6–57.9	250	154.1
Pasture, range, and paddock manure	N_2O	29.2	150	152.8
Indirect soil emissions	N_2O	54.6–74.0	50	61.2

demonstrates that the highest uncertainties are principally associated with the emission factors, and that the overall uncertainties are also rather high.

15.9 CO, CH_4, N_2O, and NO_x Emissions from Biomass Burning (Burning of Slash in Forests)

Non-CO_2 emissions from burning originate from onsite burning of slash on FL. Emissions from this burning are not significant and are only estimated for the sake of completeness. Note that CO_2 emissions due to this burning are accounted for in the biomass pool, because the stock-change method covers all carbon stock changes. Also note that, theoretically, the CO_2 emission estimates include the carbon of CO and CH_4, too; however, these gases are nevertheless estimated (complying with the methodology of the GPG for LULUCF, IPCC, 2003) because their global warming potential is much higher than that of CO_2.

The estimation methodology is based on the method suggested by IPCC (2003), Equation 3.2.19, as well as by the IPCC (1997) Guidelines. Carbon released is estimated using harvest statistics (m^3 of wood removed from forest, from which the amount of slash was calculated using average values by species, developed in a former country-wide specific project for statistical purposes). In addition, expert judgment was applied with respect to the fraction of slash burnt on site (0.2), and to the fraction that oxidized on site (0.9). Finally, the IPCC default values were used for the carbon fraction of harvested wood (as above). The product of these values is multiplied first by default emission ratios by gas: 0.012 for CH_4, 0.06 for CO, 0.007 for N_2O, and 0.121 for NO_x. Then, for nitrogen compounds a general default value of 0.01 is applied to yield the total amount of nitrogen (N) released. Finally, the products obtained are multiplied by the appropriate molecular weight ratios, which are the following: 16/12 for CH_4, 28/12 for CO, 44/28 for N_2O, and 46/14 for NO_x The estimated emissions are included in the total emission estimates below.

15.10 Net Emissions of the Terrestrial Biosphere

The net of all the above emissions and removals highly depends on how the "biosphere" is defined. Clearly, almost all major elements of the terrestrial biosphere are managed; therefore, if "management" is the only criterion, then the "unmanaged" biosphere has a very tiny GHG flux, and the "managed" biosphere is a net emitter (cf. Table 14.2 in this book). The majority of the emissions (expressed in CO_2-eq.) is released by agricultural soils (N_2O and CO_2), followed by CH_4 emissions from enteric fermentation and N_2O emissions from manure management. Trends of total net emissions are shown on Fig. 15.6.

If, however, these rather intensive sources are excluded, and only "more natural" ecosystems such as forests are concerned, then the "biosphere" is a net sink, and by far the largest amounts of CO_2 are removed from the atmosphere by forests.

Irrespective of how "biosphere" is defined, both removals and emissions have rather high interannual variability. Also, the rate of emissions and removals heavily depends not only on natural factors, but also on human interventions. Therefore, in order that relevant fluxes can be captured, these interventions must continuously be monitored. Finally, as the above information is rather incomplete, further research efforts are necessary to extend the estimation especially with regard to the forest soil and dead organic matter pools.

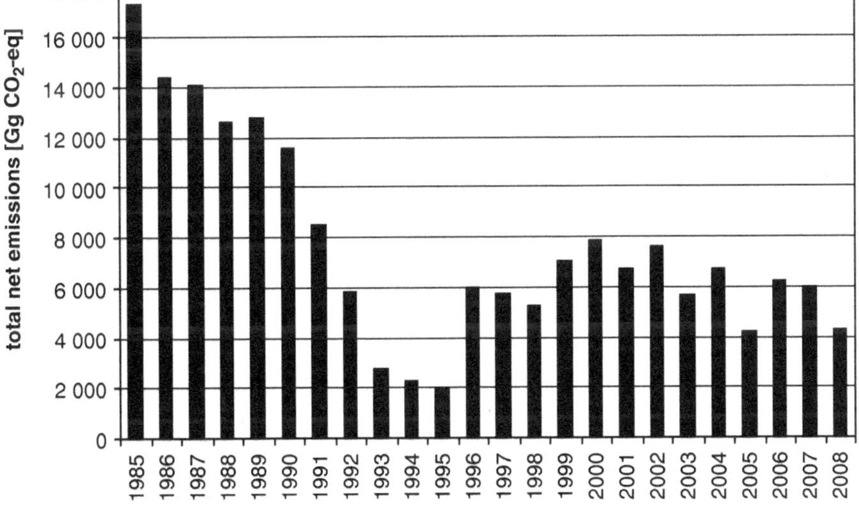

Fig. 15.6 Total net emissions from the terrestrial biosphere as managed land in Hungary

References

AERI (2010) The statistical database of the Agricultural Economics Research Institute, URL: http://www.akii.hu

Birkás M (2002) Environment-friendly and energy-econom soil management (in Hungarian). Akaprint Kiadó, Budapest, p 345

Birkás M, Kalmár T, Fenyvesi L, Földesi P (2007) Realities and beliefs in sustainable soil tillage. Cereal Res Commun 35:257–260

Borka G, Lovas K (2010) National Inventory Report for 1985–2008 – Hungary, Part Agriculture. In: Kis-Kovács, G (ed) National Inventory Report for 1985–2008 – Hungary. Hungarian Meteorological Service, Greenhouse Gas Inventory Division, Budapest, April 2010, http://unfccc.int/files/national_reports/annex_i_ghg_inventories/national_inventories_submissions/application/zip/hun_2009_nir_15apr.zip

Forgács L, Zsembeli J, Tuba G (2005) Examination of a soil protective cultivation method in the Research Institute of Karcag. In: Realizáciou poznatkov vedy a vyskumu k trvalo udrzatel'ného pol'nohospodárstvu. Michalovce, Slovakia, pp 64–68

Gyuricza Cs, Földesi P, Mikó P, Ujj A (2005) Carbon dioxide emission from arable lands. Cereal Res Commun 33:89–92

HCSO (2009) http://portal.ksh.hu/pls/ksh/docs/eng/agrar/html/tabl1_3_1.html (retrieved 15 January, 2010)

Horvath B (2006) Kohlenstoff-Akkumulation im Boden nach Neuaufforstungen: Beitrag zur Reduzierung der C-Emission in Ungarn? (C-accumulation in the soil after afforestation: contribution to C-mitigation in Hungary?). Forstarchiv 77:63–68

IPCC (1997) In: Meira Filho LG, Tréanton K, Mamaty I, Bonduki Y, Griggs DJ, Callander BA, Houghton JT (eds) Revised 1996 IPCC guidelines for national greenhouse inventories. IPCC/OECD/IEA, Paris

IPCC (2000) In: Penman J, Kruger D, Galbally I, Hiraishi T, Nyenzi B, Emmanuel S, Buendia L, Hoppaus R, Martinsen T, Meijer J, Miwa K, Tanabe K (eds) Good practice guidance and uncertainty management in national greenhouse gas inventories. IPCC/OECD/IEA, IGES, Hayama, Japan

IPCC (2003) In: Gytarsky M, Krug T, Kruger D, Pipatti R, Buendia L, Miwa K, Ngara T, Tanabe K, Wagner F, Penman J (eds) Good practice guidance for land use, land-use change and forestry. IGES, Hayama, Japan

IPCC (2006) In: Eggleston HS, Miwa K, Ngara T, Tanabe K (eds) 2006 IPCC guidelines for national greenhouse gas inventories. IGES, Hayama, Japan

Kecskés Cs (1997) Methodological changes in land-use statistics (in Hungarian). Statisztikai szemle 75:1047–1061

NFD (2010) National Forestry Database, managed by the Central Agricultural Office Forest Directorate, Budapest

Somogyi Z (2008) Greenhouse gas inventory of the Hungarian forests. Erdészeti kutatások 92:145–162

Somogyi Z (2010) Advancements in understanding and demonstrating forest carbon cycle. Reserach report for the Ministry of Agriculture, Budapest

Somogyi Z, Cienciala E, Merganic J, Merganicova J, Illes G (2005) Guidelines for monitoring and verification of carbon removals in Afforestation/Reforestation Joint Implementation projects. Improved monitoring standards for LULUCF projects. Analysis of case study / test site in Hungary. CarboInvent research report (Deliverable 8.5). (CarboInvent: Multi-Source Inventory Methods For Quantifying Carbon Stocks And Stock Changes In European Forests; Contract number EVK2-CT-2002-00157, see http://www.joanneum.at/CarboInvent), pp. 88. URL: http://www.joanneum.ac.at/carboinvent/D_8_5.pdf

Sopp L, Kolozs L (eds) (2003) Volume tables for the main Hungarian tree species. ÁESZ, Budapest, In Hungarian

Tóth E, Koós S (2006) Carbon dioxide emission measurements in a tillage experiment on chernozem soil. Cereal Res Commun 34:331–334

Zsembeli J, Kovács Gy (2007) Dynamics of CO_2-emission of the soil in conventional and reduced tillage systems. Cereal Res Commun 35:1337–1340

Zsembeli J, Tuba G, Forgács L (2005) CO_2-emission of soil in an energy saving soil cultivation system. In: Realizáciou poznatkov vedy a vyskumu k trvalo udrzatel'nému pol'nohospodárstvu. Michalovce, Slovakia, pp 59–63

Zsembeli J, Tuba G, Kovács Gy (2006) Development and extension of CO_2-emission measurements for different soil surfaces. Cereal Res Commun 34:359–362

Chapter 16
Energy Production, Industrial Processes, and Waste Management*

Klára Tarczay, Edit Nagy, and Gábor Kis-Kovács

Abstract Pursuant to the United Nations Framework Convention on Climate Change, Hungary, as a Party of the Convention, prepares annual inventories of greenhouse gas (GHG) emissions. This chapter discusses the nonbiospheric anthropogenic sources of the Hungarian greenhouse gas inventory. The emission estimations are calculated – following the mandatory methodology compiled by the Intergovernmental Panel on Climate Change – using specific emission factors and activity data. For the most important sources, it is recommended to report more accurate estimation; therefore, country- or plant-specific emission factors are developed and presented in this chapter.

The largest source of GHG in the Hungarian inventory is the combustion and handling of energy carriers, namely the *Energy* sector; the related emission was 55,476 Gg in 2008 expressed in CO_2-equivalent, which covers 76% of the total GHG emission of the country. The *Industrial Processes* and the *Solvents and other product use* sectors emitted together 5,154 Gg of GHGs, while the *Waste* sector represented the smallest source with 3,725 Gg (5%). The chapter gives a general idea of the main processes and methodologies used for the emission calculations of these sectors.

Keywords Greenhouse gas inventory • Energy sector • Industrial processes sector • Solvents • Waste sector

---

*Citation: Tarczay, K., Nagy, E., Kis-Kovács, G., 2010: Greenhouse gas emissions and removals in Hungary based on IPCC methodology – Energy production, industrial processes, and waste management. In: Atmospheric Greenhouse Gases: The Hungarian Perspective (Ed.: Haszpra, L.), pp. 367–386.

K. Tarczay (✉), E. Nagy and G. Kis-Kovács
Hungarian Meteorological Service, H-1525 Budapest, P.O. Box 38, Hungary
email: tarczay.k@met.hu

L. Haszpra (ed.), *Atmospheric Greenhouse Gases: The Hungarian Perspective*,
DOI 10.1007/978-90-481-9950-1_16, © Springer Science+Business Media B.V. 2011

16.1 Introduction

This chapter summarizes the most important methods of the emission calculations applied for the nonbiospheric anthropogenic sources in Hungary, and the results of these calculations. The emission figures published in this chapter refer to 2008 primarily, but trends from 1985 to 2008 are also shown in the graphs.

In the following sections, detailed descriptions are presented about the energy, industry, and waste sectors of the national greenhouse gas (GHG) inventory. These sectors covered almost 90% of the national total greenhouse gas emissions in 2008, which is illustrated by Fig. 16.1. The main source of CO_2 in the inventory is the *Energy* sector. CH_4 emerges from waste in the highest degree. The importance of the *Industrial process* sector has been decreasing since the base year (average of 1985–1987 for CO_2, CH_4, and N_2O; 1995 for F-gases); however, the so-called F-gases (PFCs, HFCs, and SF_6), which are emitted only by industrial processes, show increasing tendency.

16.2 *Energy* Sector

16.2.1 Methodology of Emission Estimates in the Energy Sector

The *Energy* sector covers emissions from combustion processes and fuel-related fugitive emissions from exploration, transmission, distribution, and conversion of primary energy sources.

Calculations of emission from fuel combustion are based on energy consumption data, since CO_2 emission mainly depends on carbon content of fuels (see Table 16.1). This method is quite reliable in contrast with emission estimates of other GHGs in this sector, and also to methods of other sectors, where emissions

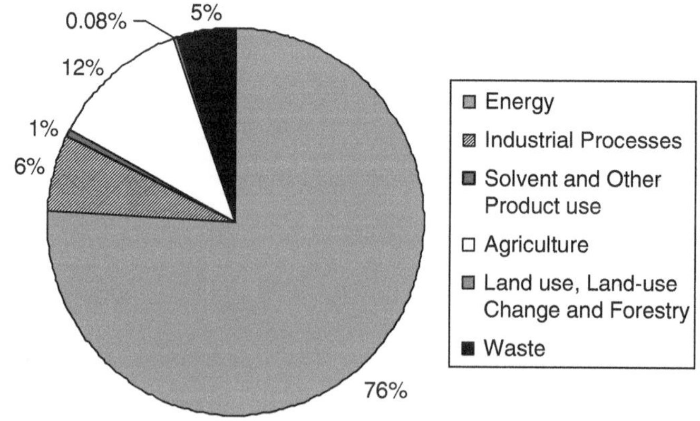

Fig. 16.1 Sectoral proportion of total national anthropogenic emissions in 2008 (excluding CO_2 from Land Use, Land-Use Change, and Forestry)

Table 16.1 Emission and oxidation factors of selected fuel types used in the energy industries

Fuel type	Emission factor ($t\ CO_2\ TJ^{-1}$)	Oxidation factor
Coking coal	94.6	0.98
Other bituminous coal (hard coal)	95.71[a]	0.98
Lignite (brown coal and lignite)	107.5[a]	0.977[a]
Brown coal briquette	94.6	0.98
Coke oven/gas coke	108.17	0.98
Coke oven gas	43.49[a]	0.98
Crude oil	73.33[a]	0.99
NGL	63.07	0.99
Gasoline	69.3	0.99
Jet kerosene	71.5	0.99
Gas/diesel oil	74.06[a]	0.99
Residual fuel oil	81.95[a]	0.99
LPG	63.07	0.99
Bitumen	80.67	0.99
Petroleum coke	100.83	0.99
Other oil	80.5[a]	0.99
Natural gas	56.1	0.995

[a] Country- or plant-specific factors

are more sensitive to technology/source-specific calculations (Rypdal and Winiwarter 2001). Fuel-related fugitive emissions can be estimated only loosely and the bias of estimation from the real value can be an order of magnitude.

It should be noted that CO_2 emission from biomass burning is not included in the *Energy* sector; it is mentioned among "memo items"[1] in the inventory. However, waste incineration for energy purposes is accounted in this sector.

Emissions in the *Energy* sector are estimated with two approaches (reference and sectoral) according to the instructions of the guidelines (IPCC 1997, 2000). The reference approach is based on national energy balance: production, import, export, stock changes, and international bunkers[2]. It is necessary to use calorific values to harmonize information across all fuels. In our case, the apparent consumption of the energy balance components mentioned is already given in calorific value, which makes the calculations easier, because default emission factors, which are available in the IPCC guidelines (IPCC 1997, 2000) with uncertainty range, are developed for this kind of activity data.

[1]Emissions from combustion of biomass and international bunkers are excluded from the national total GHG emissions, but must be reported separately. These sources can be found under the category "memo items" in the national inventory reports.

[2]Methodology for accounting the international bunkers (aviation, navigation) is not adopted by COP yet (FCCC/SBSTA/2009/L.15). The current calculation is based on fuel sold in national territories and it is indicated under "memo items" in the inventory, separated from other fuel consumptions.

The amount and carbon content of long-lived materials manufactured from fuel are subtracted from apparent consumption, because they store their carbon contents for several years (depending on the nature of use), and in the end of their life cycle they appear in the *Waste* sector. Emissions from feedstocks and non-energy use of fuels are taken into account in the *Industrial Processes* sector.

Usually, the combustion process is not complete in reality, so small amount of carbon must be discounted – which means a multiplication with an oxidation factor (see Table 16.1) – to consider the unoxidized part of fuel. This factor is technology- and fuel-specific; it is known fairly accurately for gaseous fuels and oils. It is more dependent on combustion conditions for coal and other solid fuels, and can vary by several percent.

It is relatively easy to use the reference approach, which gives good results for CO_2 emissions from fuel combustion, and it is internationally transparent, but it has two disadvantages, namely stationary emissions cannot be distinguished from mobile combustion emissions, and methane and nitrous oxide emissions cannot be estimated with aggregated activity data.

Sectoral approach allocates the emissions by source category and allows to use more technology-specific information. This approach is needed for atmospheric monitoring and abatement policy discussions. However, the source categorical results are more uncertain than total emission because the distribution of fuel consumption can have higher statistical uncertainty.

Estimates of the two approaches are not exactly the same, but the differences between them should not be significant. For national totals, the sectoral approach is taken into account, but for verification the results of the reference approach are also reported in the inventory with comments on the differences between the two approaches.

The main source of activity data for these calculations in our case is the energy balance of Hungary including the fuel balance for each fuel type and fuel consumption for each subcategory prepared by Energy Efficiency, Environment and Energy Information Agency Non-Profit Company (Energy Centre 1986–2010).

16.2.2 Background of the Trends in the Energy Sector

The trend of energy demand and associated emissions (see Chapter 14) depend on joint impact of several factors, such as the changes in economical structure, weather conditions during the heating period, growing energy demand of transport, improvement of energy efficiency in the *Manufacturing industries* and *Energy industries* categories, fuel-mix changes, growing energy prices, energy and environmental policies.

Recently, one of the most favorable environmental policies was the fuel-mix change (see Fig. 16.2) in all source categories of the *Energy* sector. If we compare the specific emission factors (see Table 16.1) of different fuel types, it is obvious that significant emission reduction can be achieved in this way without structural changes in the energy production. The results of these changes can be seen in

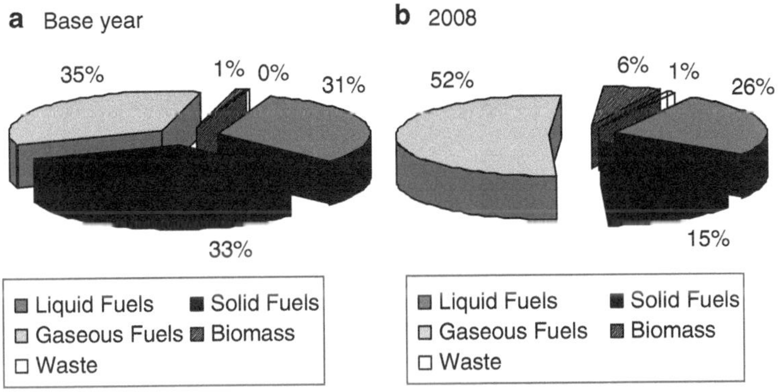

Fig. 16.2 Share of the combusted fuel types by energy produced (Base Year [1985–1987] – *left*, 2008 – *right*)

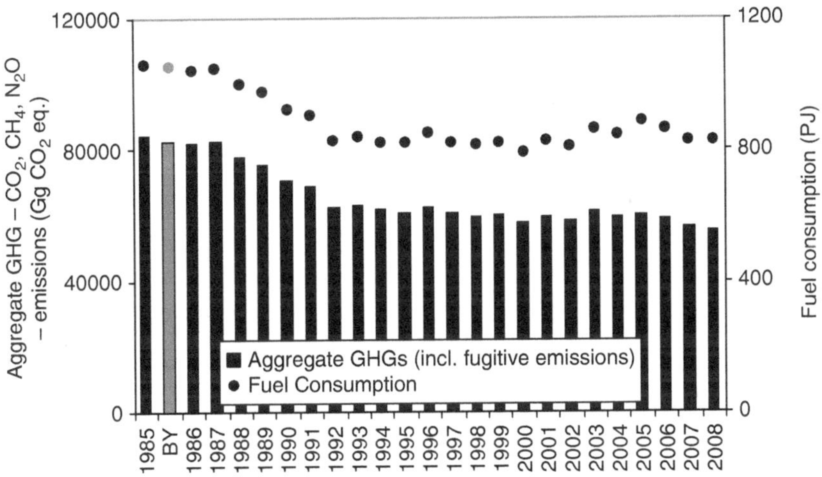

Fig. 16.3 Emission and fuel consumption trends in the *Energy* sector (BY – Base Year [1985–1987])

Fig. 16.3, where the total GHG emission of the sector does not follow the trend of total energy consumption from 1992.

The *Energy* sector emitted 55,476 Gg CO_2-eq,[3] which is equal to 76% of total GHG emissions (excluding CO_2 from LULUCF) in Hungary (see Fig. 16.1), and stationary combustion was responsible for 73% of the GHG emissions from the

[3]For the purpose of official submissions of national GHG inventories under the UNFCCC, Annex I Parties are required to use the global warming potential (GWP) values provided by the IPCC in its Second Assessment Report based on the effects of GHGs over a 100-year time horizon, as it was decided by the Conference of the Parties at its 5th session. Under the Kyoto Protocol, the Conference of the Parties also decided (decision 2/CP.3) that the values of GWP calculated for the IPCC Second Assessment Report are to be used for converting the various greenhouse gas emissions into comparable CO_2 equivalents.

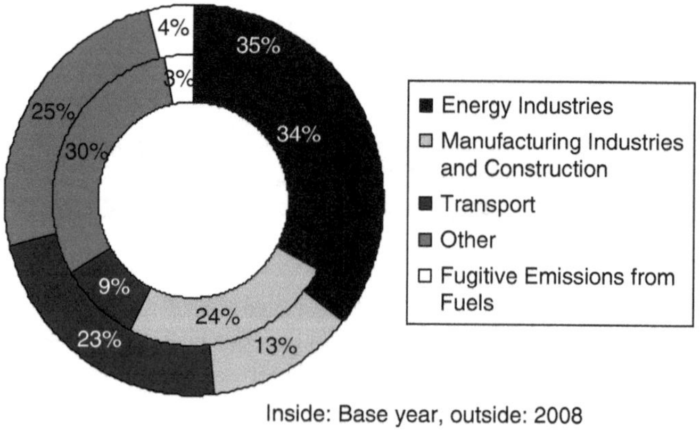

Fig. 16.4 Shares of GHG emissions (CO_2-eq) in the *Energy* sector by source categories (inside: base year [1985–1987], outside: 2008)

Energy sector. 49% of these stationary emissions were associated with combustion in *Energy industries*, mainly by power plants and refineries. Mobile combustion (road and other traffic) caused 23% of the emissions in the *Energy* sector with gradually increasing tendency (see Fig. 16.4). These figures are in line with the findings of IPCC for developed countries (IPCC 2006).

Further examining the rank of source categories, it can be found that the *Other* category with 25% was the second largest emitter. The greatest final energy consumption including electricity – its emission is accounted in the *Energy industries* category – was associated with *Residential* and *Commercial/Institutional* categories of *Other* category; so the highest emission reduction can be realized in these categories. Because of these facts, the cost-effective emission reduction potential of buildings was also a very important question among mitigation options in the Fourth Assessment Report of the IPCC (IPCC 2007).

16.2.3 Energy Industries

The *Energy industries* category includes facilities generating electricity, district heating stations, oil refineries, and coking and briquetting plants. On an overall level, they are the largest energy consumers and the largest emitters (see Fig. 16.3). In 2008, 36% of the domestic energy consumption, of which 92% was of fossil origin (including wastes), was used by energy industries. The total fuel consumption of this source category has a strong fluctuation, but it was lower by 14% in 2008 compared to the 1985–1987 average. Within this, the consumption of liquid and solid fuels decreased significantly. In contrast, the consumption of natural gas increased to a small extent. The biomass use has become more and more important due to the so-called co-burning in power plants, and by now it has overtaken the consumption of liquid fuels in amount. This change is induced partly by the

European Union Emission Trading Scheme (EU ETS), which is based on Directive 2003/87/EC and entered into force on October 25, 2003. This law came into force in the Hungarian legal system in 2005 (Act XV of 2005 on trading with greenhouse gas allowances, http://www.kozlonyok.hu/nkonlinc/MKPDF/2005/MK54.pdf).

The Act XV of 2005 appoints which installation has to join the EU ETS. Establishments emitting more than 500 kt CO_2 year^{-1} are required to get the calorific value, the carbon content, and oxidation factor of used fuel measured in an accredited laboratory. With the evaluation of the measurements, it is possible to define more relevant emission factors (Table 16.1) that suit better to the Hungarian conditions, and can be used instead of IPCC default emission factors. Nowadays, the calculations of emission in the *Energy industries* category consider these measurements improving the quality of the inventory.

16.2.4 Manufacturing Industries and Construction

This source category covers emissions from combustion of fuels in the industrial sector. Since the EU ETS data are available, more accurate emission estimates can be carried out with measured and verified carbon content of fuels. For solid fuels, we calculate the emission from plant-specific information in categories, for which the EU ETS covers the whole sector. In other cases, the reported data are used only for QA/QC activities or to appreciate the uncertainty of activity data and emission factors.

16.2.5 Transport Activities

Transport category covers all emission from fuels used for transportation purposes. It should be noted that emission estimates are bound to the principle of political responsibility (allocation according to fuel sale in national territory); this should be taken into account in comparisons with other emission inventories or atmospheric modeling.

CO_2 emission from transport is calculated with default IPCC emission and oxidation factors. Other GHG emissions from railways and navigation estimated with country-specific factors chosen by experts. Calculation of CH_4 and N_2O emissions from road transport considers the different control technology types of the vehicle fleet, but uses default IPCC emission factors.

The composition of the national passenger car fleet underwent considerable changes after 1990. The proportion of Eastern European cars characterized by high fuel consumption and high methane emission has decreased; in 2008, about 80% of the gasoline fueled vehicles were more advanced cars, which fulfilled the NO_x reduction commitments of the EU, but as a side effect they emitted more nitrous oxide than the former types.

Electrification of the railways in Hungary decreased the solid fuel consumption by 99.5% compared to the base year (average of 1985–1987). Today, there are only few lines where steam engines provide occasional nonscheduled service; diesel engines are being taken out at the moment. The switch to electricity has allocated the emission from the *Transport* category to the *Energy industries* one. In spite of this change, in contrast to the other sectors described, transport consumption and emission show a rising overall tendency caused by road transport (see Fig. 16.5).

International aviation and navigation are excluded and reported separately under "memo items."

16.2.6 The Other Source Category

The source category called *Other* is a collective noun for the following categories: *Residential, Institutional, Commercial,* and *Agriculture, Forestry and Fisheries.* CO_2 emissions are calculated mostly using default emission and oxidation factors. The only exception is the emission from Hungarian lignite because measurements of its carbon content are available for the single active mine from the EU ETS database. Other GHGs are estimated with factors taken from international literature chosen by energy experts. Since combustion appliances in these categories are relatively small, presumably their methane and nitrous oxide emissions are higher compared to the emissions of autoproducer or public power plants, where the combustion process is controlled to meet the standards.

Since more than 60% of the fuel consumption in the *Other* category was related to the residential subcategory, the fuel structure was influenced principally by changes in this subcategory. In contrast to the significant reduction of coal and oil

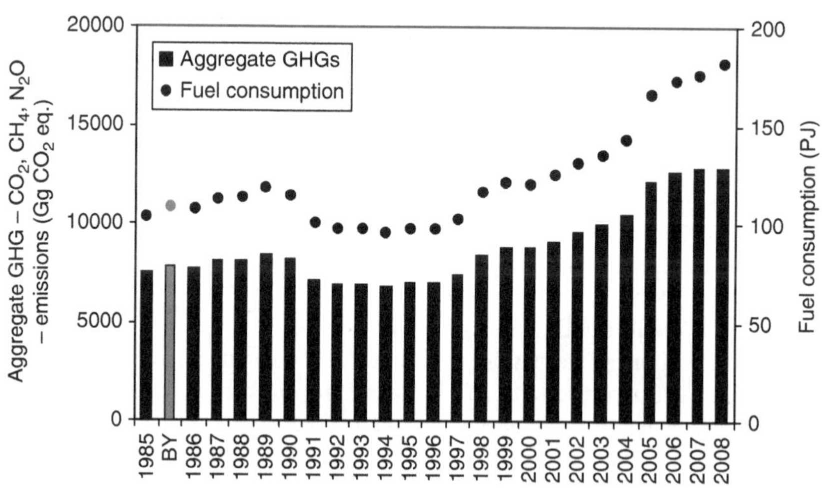

Fig. 16.5 Trends of fuel consumption and emissions in the *Transport* source category

consumption, the quantity of natural gas used has increased dynamically since 1985. During the period 1985–2008, natural gas pipelines' length doubled, and the number of households supplied with natural gas has been increasing continuously. The population switched from coal to natural gas combustion. Houschold heating oil was completely replaced by LPG during the last years of the period analyzed.

As the dominant fuel is natural gas in the *Other* category, the following basic statistical data will help to get acquainted with the Hungarian situation (HCSO 2009). *Residential* and *Commercial/Institutional* consumption nowadays represents 58% of the total quantity of natural gas used (transported by pipelines). 91.1% of the settlements were supplied with this facility in 2008. Some 84% of the households use natural gas for heating purpose.

Heating degree day (HDD) is a quantitative index, which reflects the demand for energy to heat houses and businesses. This index is derived from daily temperature observations. The following equation was used to determine HDD in the inventory[4]:

$$HDD = \sum_{i} (18^{\circ}\mathrm{C} - T_{mean,i}) \text{ if } \mathrm{T}_{\mathrm{mean,i}} \leq 15^{\circ}\mathrm{C},$$

where i is the number of the day, while $\mathrm{T}_{\mathrm{mean,i}}$ is the daily mean temperature on the ith day of the year.

Fig. 16.6 illustrates the relationship between residential fuel consumption and HDD for the period 2000–2008.

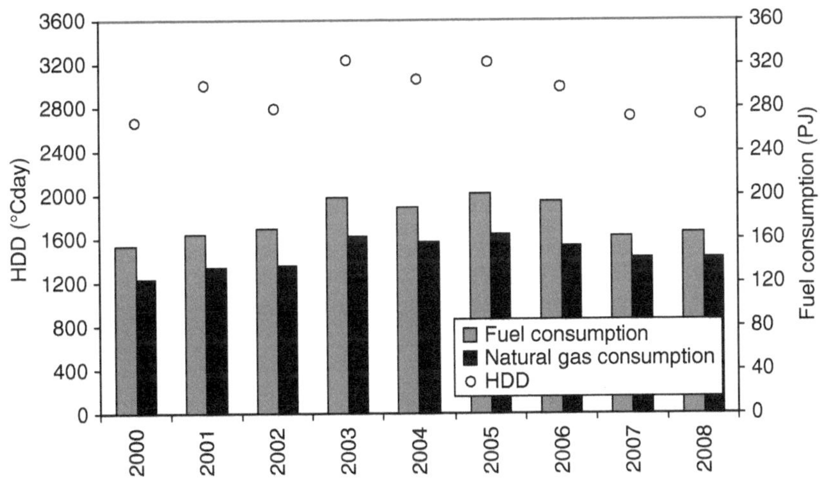

Fig. 16.6 Comparison of residential fuel consumption and heating degree day (HDD) between 2000 and 2008

[4]The inside temperature is 18°C and the base temperature (the outside temperature above which a building needs no heating) is 15°C in our calculation following the standard European methodology (EC 2007).

16.2.7 Fugitve Emissions from Fuels

The source category *Fugitive emissions from fuels* includes CO_2, CH_4, and N_2O emissions released during coal mining and handling, and also oil and natural gas related activities (exploration, production, processing, distribution, and transmission).

In Hungary, both underground and surface coal mines are present. Although underground mining was the predominant form in the past, it represented only 16% in the mass of the coal mined in 2008. Drastic reduction in coal production was observed between 1987 and 1988, as well as between 1989 and 1990. Underground mining continues to decrease in both relative and absolute terms.

In the past, oil production and processing was an important sector in Hungary, but the importance of the production is declining as the reserves are running out. Gas mining shows similar tendencies, although the reduction is less intensive. At the same time, natural gas uses show a significant increase covering with sharply growing import. The distribution loss from gas transport represents the highest proportion in the emissions.

Emission factors were taken into consideration according to the information from Mining Bureau of Hungary and measurement data from mines. In 2008, there was only one operating underground mine with 0.93 m^3 ton^{-1} in situ methane content of coal. Lignite is mined only in surface mines; where – based on the measurement data (REKK 2004) – methane is not emitted during mining activity, since the Hungarian lignite is relatively young in the coalification (NCV is under 10 MJ kg^{-1}). In case of postmining activities, the emission factor was calculated to be 10% of that of the mining activities according to the IPCC 2000 Guidance (IPCC 2000).

Activity and consumption data related to extraction and primary handling of oil and natural gas were taken from the Energy Statistics Yearbooks – prepared by the Energy Centre – for the period 1985–2008 (Energy Centre 1986–2010). In addition, data for the same period were used from Statistical Yearbooks of Hungary published by the Hungarian Central Statistics Office (HCSO 1986–2009) and from production companies. The technologies used in Hungary are entirely based on state-of-the-art equipment; therefore, the use of the specific emission factors for Eastern Europe, which are high and associated with great uncertainty, is not justifiable. Since we do not have own measurements, it was decided – on the basis of the data available from the production companies – that the Canadian calculation presented in the Background Papers published by IPCC (IPCC 2002) would be used. Hungarian data for the activities in this calculation were determined and multiplied by the specific emission factors provided. Flaring was estimated on the basis of detailed production data obtained from oil and gas companies and using default emission factors of the IPCC 2006 guidelines (IPCC 2006).

In Hungary, and especially in the Hungarian Great Plain, subsurface water and deep wells drilled for various purposes contain varying quantities of methane. Upon the abstraction of such waters (as drinking and/or as thermal water), methane is also abstracted and released into the atmosphere. According to a previous expert estimate, the annual quantity of methane released from wells is approximately 20 Gg.

We believe that this item should also be included in the methane emissions for the sake of completeness. However, it does not have an appropriate "slot" in the inventory.

16.3 *Industrial Processes* and *Solvent and Other Product Use* Sectors

16.3.1 Emissions from Industrial Processes

Industrial Processes sector is composed of a variety of industrial activities, which are not related to *Energy* sector. The main sources are the industrial production processes, which chemically transform materials. During these technological processes, several different greenhouse gases, such as carbon dioxide, methane, and nitrous oxide, can arise. Different HFCs (hydrofluorocarbons), PFCs (perfluorocarbons), and SF_6 (sulfur hexafluoride) are also consumed as alternatives to ozone-depleting substances in various applications. The related emissions are also included in the *Industrial Processes* sector. Emissions from this sector (Fig. 16.7) comprise the following source categories: *Mineral Products, Chemical Industry, Metal Production, Paper and Food Production, Production and Consumption of Halocarbons and SF_6.* The major processes and main sources (Fig. 16.8) are cement, ammonia, nitric acid, glass, brick and ceramics, iron and steel production. In the *Solvent and Other Product Use* sector, emissions from paint and similar material uses and technologies, which are related to the use of N_2O, are calculated.

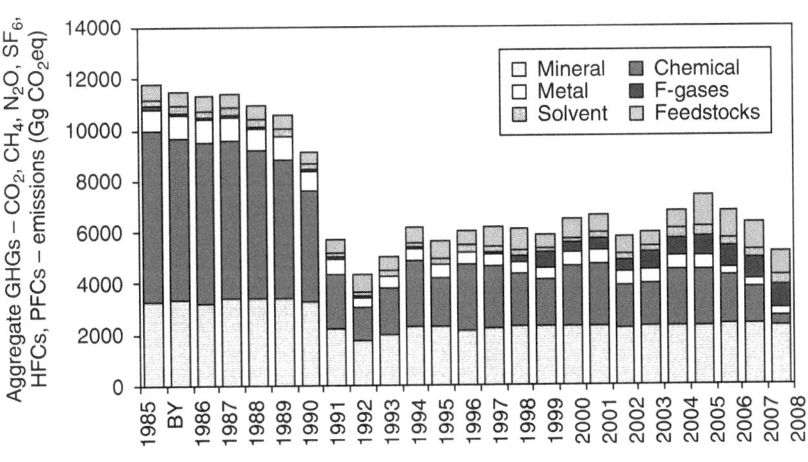

Fig. 16.7 Greenhouse gas (GHG) emissions from the *Industrial Processes* sector

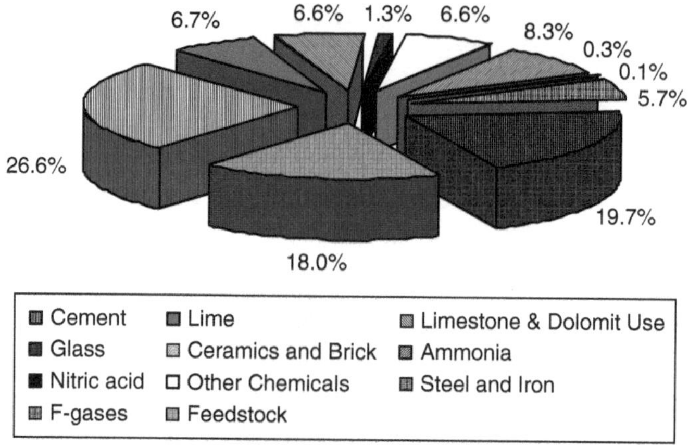

Fig. 16.8 Main sources in the *Industrial Processes* sector in 2008 (CO_2-eq)

The *Industrial Processes* and *Solvents and other product use* sectors emitted together 5,154 Gg of GHGs in CO_2-equvivalent (7%) in 2008. The following sections explain the basis of the calculations.

16.3.2 Mineral Products

Cement is a basic building material, which was already used by the Romans (Hill 1984). Cement factories contribute by 5% to the global CO_2 emission (Worrell et al. 2001). Approximately 50% of the global cement industry's CO_2 emission originates from processes of decomposition of limestone (Hendriks et al. 1999), the remainder part comes from fuel combustion for heat generation to the technological process. In Hungary, this source is responsible for about 30% of emissions in the *Industrial Processes* sector.

The product's raw material is limestone ($CaCO_3$), which is ground into raw meal in cement manufactures. Ground raw meal is fed into rotating kiln and is calcinated by at least 1,450°C to produce clinker and CO_2 as a by-product. Clinker is ground usually with the addition of different auxiliary materials like by-products or wastes from other industrial processes (e.g., fly ash, blast furnace slag, FGD-gypsum[5]) to produce cement. Substituting the amount of clinker in cement with auxiliary materials is one of the most effective ways to reduce emission; therefore, only 510 kg CO_2 per ton of clinker was released into the atmosphere during the production in Hungary in 2008.

[5]FGD-Gypsum is a synthetic product derived from flue gas desulfurization (FGD) systems at electric power plants.

16.3.3 Chemical Industry

The main sources in chemical industry of Hungary are ammonia (NH_3) and nitric acid (HNO_3) productions, where carbon dioxide and nitrous oxide are emitted. Traditional ammonia production uses natural gas, the carbon content of which is released in the form of carbon dioxide. Here, only emission from natural gas used as raw material is calculated, and emissions from combustion are taken into consideration under the *Energy* sector. Nitric acid is produced by oxidizing ammonia. The tail gas of the process contains N_2O and NO_x. In order to control the emission, NO_x is reduced to nitrogen using natural gas and the carbon content of natural gas is emitted as carbon dioxide (Pérez-Ramírez et al. 2003).

By 2008, this sub-sector underwent significant changes. Its emission decreased by 90% compared to the base year, which caused 5,639 kt CO_2 emission reduction, due to several reasons: principally, the production was significantly decreasing; in parallel, several factories were closed down. Due to the environmental investments in nitric acid production, the nitrous oxide content of tail gas considerably decreased. This investment was realized in the frame of Joint Implementation Project[6] in a new acid plant, which started the production in 2007 with taking out the out-of-date technologies. Selective catalytic reduction system (EnviNOx) for purifying tail gas from NO_x and N_2O was developed, where apart from ammonia small amount of natural gas is added to the tail gas to achieve better efficiency (Kovács et al. 2009).

16.3.4 Consumption of Halocarbons and SF_6

Since the mid-1990s, Hungary's F-gas consumption has been increasing (Fig. 16.9) due to the Montreal Protocol signed in 1987, which controls the phase-out of ozone-depleting substances' production and use. F-gases have been applying as substitutes for chlorofluorocarbons (CFCs), hydrochlorofluorocarbons (HCFCs), and halons to a growing extent. These materials are harmless for the ozone layer, but they increase the greenhouse effect because their global warming potential (GWP) is extremely high and their residence time is very long (IPCC 1996). F-gases are exclusively anthropogenic gases. They are not produced in Hungary, they are only imported.

[6]The mechanism known as "joint implementation," defined in Article 6 of the Kyoto Protocol, allows a country with an emission reduction or limitation commitment under the Kyoto Protocol (Annex B Party) to earn emission reduction units from an emission-reduction or emission removal project in another Annex B Party, each equivalent to one ton of CO_2, which can be counted towards meeting its Kyoto target. Joint implementation offers Parties a flexible and cost-efficient means of fulfilling a part of their Kyoto commitments, while the host Party benefits from foreign investment and technology transfer.

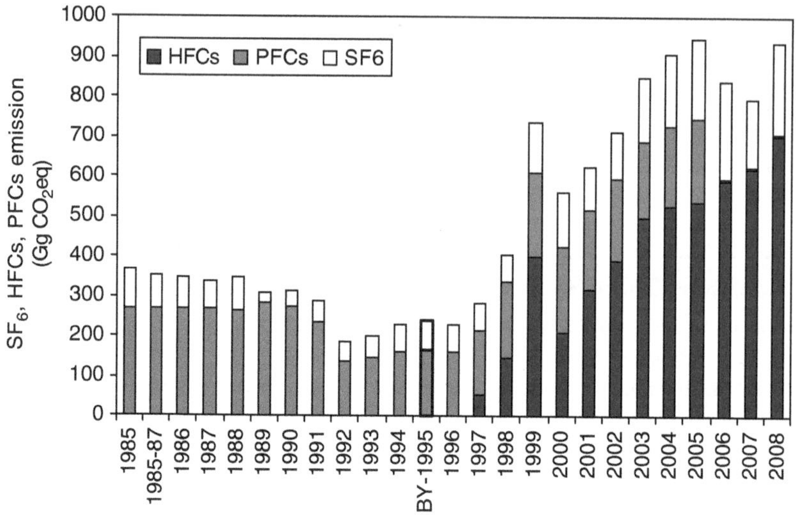

Fig. 16.9 Trends of F-gases (in accordance with Article 3.8 of the Kyoto Protocol, Hungary has chosen the year 1995 as the base year for the emissions of F-gases)

HFCs are used in household and commercial refrigeration equipments, in the production of foams used in construction/insulation industry, for cushioning and packaging, by the automotive industry and furniture manufacturers, for medical and technical sprays (as propellant gas). PFCs are applied as solvents or as an ingredient of cooling mixes, but they are rarely used.

SF_6 is also imported and is mainly used as an insulation gas in electrical switchboards. Furthermore, it is used as intermediate gas in double-glass heat insulation windows and production of optical bodies.

Activity data are obtained directly from the factories or importers. The applied methodology for each subcategory depends on the quality of the data and the scope.

HFCs are used for the production of both soft and hard foam. Since HFCs used for open-cell foam blowing are released immediately, all of the emissions will occur in the foam-manufacturing country.

According to the IPCC (1997; 2006) methodology, emissions from closed-cell foam occur in the following way: first-year losses from manufacturer – assuming 10% of the original HFC charge; these emissions occur where the product is manufactured. Then, closed-cell foam will lose a fraction of its initial charge each year (in situ losses from foam use) – assuming 4.5% of the original HFC charge per year – until decommissioning. Since we had no information about the decommissioning amount, it was assumed that all chemicals not emitted in manufacturing are emitted over the lifetime of the foam, in our case during 20 years.

The subcategory of *Refrigeration and Air Conditioning Equipment* includes the emissions related to domestic refrigerators and freezers, food refrigerators in shops, commercial and household air conditioners, refrigerated storages. Fugitive assembly emission does not occur when the equipments are filled because the system of filling is a closed one. Emission data, export–import data, refilled amount of domestic refrigerators are received directly from the manufacturers. In case of commercial and industrial equipments, distributors or trading companies were contacted to get information on the quantities used for filling new refrigerators and for refilling existing ones. For certain operators, the filling/refilling ratio was determined by estimation taking into account their activities. This refilled amount was taken as emission.

Nowadays, isobutane, also known as methylpropane (R600a), is used predominantly as refrigerant instead of F-gases. Its popularity has been significantly increased since 1994.

Most aerosol packages (metered dose inhalers, personal care products, household and industrial products) contain some kind of propellants. In a small fraction, HFCs or PFCs are used. Emission from aerosol packages occurs shortly after production, and all the initial charge escapes within the first year.

Emissions from primary aluminum production is not related directly to this subcategory, but it should be noted that PFCs like tetrafluoromethane, also known as carbon tetrafluoride (CF_4), and hexafluoroethane (C_2F_6) are emitted during the alumina electrolysis from cryolite as a result of anode effect when aluminum oxide concentration is low in the electrolyte of the reduction cell.

16.3.5 Solvents and Other Product Use

Paints and similar materials (e.g., lacquers, kits, glues) used by various industries and households contain organic solvents. During the surface treatment of objects, their solvent contents are evaporating. The emitted nonmethane volatile organic compound (NMVOC) and resultant CO_2 turn up in the inventory. Emissions, which are related to use of N_2O like in dry cleaning or metal degreasing, are also accounted here. N_2O is used for anesthesia and analgesia in Hungary. It is also filled into whipped cream canisters as aerosol propellant. N_2O leaves the human body almost unchanged – less than 0.004% is metabolized in humans (Hong et al. 1980) – therefore, the emission is equal to the total consumption of the last three activities mentioned.

16.3.6 Emissions from Feedstocks and Non-energy Use of Fuels

This category is created for calculating carbon dioxide emissions from fuels used as feedstocks or for other Non-energy purposes. The CO_2 emission arise from oxidization during use; however, methane emission is expected to be minor or not to occur at all.

It is an aggregated category, because the real consumers of these fuels (mostly oil products) are unknown. These kinds of oil products are widely used. Just a few examples: paraffin waxes are used for candles, corrugated boxes, paper coating, board sizing, adhesives, food production, packaging; lubricants are consumed in transportation and industry; white spirit, kerosene, some aromatics are applied as solvent, e.g., for surface coating (paint) and dry cleaning.

The amount of carbon dioxide released is estimated from the carbon content of fuels and fraction of carbon not stored based on figures provided by IPCC Guidelines (1997).

Bitumen or asphalt for road paving and roofing is taken into account in the appropriate industrial subcategory.

16.4 *Waste* Sector

The *Waste* sector contributes by around 5% to total national GHG emissions, and comprises the following source categories: *Solid waste disposal on land* (79%), *Wastewater handling* (19%), and *Waste incineration* (2%). The relative order of magnitude and trends of emissions are summarized in Fig. 16.10.

16.4.1 Solid Waste Disposal in Landfills

In case of a managed disposal, the waste is disposed in landfills (solid waste disposal sites, SWDS) where it is compacted and covered. Under these circumstances, anaerobic degradation occurs, during which methane and carbon dioxide is emitted. CO_2 generated in landfills is of biogenic origin and thus it is excluded from the inventory. Under the conditions prevailing in landfills, CO_2 generated from

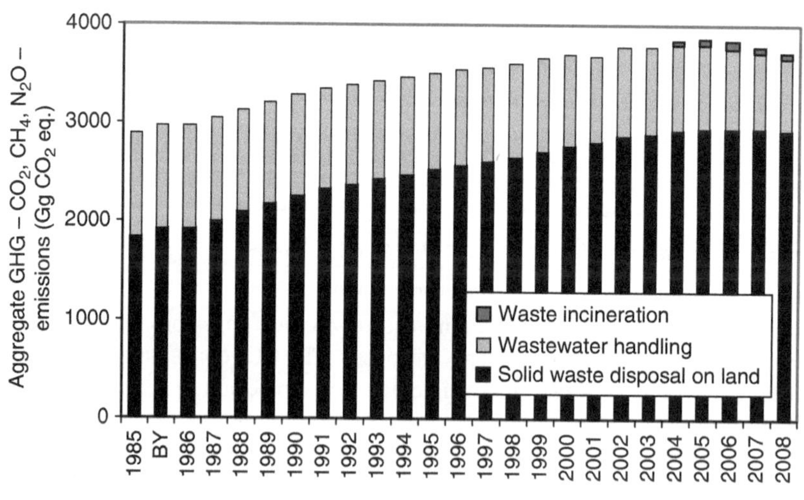

Fig. 16.10 Trend of emissions of the different categories in the *Waste* sector

waste containing carbon of fossil origin is insignificant. In advanced disposal sites, the generated methane is recovered by incineration or flaring. Degradation requires several decades and occurs at varying rates. Therefore, a bit more complex method, a first-order decay (FOD) model, is used for estimating CH_4 emissions, which produces a time-dependent emission profile acknowledging the experience that CH_4 is emitted over a long period of time rather than instantaneously. In a first-order reaction, the amount of product (methane emission, ultimately) is proportional to the amount of reactive material (the mass of degradable organic carbon deposited). This is expressed by the basic equation of the IPCC Waste Model (IPCC 2006):

$$\text{DDOCm decomp}_T = -\text{DDOCma}_{T-1} \cdot (1 - e^{-k})$$

where T = inventory year; DDOCm = mass of decomposable organic carbon deposited; DDOCma_{T-1} = DDOCm accumulated in the SWDS at the end of year (T−1), Gg; DDOCm decomp_T = DDOCm decomposed in the SWDS in year T, Gg; k = reaction constant, $k = \ln(2)/t_{1/2}$, $t_{1/2}$ = half-life time

It is important to note that in a given year, the amount of the decomposed carbon, which is closely related to the emitted methane, depends only on the total mass of decomposing material at the disposal site and not on the year of deposition. The accumulation of decomposable organic matter is influenced by the amount of waste deposited year by year, its composition, and the rate of decay. The latter parameter, the reaction rate in the equation above, can be expressed as a function of half-life, i.e., the time for the organic carbon in waste to decay to half of its initial mass. Under the climatic conditions characteristic in Hungary, the half-life of food waste is around 10 years, whereas for wood 35 years are needed for decomposition. Therefore, the FOD method requires a quite long time series of activity data (amount and composition of waste). The default first year in the IPCC Waste Model is 1950. As the eldest data, which can be found in statistical publications, are for 1975, extrapolation had to be made. As regards waste composition, statistics only exist for the waste collected in Budapest, capital of Hungary, and only from 1980. For the FOD method, the default values in the IPCC Waste Model were used for the year of 1950, but the measured values for 1980, and interpolation was carried out between these two dates.

Both international and national studies suggested that the mass of municipal solid waste increased hardly, while waste volumes grew drastically all over the world (Barótfi 2000). These changes are attributable to the increasing amounts of paper and plastic in the packaging sector. Data from the regular analysis of waste composition from Budapest show similar pattern: the share of easily degradable food and garden waste decreased from 41 to 23 mass percent, while that of plastics increased to fourfold in the last 20 years (1988–2008). As plastics do not degrade easily, the growth of emissions has stopped recently. Due to the Hungarian waste policy, the amount of disposed waste is expected to decrease in the near future, which will lead to further emission reduction in this source category.

16.4.2 Wastewater Handling

This section covers CH_4 and N_2O emissions generated during municipal and industrial wastewater treatment. While estimating the methane emission of wastewater handling, the first thing is to determine the total organic waste to be handled. For domestic wastewater, it can be calculated by multiplying the population of the country by the IPCC default value of biochemical oxygen demand (BOD_5) that is 60 g $person^{-1}$ day^{-1}. Total organic waste from industrial wastewater can be estimated by using the default chemical oxygen demand (COD) values per wastewater output, which is a good measure of organic compounds in water. Special emphasis should be given to industries with high COD output, e.g., food and beverage, paper and pulp, and chemical industry. For the calculation of the emission factor, the default maximum CH_4 producing capacities of 0.25 kg CH_4 $(kg\ COD)^{-1}$ and 0.6 kg CH_4 $(kg\ BOD)^{-1}$ are used for industrial and domestic wastewater, respectively. As methane is generated only in anaerobic condition, the fraction of wastewater treated anaerobically is another important parameter. Usually, collected wastewater undergoes aerobic treatment in plants; therefore, the share of people not connected to sewerage system determine the emission to a great extent. In case of septic systems or any other domestic means (no connection to public sewerage system), it can be assumed that half of the organic waste settles, and methane is generated. In the 1980s, the portion of population connected to public sewerage system was less than 40%, while in 2008 it was around 71%. This development led to a significant decrease in emissions, since from centralized treatment plants hardly any methane emission arises. However, it must be noted that the percentage of dwellings connected to the public sewerage systems is still below the Central European average.

Nitrous oxide emission from domestic wastewater effluent was estimated based on protein consumption data. On average, 37–38 kg protein is consumed per person annually, which leads to 0.6–0.7 Gg nitrous oxide emission in Hungary.

16.4.3 Waste Incineration

This source category covers only the emission from thermal waste treatment without energy recovery. During waste incineration, mainly CO_2 is emitted of which only the fossil part contributes to the total emissions in the inventory. Nevertheless, 90–97% of all incinerated waste in this source category was hazardous waste with high fossil carbon content. Methane emission is insignificant and N_2O generation is also minimal.

Emissions from waste incineration for energy purposes are accounted for in the *Energy* sector and waste is treated as a "normal" fuel, except for that the amount of waste from waste statistics has to be converted from mass to energy units, and the fossil carbon fraction has to be determined. Waste incineration has become increasingly popular among different industrial facilities: cement factories incinerate large amount of waste of fossil origin (plastics, rubber, etc. with relatively high energy content 26–28 GJ ton^{-1}), whereas other industrial facilities incinerate predominantly waste of biogenic origin, mostly wood waste; therefore, their CO_2 emissions do not contribute to the national total. From municipal solid waste 8–10 GJ energy per ton can be recovered.

References

Barótfi I (2000) Környezettechnika, Mezőgazda Kiadó. Available online at: http://www.tankonyvtar.hu/konyvek/kornyezettechnika/kornyezettechnika-eloszo-081029-1

EC (European Commission) (2007) Panorama of energy – energy Statistics to Support EU Policies and Solutions. Office for Official Publications of the European Communities, Luxembourg. ISBN 92-79-03894-X

Energy Centre (1986–2010). Energiagazdálkodási Statisztikai Évkönyv, 1985–2010 (Energy Statistics Yearbook, 1985–2008). Budapest

HCSO (Hungarian Central Statistical Office) (1986–2009) Magyar statisztikai évkönyv 1985–2008 (Statistical yearbook of Hungary 1985–2008), Budapest.

HCSO (Hungarian Central Statistical Office) (2009) A települések infrastrukturális ellátottsága, 2008 (Infrastructural supply of settlements, 2008). Statisztikai tükör, III., No. 154

Hendriks CA, Worrell E, de Jager D, Blok K, Riemer P (1999) Emission reduction of greenhouse gases from the cement industry. International Energy Agency Greenhouse Gas R&D Programme, Cheltenham, UK, Report Number PH3/7, pp 25–49

Hill D (1984) A history of engineering in classical and medieval times. Routledge, p 106

Hong K, Trudell JR, O'Neil JR, Cohen EN (1980) Metabolism of nitrous oxide by human and rat intestinal contents. Anesthesiology 52:16–19

IPCC (Intergovernmental Panel on Climate Change) (1996) In: Houghton JT, Filho LGM, Callander BA, Harris N, Kattenberg A, Maskell K (eds) Climate Change 1995: the science of climate change. Cambridge University Press, Cambridge, UK

IPCC (Intergovernmental Panel on Climate Change) (1997) Revised 1996 IPCC guidelines for national greenhouse gas inventories. Intergovernmental Panel on Climate Change, Organisation for Economic Cooperation and Development, and International Energy Agency. (IPCC/OECD/IEA), UK Meteorological Office, Bracknell. Available online at: http://www.ipcc-nggip.iges.or.jp/public/gl/invs1.htm

IPCC (Intergovernmental Panel on Climate Change) (2000) Good practice guidance and uncertainty management in national greenhouse gas inventories. Intergovernmental Panel on Climate Change National Greenhouse Gas Inventories Programme, Institute for Global Environmental Strategies, Japan. Available online at: http://www.ipcc-nggip.iges.or.jp/public/gp/english/

IPCC (Intergovernmental Panel on Climate Change), Background Papers, (2002) IPCC expert meetings on good practice guidance and uncertainty management in national greenhouse gas inventories, p. 112, Japan

IPCC (Intergovernmental Panel on Climate Change) (2006) 2006 IPCC guidelines for national greenhouse gas inventories. In: Eggleston HS, Buendia L, Miwa K, Ngara T, Tanabe K (eds) Intergovernmental Panel on Climate Change National Greenhouse Gas Inventories Programme. Institute for Global Environmental Strategies, Japan

IPCC (Intergovernmental Panel on Climate Change) (2007) In: Metz B, Davidson O, Bosch P, Dave R, Meyer L (eds) Climate Change 2007: Mitigation, Contribution of Working Group III to the Fourth Assessment Report of the Intergovernmental Panel on Climate Change. Cambridge University Press, Cambridge, United Kingdom and New York

Kovács Zs, Kakucs O, Lakó J, Rédey Á, Fülöp T, Yuzhakova T, Utasi A, Domokos E (2009) Environmentally friendly industrial technologies; Case Studies. E-tudomány, 2009/4. E-tudomány

Egyesület, Budapest; Available only online: http://www.e-tudomany.hu/etudomany/web/uploaded_files/20090408.pdf
Pérez-Ramírez J, Kapteijna F, Schöffelb K, Moulijn JA (2003) Formation and control of N_2O in nitric acid production: where do we stand today? Appl Catal, B 44:117–151
REKK (Regionális Energiagazdasági Kutatóközpont – Regional Centre for Energy Policy Research) (2004) Magyarország üvegházgáz kibocsátásainak előrejelzése 2012-ig a jelentős kibocsátó ágazatok közgazdasági kutatása alapján (Projection of greenhouse gas emission in Hungary until 2012 based on economical research of significant emitters). Research paper for the Ministry for Environment and Water, Corvinus University, Budapest – Regional Centre for Energy Policy Research, Budapest, Hungary
Rypdal K, Winiwarter W (2001) Uncertainties in greenhouse gas emission inventories – evaluation, comparability and implications. Environ Sci Policy 4:107–116
Worrell E, Price L, Hendricks CA, Ozawa Meida L (2001) Carbon dioxide emissions from the global cement industry. Ann Rev Energy Environ 26:303–329

Index

T

U

V

FSC
www.fsc.org
MIX
Papier | Fördert gute Waldnutzung
FSC® C083411

Kolibri 360 GmbH
Ferdinand-Jühlke-Straße 7
99095 Erfurt, Deutschland
produktsicherheit@kolibri360.de